Global Change and Caribbean Vulnerability

Global Change and Caribbean Vulnerability

Environment, Economy and Society at Risk

EDITED BY

Duncan McGregor
David Dodman
David Barker

University of the West Indies Press
Jamaica • Barbados • Trinidad and Tobago

University of the West Indies Press
7A Gibraltar Hall Road Mona
Kingston 7 Jamaica
www.uwipress.com

CATALOGUING IN PUBLICATION DATA

Global change and Caribbean vulnerability: environment, economy and society at risk / edited by Duncan McGregor, David Dodman, David Barker.

p. cm.

Papers presented to the international conference: Global Change and Caribbean Vulnerability, held at the Mona campus, University of the West Indies in July 2006.

Includes bibliographical references.

ISBN: 978-976-640-221-1

1. Sustainable development – Caribbean, English-speaking. 2. Natural disasters – Jamaica. 3. Natural disasters – Caribbean, English-speaking. 4. Food supply – Jamaica. 5. Urbanization – Environmental aspects – Caribbean, English-speaking. 1. McGregor, Duncan F.M. II. Dodman, David. III. Barker, David, 1947–

GE160.C27G48 2006 363.7'009729

Cover design by Robert Harris.

Printed in the United States of America

Contents

Foreword

I consider it an honour to have been asked to write the foreword for this book, which represents the compilation of the contributions made at the International Conference *Global Change and Caribbean Vulnerability*, held at the Mona campus of the University of the West Indies in July 2006.

As small nation states, the islands of the Caribbean are vulnerable on many different fronts. Although one focus of the conference was global climate change, there are other areas – for example, international economic arrangements – that also serve to highlight the vulnerability of small Caribbean nation states. Such vulnerability has resulted from the effects of the phenomenon of "globalization". Although often not defined in precise terms, global changes are nonetheless of significance, in terms of adverse impact on small economies which face new competition in areas of trade where they once received special "treatment".

In this foreword I wish to assess the impact of global changes on the natural environment, as well as in terms of social, economic and trade issues. In making this assessment, it is useful for us to consider three levels of interaction: first between countries in international economic fora; second between governments and their citizens, in particular those in the lower socio-economic cohorts; and finally between various special interest groups within individual countries.

In terms of the environment, the early years of this new century have clearly demonstrated the vulnerability of the Caribbean region to natural disasters, in particular hurricanes. Countries such as Grenada and Barbados, and even certain states in the United States which are unaccustomed to damage caused by hurricanes, have been affected. While all of the population feels the adverse effects when such natural disasters strike, the poorest groups are those who suffer most in a variety of ways.

Hence, damage to infrastructure and to economic activity has the most serious effects on those with limited or no savings, limited economic options and limited ability to recover. For example, very few Caribbean small farmers have the protection of crop insurance. Therefore, when their crops are damaged or, worse yet, totally destroyed, a whole segment of the previously employed labour force has no option but to become "wards of the state". Similarly, when there is damage to the houses of the poor, only a small percentage has the protection of insurance coverage. In many instances, even if the householders were willing to purchase such protection, it is not available based either on the assessed vulnerability of their locations or the structure of the housing units.

When a major disaster strikes a vulnerable state, the negative effects are multifaceted. To begin, there is the distinct possibility of loss of lives, the impact on economic activity, and the destruction of social and physical infrastructure. In the majority of instances, and in the absence of the private "safety nets" of crop or home insurance, the expectation is that the state will not only have the responsibility to restore public infrastructure, but also to contribute to the rehabilitation of individual households.

On the economic front, global change has, over the last decades, brought about a reduction/elimination of special tariff barriers and subsidies. As such, local producers in poor countries, whether of manufactured goods or agricultural commodities, have been forced to face increased competition from imports. In many instances, the relative scales of activity, as well as differences in the level of technology applied, make such competition lopsided. This has become most evident with the planned, rapid expansion of China's export thrust. Perhaps most importantly, the ability of home governments to "protect" domestic producers against external competition has been eroded over time, as international trade agreements have made reciprocity a fundamental requirement for participation. At best, regardless of how just is their cause, developing countries can only hope to reduce the rate at which special trade arrangements are dismantled.

For persons from low-income groups, the challenge of addressing the negative repercussions of global changes often seems overwhelming. On the one hand, what had been a predictable way of life – for example, producing agricultural commodities for export to the European Union – is now being challenged as the subsidies that facilitated such exports have been systematically reduced. If, in addition to this challenge, the same farmers and producers are faced with destruction of their crops and housing units by natural disasters, it is not difficult to imagine that their confidence in the future is being undermined.

The fundamental question which a collection of papers such as this one must seek to answer is as follows: How can these small island states not only survive these challenges, but improve the socio-economic situation of the most disadvantaged groups in an environment of change thus making them more capable of dealing with global changes? The fact is that while the state must lead the fight, there has to be a coordinated effort, with civil society, the NGO community and the individuals themselves all playing roles.

We must first accept that most of these changes are inevitable, and while diplomatic interventions and negotiations have value and must be pursued, at best, they can only slow the pace of adjustments. Against this backdrop, in each country there needs to be a clinical assessment of whether the economic activities that once dominated can be raised to higher levels of efficiency so as to meet competition. This competition must be seen as global, regardless of whether production is for the domestic market or for export. Explicit reference is made of competition in the domestic market which once represented an "enclave of protection" determined by national governments. As has been stated previously, this ability to protect domestic producers is rapidly disappearing with a growing consensus to remove tariff barriers.

Assessment of which sectors have the potential to survive and prosper in an increasingly competitive environment is always difficult. Abandoning traditional economic activities and a way of life is not a decision which is easily made, either by governments or by those involved in a particular sector. A natural tendency is to assume that steps can be taken to preserve the status quo – either by appealing to the goodwill of the rich countries or by lobbying at home for special action to be taken by national governments, if not by tariffs then by production subsidies.

A major challenge in such instances, when it is apparent that changes are inevitable, is in identifying how the workers/producers in these sectors, who throughout their lives have been productive members of society, can be "transformed" and "transferred" to another area of activity. This "transformation" transcends a question of identifying a suitable, alternative economic activity. For example, it is often difficult to convert someone who, for all his or her working life, has been a banana farmer, into a producer of some other commodity.

This challenge is not unique to the Caribbean. In the United States, for example, the economic bases of several regions have been radically changed over the latter half of the twentieth century. The economic landscape of many industrial towns in the northeast seaboard has been altered and in few instances have these changes been managed flawlessly. One disturbing phenomenon, which has received too little analytic assessment, has been the social impact on households where the members formerly held jobs in the sectors of economy which have either declined or been totally abandoned.

The second major challenge is the need to systematically increase the technical capabilities and knowledge set of the general population. What is clearly established is that poverty is the inevitable result of the continued application of low skills and limited knowledge to any task. Hence, raising the standard of living can only be achieved, and maintained on a sustained basis, if the level of training of the population is correspondingly improved in a systematic manner.

The third issue that arises, particularly in combating devastation of natural hazards, is that some of the cultural habits of the lowest income groups have to be confronted. As such, in many instances where well-meaning organizations seek to protect the "rights" of disadvantaged groups, it is sometimes the case that preservation of these "rights" is detrimental to the welfare of those for whom protection is being sought.

A specific example illustrates the point. A major reason for loss of lives and widespread devastation following a hurricane or major flood in the Caribbean is the fact that many housing units were illegally built in locations where no construction should have been permitted and no official permission had been granted. Such locations are sometimes referred to as "informal settlements". No long term solution lies in seeking to provide legal status to such settlers. Rather, the state, in collaboration with community groups, must systematically address the issue of relocation and ensure that the new sites chosen for relocation activities must meet the most rigid environmental requirements.

As we assess the various ways in which the challenges of global change can be met, a realization is that a coordinated response covering various academic disciplines, as well as interest groups, provides the greatest chance for minimizing the negative effects and at the same time maximizing the potential positive results. As such, the profession of engineering will need to take greater cognizance of the advice of the environmentalists and physical planners, not only in terms of structures to be erected but also the areas where these constructions should be permitted. The same principle applies to farming activities. We must, at the same time, re-examine the basic principles of our education system. It is now clear that there is need for a greater recognition of the potential negative impact on the economy and society of the abuse of the environment, and this awareness must be integrally woven into the formal school curriculum.

There is no question that the changes that have taken place have represented major challenges to Caribbean nation states. However, they also have provided us with an opportunity to analytically assess our resources and determine how various sectors of the society can combine, not simply to combat the challenges, but to move these societies forward in this increasingly competitive world.

On a positive note, a recent initiative taken by most of the countries of the Caribbean, assisted by the World Bank, demonstrates a proactive response to the negative effects of natural disasters. Eighteen states have joined together to establish an insurance fund which will provide immediate support when a participating member country is hit by a hurricane or earthquake. The insurance payments will be immediate, thus providing the authorities with the ability to respond to needs, whether these are for emergency relief, repairs to vital physical infrastructure or even to replace revenues lost as a result of reduced economic activity. Work is underway to assess the actuarial implications of adding crop insurance and damage resulting from floods to this facility.

A publication such as this, by bringing together academics from difficult fields as well as professionals and state technocrats, provides a unique opportunity for us to aggregate knowledge and maximize the benefits of interdisciplinary research and solutions. The contents of this publication will provide, not only to the participants, but also to researchers, policy makers and the interested lay observer, much on which to reflect and act.

Omar Davies

Preface

This volume is the fourth in a series of books arising out of the British-Caribbean Geography Seminar (BCGS) Series. The chapters have been selected from among the thirty-two papers presented at the fifth BCGS held at the Mona campus of the University of the West Indies in July 2006. The papers have been suitably revised and edited for publication. The theme of the meeting and the title of the book, *Global Change and Caribbean Vulnerability: Environment, Economy and Society at Risk* focuses on the impact of global change on human and natural systems in the Caribbean.

The seminar series was inspired originally by the successful "Overseas Seminars" promoted by the then Institute of British Geographers during the 1970s and 1980s. The BCGS has gone from strength to strength and is the only seminar to last into the new century. This is a fact recognized by the Royal Geographical Society/Institute of British Geographers, as it continues to lend its prestigious support to the series. Past meetings also have received funding support from the Commonwealth Foundation and the British Council.

A feature of these meetings has been the very high proportion of the papers presented which are successfully revised and converted into peer-reviewed publications. The cumulative editorial effort by those involved has been prodigious. From the outset, it was felt strongly that the lasting value of an academic meeting can only be measured by its scholarly output. The task of moving from conceptualizing and organizing a meeting to publishing a volume is daunting, especially because academic conferences have evolved considerably from the days when papers were prepared and written prior to a conference and often formally read out to the audience. PowerPoint presentations have made conference "papers" much more visually interesting, even entertaining. To translate a slide presentation into a scholarly article, however, requires a serious commitment from an author, and determination, patience and fortitude from the editors.

Thus, as in the previous volumes, the chapters here have been subjected to a rigorous peer-review process. Authors of conference papers selected for the volume were invited to submit manuscripts based on their presentations, and these were reviewed by the editors. Changes were then requested for each of submitted chapters to ensure quality and consistency in the collection, as well as to help elucidate the main themes of the book. Next, the publishers engaged three independent, external reviewers who made general comments on the collection and specific comments on each chapter. Both authors and editors then went through another round of changes, trimming, extending and generally fine-tuning the collection. We feel confident that this rigorous two-step review process has produced another volume of high quality papers.

An overarching theme in each of the five BCGS meetings has been empirical research connecting environmental issues and development problems in the Caribbean region. The inaugural BCGS was held in 1992 at Mona campus, and its theme of Environment and Development was captioned in the title of the subsequent edited volume *Environment and Development in Small Island States: The Caribbean* (edited by Barker and McGregor, 1995). The book was published by the fledgling Press at the University of the West Indies, and we are indebted to them for their continued support which has resulted in four books in the series. The first BCGS was held in the same year as the landmark Rio Summit, and, not surprisingly, geographical research in the Caribbean region during the 1990s often addressed aspects of sustainable development, in the context of vulnerable small islands and the fragile connections between their environmental and human systems. Such concepts and ideas were captured in the second BCGS, held in 1995 at Royal Holloway and the third BCGS, which returned to Mona in 1998. These meetings resulted in the subsequent UWI Press publications *Resource Sustainability and Caribbean Development* (edited by McGregor, Barker and Sally Lloyd Evans, 1998) and *Resources, Planning and Environmental Management in a Changing Caribbean* (edited by Barker and McGregor, 2003). Other papers from these meetings have appeared as special issues of *Caribbean Geography*: volume 3 (4) in 1992 and followed by volume 7 (1) in 1996, then two special issues, volume 9 (2) in 1998 and volume 10 (1) in 1999. The fourth BCGS was held at San Ignacio in Belize in 2002, in conjunction with the Fourth International Belize Conference, and a selection of papers appeared in 2003 in *Caribbean Geography* volume 13 (1–2).

The aims of the fifth BCGS were similar to the earlier meetings: to bring together geographers and researchers from cognate disciplines to exchange ideas and present their research; to provide a platform for younger researchers to present their work to a professional audience, perhaps for the first time; and to focus on Caribbean environments. It involved over fifty participants and the thirty-two papers presented ranged across the geographical spectrum from regional and sub-regional climate prediction to the social impacts of environmental change and economic globalization. Once again, the meeting had a truly international flavour, including participants from eight Caribbean territories (Jamaica, Barbados, Antigua, St Lucia, Trinidad, Guyana, Haiti and Suriname) and from the United Kingdom, the Netherlands, the United States and Canada. A particular highlight on this occasion was the opening address by Dr Omar Davies, MP, then Jamaica's minister of finance and planning, who, using his background in geography and planning, reflected on economic and social vulnerability in small island states such as Jamaica.

Duncan McGregor
David Dodman
David Barker

Acknowledgements

We would like to thank a number of institutions and individuals for their assistance and support in facilitating the fifth British-Caribbean Geography Seminar, held at the Mona campus of the University of the West Indies in July 2006. The Royal Geographical Society/Institute of British Geographers kindly provided pump-priming funds for the seminar. Duncan McGregor thanks the Royal Society (Conference Grants Fund) and the Department of Geography, Royal Holloway, University of London, for covering the costs of his attendance. We are particularly indebted to Dr David Howard (Institute of Geography, University of Edinburgh) for organizing a stimulating session on urban vulnerabilities in the Caribbean. The administrative staff and geography postgraduates of the Department of Geography and Geology at Mona are thanked for their invaluable assistance, especially during registration and throughout the week of the meeting. Lisa Williams, map curator, Department of Geography and Geology, is thanked for her help in organizing the book exhibition and map displays. UWI (Mona) and Community Co-operative Credit Union is thanked for its generous sponsorship which facilitated an excellent reception at the Mona Visitors' Lodge after the opening ceremony. Most of the maps and diagrams have been drawn or redrawn by Jenny Kynaston, Geography Department, Royal Holloway, University of London. The Office of Graduate Studies and Research at the Mona campus of the University of the West Indies is gratefully acknowledged for the award of a grant towards the cost of publishing this volume.

Duncan McGregor
David Dodman
David Barker

PART 1

Caribbean Vulnerability and Global Change

CHAPTER 1

Caribbean Vulnerability and Global Change

Contemporary Perspectives

DAVID BARKER, DAVID DODMAN AND DUNCAN McGREGOR

The Caribbean region is confronted by global change on many fronts. Global warming and climate change comprise the environmental engine, while globalization and population growth are the fuel that drives changes in the economy, society and landscapes. Caribbean island nations are vulnerable to these global changes in complex and interconnected ways. This volume reports on current research into the interactions between global physical and human forces and peoples' vulnerability to these forces in the Caribbean.

The volume contains seventeen substantive chapters organized thematically into five parts: Caribbean Vulnerability and Global Change, Managing Vulnerable Environments, Vulnerability and Domestic Food Supply, Urban Vulnerability and Urban Change, and Synthesis. The individual focus of chapters varies, with a range from Caribbean-wide discussion, a majority of case studies from Jamaica, and case study material from Guyana, Trinidad and Tobago, Curaçao, Suriname and the Dominican Republic (the three latter from outside the anglophone Caribbean). Additionally, however, all contributors were asked to extend their discussions beyond their specific case studies to the wider Caribbean. The editors have further attempted to broaden the empirical base of the collection within a framework of cognate Caribbean issues and research, in both this introductory chapter and in the concluding synthesis. Geography, geology, sociology, social anthropology, environmental management and national park management are the professional disciplines represented by the authors, and the work of several postgraduates is also included in the volume.

This is the fourth volume in a series that highlights the fulcrum of environmental and developmental research in the Caribbean. In the earlier volumes, current geographical research was incorporated into themes such as sustainable development, resource sustainability, small islands and development, and the omnipresent threat of environmental hazards such as tropical storms and hurricanes (Barker and McGregor 1995; McGregor, Barker and Lloyd Evans 1998; Barker and McGregor 2003). However, from the late 1990s, the impact of economic globalization increased the pace of change and introduced new challenges for Caribbean people and nations. Trade liberalization and the global reach of multinationals have compounded the vulnerability of island economies and people's livelihoods, especially in the field of agriculture. More recently, there has been another major reconfiguration of the interrelated themes of environment and development, as climate change has captured centre stage and its implications for the Caribbean have started to be debated and researched. In this volume, we bring the themes of environment and development together in a fresh way, bracketing climate change and economic globalization as two major external forces driving global change, and highlighting the reality of the Caribbean region's vulnerability to them.

In the discourse on vulnerability, scientists and social scientists often use different definitions and different perspectives. For example, in the social sciences, social vulnerability may refer to the capacity or inability of individuals, organizations or society to cope with the stressors or changes to which they are exposed (Allen 2003). In these cases their vulnerability is a function of social, economic, cultural and political processes. Scientists, on the other hand, tend to view vulnerability in terms of the likelihood of occurrence of physical environmental events (especially natural hazards) and emphasize the concept of risk assessment (Smith and Petley 2009). It is of note that the concept of vulnerability was introduced into the disaster literature by geographers working in developing countries, specifically to emphasize the connection between natural hazardous events and the poor socio-economic conditions that are often the root cause of disasters (Blaikie et al. 1994).

Chambers (1989) argues that vulnerability has two characteristics, an internal and an external component. This framework is useful in the context of this volume's theme because it allows us to conceptualize twin forces of climate change and economic globalization (global change) as two sources of external shocks and stresses. The internal components of vulnerability are considered to be the ability (or inability) to adjust to, or cope with, these external forces. Thus, this volume uses the conceptual lens of vulnerability to focus attention on some of the environmental and developmental impacts of climate change and economic globalization, and on the adaptive capacity and resilience of individuals, communities and nations in the Caribbean region. Hence, the theme and the title of this book: *Global Change and Caribbean Vulnerability: Environment, Economy and Society at Risk?*

Perhaps the most significant global change factor influencing Caribbean vulnerability is the threat posed by climate change. The Intergovernmental Panel on Climate Change (IPCC) Fourth Assessment Report (2007) states that the warming of the earth's climate system is unequivocal, many natural systems are being affected by regional climate changes and the IPCC report suggests that it is "very likely" that most of the observed warming is due to an increase in anthropogenic greenhouse gas concentrations. These changes are expected to have major impacts for agriculture, forestry and ecosystems; water resources; human health; and industry, settlement and society (IPCC 2007).

The IPCC report also took the important step of linking climate change and development pathways, adding significant weight to our conceptualization of Global Change and Caribbean Vulnerability as a critical current research arena. The report urges, for example, that it is necessary to consider projections of social and economic changes at the same time as projections of climate change (IPCC 2007, 19). Elsewhere it emphasizes the reciprocal nature of the relationship between sustainable development and climate change; sustainable development can reduce vulnerability to climate change while climate change itself can impede the ability of nations to achieve sustainable development (IPCC 2007, 20).

One of the likely effects of climate change in the Caribbean is the potential threat posed by an increase in the intensity and severity of hurricanes (Emanuel 2005). A portent may be the record losses experienced as a result of the 2004 and 2005 hurricane seasons in the Caribbean Basin and Gulf of Mexico (Boruff and Cutter 2007). Boruff and Cutter report that Hurricane Wilma caused losses estimated at US$700 million in Cuba while losses in Grenada by Hurricane Ivan were estimated at US$3 billion (2007, 25). Hurricane Ivan also devastated Jamaica and the Cayman Islands, but its impact on Grenada was utterly catastrophic; the damage in monetary value and the disruption to the economy and people's lives was far greater than the disastrous impact of Hurricane Gilbert on Jamaica in 1988 (Barker and Miller 1990).

In discussing the Caribbean research agenda, McGregor and Barker (2003) alluded to the fact that the environmental and developmental challenges facing Caribbean islands (classed as small island developing states [SIDS]) are quite different from those facing larger developing countries like Brazil, Kenya and India. The same characteristics that distinguish SIDS from larger countries (small tropical islands, small populations and economies, limited natural resources) also makes them especially vulnerable to the effects of climate change, sea-level rise and extreme events (Briguglio 1995; Pelling and Uitto 2001). Sea-level rise is expected to exacerbate inundation, storm surge, erosion and other coastal hazards; water resources are likely to be seriously compromised; coral reefs, fisheries and other marine-based resources are likely to be heavily impacted; agriculture is likely to be adversely affected; and there are likely to be negative consequences on tourism (Mimura et al. 2007). The nations of the Caribbean face particular stresses from climate change due to their heavy reliance on coastal areas for their main economic and industrial activities; this is a theme we return to later.

Recent research also suggests that urban areas face distinctive challenges from climate change (Huq et al. 2007; Bicknell, Dodman and Satterthwaite et al. 2009). This is an important issue for the Caribbean because it is one of the world's most highly urbanized regions, with more than 65 per cent of the total population living in towns and cities (Potter 2000). The implications of global environmental change for urban areas are significant because of the scale of the population at risk (more than half the world's population now lives in urban areas), the economic costs without adaptation, and the high levels of vulnerability of urban populations (particularly low-income urban residents) to climate change (Satterthwaite et al. 2007). Simultaneously, the concentration of populations and economic activities means that cities have an important role to play in the reduction of greenhouse gas emissions and the prevention of catastrophic climate change (Dodman 2009). This is an area requiring increased attention from geographers and other researchers whose expertise incorporates both the physical and the human sciences. While climate change will generate new physical threats for towns and cities, it will also have a disproportionate impact on the urban poor and other vulnerable groups (Dodman and Satterthwaite 2008).

Climate shocks can affect livelihoods in many ways: wiping out crops, reducing opportunities for employment, pushing up the price of food and destroying property. Although wealthy countries and households can manage these shocks, poorer countries and individuals have many fewer options (UNDP 2007). Poorer communities are especially vulnerable because of their more limited adaptive capacities and their greater dependence on climate-sensitive resources such as local water and food supplies (Wilbanks et al. 2007). Climate change in the Caribbean threatens both environmental and developmental objectives, and climate change adaptation needs to be incorporated more thoroughly throughout the main sectors in the region. The mutually reinforcing effects of climate shocks and economic globalization are brought into sharp focus in relation to domestic food production, small farmers' livelihoods, and associated broader issues of national food security and high levels of food importation. Geographical research on these themes is gathering momentum in the region, and this is reflected throughout this volume, particularly in the section "Vulnerability and Domestic Food Supply".

One of the major issues underlying contemporary research into both physical and human factors contributing to change is the shifting nature of the Caribbean climate. Douglas Gamble (chapter 2) reviews observed climate trends in the Caribbean and notes that the greatest limitation in the use of global climate models is that their relatively coarse resolution makes it difficult to predict climate trends for individual islands. Gamble argues that this renders assessment of Caribbean vulnerability incomplete, and thus undermines realistic planning and mitigation strategies for climate change. Gamble takes forward the development of an appropriate climatic framework through an analysis of the widely observed Caribbean mid-summer drought.

Managing Vulnerable Environments

Managing vulnerable environments is a major challenge facing Caribbean governments, land owners and individual land users alike. Natural resources not only impose limits on a wide range of human activities, but they themselves are prejudiced by the interaction of changing climates and population pressure. This part examines a range of vulnerable Caribbean environments, from the coastal zone to hillside and mountain environments. First, we introduce the individual chapters in the section and then draw some inferences from them relevant to the wider Caribbean.

Starting in the coastal zone, Deborah-Ann Rowe, Shakira Khan and Edward Robinson examine the implications for coastal zone management of extreme climatic and seismically induced events (chapter 3). The presence of extensive fields of large boulders and debris ridges around Jamaica's coasts attests to the power of hurricane storm waves and possible tsunami events. Rowe et al. relate these features to specific events, both recent and historical. The spatial extent of these features delimits coastal areas within which development should be considered inappropriate, yet the authors point out that tourist-related coastal construction continues in these vulnerable locations.

Patrick Williams and Linda Johnson-Bhola continue the coastal theme by examining the problem of flooding in Guyana's coastal zone (chapter 4). They ascribe recent floods east of Georgetown to a combination of human and environmental factors. Poor maintenance of coastal defences and inadequate land use planning combine to increase people's vulnerability to flooding. Extensive surveys in coastal communities reveal socio-economic and health-related impacts of flooding on these communities, and point to relative inadequacies in government response. Responses from householders enable the authors to suggest structural approaches to reducing vulnerability.

Mick Day and Sean Chenoweth examine the implications of environmental change on karst (limestone) landscapes in the Caribbean (chapter 5). Covering more than half the total land area of the insular Caribbean, karst landscapes are vulnerable because of their unique hydrology. This is particularly notable in areas prone to flooding or in areas where there has been excessive abstraction from the groundwater aquifer. The authors illustrate the sensitivity of karst to human interference through examples from Jamaica and elsewhere. Day and Chenoweth conclude by pointing to the urgent need for more research into the environmental response of Caribbean karst lands to climatic and anthropic change.

Elizabeth Thomas-Hope and Adonna Jardine-Comrie (chapter 6) consider the question of environmental valuation in response to ever-increasing tourism. Using the Jamaican tourist industry as a case study, they investigate the environmental implications of continuing tourism development. Central to their argument is the increasing vulnerability of the physical environment in the face of pressures created by resource use. Increased CO_2 emissions, water pollution and effluent are just some of the pressures on the natural environment which require careful evaluation, particularly when set against the consequences of deforestation. The authors argue that only when appropriate payments (in cash or in kind) are made by user industries, such as tourism, can environmental vulnerability be managed appropriately.

Kemi George (chapter 7) considers the case of biodiversity management in Cockpit Country, Jamaica, using epistemic communities – knowledge-based networks of individuals – as the key actors in developing effective policy for biodiversity management. The bauxite industry rep-resents the most significant anthropogenic environmental threat to Cockpit Country, while agricultural degradation is currently the most significant issue related to the effects of that industry on land management. He concludes that epistemic community members adopt different frames for communicating "internally" and with policy makers, emphasizing the ecological value and the economic value of biodiversity management respectively. This divergence of emphasis, George argues, significantly weakens the ability of the epistemic communities to persuade policy makers to adopt environmental friendly approaches, not least due to their imperfect employment of cost-benefit analysis.

In the final chapter in this section (chapter 8), Susan Otuokon and Shauna-Lee Chai present the work of the Jamaica Conservation and

Development Trust, and other organizations, in managing the Blue and John Crow Mountains National Park, from the perspective of coping with global changes and their impact on vulnerable natural and social environments. They share experiences and lessons learned, making recommendations for approaches to plan, implement, monitor and evaluate adaptation strategies. Key to the success of mitigation is the maintenance of a coherent ecological buffer zone between the core national park and the surrounding agricultural landscape. This requires careful co-management with farmers to develop and implement farming techniques that provide adequate income generation alongside effective watershed protection.

The challenge of managing vulnerable environments in the Caribbean region is significant because tropical SIDS are characterized by fragile ecosystems directly and indirectly impacted by intensive human activities in relatively confined geographical spaces. The region's population distribution is such that the coastal zone is particularly vulnerable, not only in the islands, but also in the littoral regions of South and Central America. It is noteworthy that four of the top ten nations in the world with the highest share of their population living in the low-elevation coastal zone (LECZ) are in the Caribbean region: the Bahamas (with 88 per cent of its population living in the LECZ), Suriname (76 per cent), Guyana (55 per cent) and Belize (40 per cent) (McGranahan, Balk and Anderson 2007). It is worth noting additionally that this ranking excludes countries with a total population of under 100,000 people or a total area smaller than 1,000 square kilometres. If these countries were included, the Cayman Islands and Turks and Caicos Islands would be in the top five – with more than 90 per cent of their populations living in the LECZ.

Coastal vulnerability is particularly acute in Guyana, where the majority of the population live in the coastal zone. Thus, Williams and Johnson-Bhola's case study of the major 2005 flooding events in Guyana builds on earlier pioneering research by Pelling (2003a, 2003b). In order to build up a broad picture of coastal vulnerability in the region, CDERA (the Caribbean Disaster Emergency Relief Agency) has produced a series of country reports that document coastal hazards, hazard maps and vulnerability assessments as part of its Caribbean Hazard Mitigation Capacity Building Programme (http://www.cdera.org/projects/cadm/docs/). Rowe, Khan and Robinson's study of boulder arrays in coastal sites around Jamaica provides additional scientific data on the potential threats posed by storms and tsunamis, and this study helps to contribute to such regional data bases.

Caribbean coastal environments are not only vulnerable to environmental hazards but are often degraded through the impact of tourist development. For example, large areas of coastal wetlands have been irrevocably lost to tourism and resort developments in places like Antigua (Lorah 1995). Thus, the arguments regarding the need to calculate and incorporate a monetary value for "environmental services" in the coastal zone, as posed in the chapter by Thomas-Hope and Jardine-Comrie, have relevance to Caribbean tourism planners and the tourist industry. This work also contributes to the existing literature on techniques for natural resources valuation (Ulibarri and Wellman 1997).

The three other chapters in this section touch on issues relating to Caribbean conservation. Protected areas and national parks have become important tools for managing vulnerable environments and the Caribbean Natural Resources Institute (CANARI) based in Trinidad has been particularly active in advocating and monitoring various participatory and co-management approaches to protected area management in the eastern Caribbean (Krishnarayan, Geoghegan and Renard 2002; see also Conway and Lorah 1995). Otuokon and Chai's chapter on the progress and management problems in a Jamaican national park thus contributes further insights to the Caribbean experiences in protected area management. Protected area status is also relevant for a number of unique karst environments in the region. Day's work over the last two decades has systematically documented Caribbean karst and championed the idea of protecting limestone upland regions (for example, see Day 2007). The Day and Chenoweth chapter here is another milestone in this impressive literature on Caribbean karst.

Generally speaking, Caribbean environmental issues have received well-merited media attention during the last couple of decades, and the environmental lobby has become much more influential, though its influence on environmental and development policies is not easy to decipher. However, it is important in discussing the impacts of particular development projects on vulnerable environments to go beyond technical environmental impact assessments, which can be controversial. Valuable insights on the roles and tactics of environmentalists, developers and property owners, development planners and politicians can deepen our understanding of the politics of the environment and development. Yet, in the Caribbean, such studies are few and far between. One example is Sletto's (1998, 2002) research on the Nariva Swamp in Trinidad where he used a political ecology framework to analyse resource conflicts between large rice growers, and the conservationists who sided with small farmers and artesenal fishermen. In this volume, George's discussion of the controversy over bauxite mining in Cockpit Country provides another rare and valuable insight into discourses on Caribbean environmental issues. Similar studies are possible on a range of environmental management topics, ranging from the many controversial tourist resort developments at sensitive coastal sites to country-specific issues, such as the pollution caused by both large-scale and small-scale gold mining in Guyana (Roopnarine 2006).

Vulnerability and Domestic Food Supply

The three papers in this section focus on food security issues and illustrate generic problems in an era of environmental change, in the particular context of Jamaica. Again, firstly we comment on the specific chapters in the section, and then we broaden the discussion to the wider Caribbean.

Duncan McGregor, David Barker and Donovan Campbell briefly review recent climatic trends in the Caribbean Basin, with particular reference to Jamaica (chapter 9). The implications of these trends for Jamaica's agriculture are outlined. In recent years, domestic food production has been disrupted in Jamaica by the impact of successive hurricanes and drought conditions. The parish of St Elizabeth, one of the principal small farming regions in the country, which supplies both the domestic market and the tourist industry, has been particularly badly affected. A survey of small farmers in southern sections of the parish documents the varying impacts of multiple hazards on food production in the period from 2004 to 2006. Although assistance was theoretically in place, farmers reported significant problems in accessing government assistance. In this agriculturally marginal zone, a thorough review of relief structures is urgently required.

In a parallel paper, Clinton Beckford and Steve Bailey also consider small-scale farmers on Jamaica's southern coast (chapter 10). They consider the challenges faced by small farmers in competing against foreign food imports. Their surveys demonstrate that, in strengthening the capacity of local farmers to compete, access to land and access to financial support are the principal constraints. A lack of institutional support in the marketing and distribution of food exacerbates the domestic food farmer's problems. The authors propose that policies are required consistent with the principle of food sovereignty, a concept developed in the early 1990s by Via Campesina, an organization of farmers groups in Latin America and Europe, which invests domestic food farmers with control over the production and marketing of their crops. This implies a much more focused trade system, one which does not compromise local agriculture and removes barriers to small farmer viability.

Kevon Rhiney looks in more detail at the relationships between Jamaican small farmers and the tourist industry (chapter 11). He points out that relatively few contemporary studies are being made on the linkages between tourism and domestic agriculture. He argues that at present domestic agriculture is underdeveloped, having been in progressive decline over the past thirty years. He analyses the relationship between farmers (including agricultural co-operatives), and highlights the advantages of co-operation in the face of competition from imported produce and the quality demands of the tourist industry. The increasing exposure of tourists to Jamaican cuisine provides an opportunity for the development of niche markets.

While the empirical studies in this section focus on Jamaica, the issues that are raised have implications for food security and food supply across the region. All Caribbean islands are prone to tropical storms which, from a regional perspective, inflict millions of dollars of damage annually to agriculture. Specific events, like Hurricane Georges in 1998 and Hurricane Ivan in 2004, were catastrophic in their impacts on the agricultural sectors for the islands they struck. While aggregate national data on agricultural sector damage are generated after such disasters, detailed local studies of hazard impacts on rural people's livelihoods, and their ability to cope with such disasters, as reported here by McGregor, Barker and Campbell, are generally absent from the Caribbean literature. Yet detailed community-level empirical research is needed to inform policy makers about local impacts to allow them to mobilize recovery efforts. For example, the magnitude of Hurricane

Ivan's catastrophic impact on Grenada was quite unprecedented in the region's post-independence hurricane experience. Ivan destroyed, among other things, the entire banana crop and 60 per cent of the country's nutmeg trees. Both were important export crops produced mainly by small farmers who depended on them as major income sources (Brierley 1988). Bananas can be replanted and harvested within twelve months, posing medium-term problems for vulnerability and recovery, but the slow recovery of nutmeg (up to ten years for new trees to bear) means vulnerability and adjustment can have long term implications. It will be a long time before Grenada regains its status as the world's second largest nutmeg exporter and its small farmers are able, once more, to rely on nutmeg as a dependable income source.

Beckford and Bailey's concerns about the impact of food imports in Jamaica are also relevant across the region because all territories have been impacted by liberalization of the food trade (Thomas-Hope and Jardine-Comrie 2007). Only Guyana and Belize, in the anglophone Caribbean, are net exporters of food, and the gap between food imports and agricultural exports for other territories is over 60 per cent – over 80 per cent for Antigua and St Kitts/Nevis (Deep Ford and Rawlins 2007). The two latter countries have made the transition from export agriculture to tourism, thus heightening their vulnerability to food imports in order to satisfy tourist demands. In turn, food import dependence for tourist islands underscores the regional significance of potential linkages between tourism and agriculture. Momsen's (1998) insightful paper on the topic was an important signal of emerging trends, and Rhiney's chapter, in this volume, provides new data on the articulation of linkages between tourism and agriculture in Jamaica. Clearly there is a need for further, detailed empirical case studies from other tourist islands to compare country experiences and help formulate appropriate policy support in this important field.

Urban Vulnerability and Urban Change

It has been known for some time (for example, see Potter 1995) that the Caribbean is characterized by a high degree of urbanization. Another characteristic is a high degree of urban primacy, with countries such as Jamaica having greater than 50 per cent of the population in a few major cities, particularly the capital. Urban areas in the region act as foci for investment (Dodman 2007, 2008) and culture (Jaffe 2008), yet also face a variety of social and environmental challenges (Dodman 2004, Jaffe 2006). The effects of this marked population concentration on the Caribbean environment are wide-ranging and widespread (for some examples, see McGregor and Potter 1997). This section adds to the case studies on urban sustainability from Barbados, St Lucia and the US Virgin Islands, reported in McGregor, Barker and Lloyd Evans (1998), and the chapters explore urban vulnerability in a time of rapid and increasing urbanization.

Colin Clarke opens this section by looking at the nature and development of urban vulnerability in Kingston, Jamaica (chapter 12). Clarke traces the nature and development of multiple deprivation in Kingston's downtown areas, based on a comparison of indicators of multiple deprivation in 1960 and in 1991, the last date for which small-scale census information is available. Clarke's analysis demonstrates not only the very significant population increase in slum/ghetto population between 1960 and 1991, but also an increasing racial ghettoization in the areas of multiple deprivation.

David Dodman continues the theme of urban vulnerability in Kingston (chapter 13), but focuses on an examination of the assets utilized by the urban poor in Waterhouse, an economically deprived community in West Kingston. These assets, which include labour, housing, skills and social networks, are used by residents in a variety of creative ways to achieve livelihoods and build resilience to vulnerability. Dodman points to the need to facilitate the industry of poor urban residents who have already devised sophisticated strategies to address their own vulnerability.

David Howard examines recent urban planning initiatives in the Dominican Republic and notes that that these have shifted from producing material infrastructural change to a greater emphasis on confronting "civil disorder" via the formation of voluntary community associations (chapter 14). Drawing on interviews with residents, activists and government employees associated with the *barrios carenciados* in the northern sector of Santo Domingo, Howard addresses the relationship between territorial violence and fear, demographic change, and urban governance in neighbourhoods increasingly associated with violent crime.

Rivke Jaffe approaches the issue of peoples' vulnerability in islands such as Jamaica and Curaçao by examining the clash between the "green" global ideological landscape of NGO and government actors attempting to preserve fragile ecosystems, as well as the contradictory idea of blighted cityscapes as experienced by residents in polluted urban areas (chapter 15). It is in this area, she argues, that ecological vulnerability merges with social vulnerability. The chapter analyses the ways in which different actors make strategic use of various media to impose their visions of Caribbean environments, and concludes by exploring the possibilities for reconciling social and ecological vulnerabilities.

In concluding this section, Hebe Verrest (chapter 16) demonstrates, through extensive household surveys, that despite contrasting socio-economic environments, levels of vulnerability and well being among poor households in Paramaribo, Suriname, and Port of Spain, Trinidad and Tobago, the two are in fact very similar. Her research shows that poor people in urban areas experience daily vulnerabilities related to the informal labour market, tenure insecurity, and socially fragmented, environmentally hazardous living conditions. These vulnerabilities are complicated by urban residents' dependence on cash income and relations with formal institutions to fulfil daily social-economic needs. Verrest's analysis uses an "asset-vulnerability" framework to explore the diversity of households' vulnerability.

These chapters, together with the earlier chapter by Williams and Johnson-Bhola, illustrate that urban vulnerability takes many forms. A fundamental factor in the vulnerability of many Caribbean primate and secondary cities lies in their being situated in the coastal zone. As summarized by Potter et al. (2004), this developed from their roots in servicing the plantation system and the development of colonial administrations. But many Caribbean cities have now spread beyond their original sites on to more vulnerable terrains. Georgetown, Guyana (Pelling 2003a; Williams and Johnson-Bhola, this volume), illustrates just one environmental problem – flooding – in the development of cities in the coastal zone. Portmore, Jamaica, is a further example of a dormitory settlement built in a potentially unsafe location, susceptible to flooding from landward and from seaward, and in a zone of potential liquefaction of soils triggered by earthquake activity (McGregor and Potter 1997).

Development of Caribbean cities has often, despite the best efforts of administration, been poorly controlled. The spectrum ranges from the strong control of the Cuban government on the development of Havana to the contrasting situation of Port-au-Prince in Haiti, where uncontrolled development has led to progressive breakdown of infrastructure, services and effective governance. The problems associated with the recent rapid and variably controlled urbanization of Caribbean cities, in societies that are already highly urbanized (Potter et al. 2004), include road congestion, air pollution, stresses on services, employment and housing deficits. Overcoming these problems requires a range of approaches such as are illustrated herein.

The need to identify economic and ethnic patterns, in order to better manage urban vulnerability, is shown by the work of Clarke and Verrest. The linkages between urban deprivation and siting within the city can be seen as critical. Clarke's detailed analysis highlights areas of multiple deprivation in Kingston, Jamaica, while Howard's ethnographic work in Santa Domingo, Dominican Republic, points to the need for closer relationships between policing and self-policing to help resolve issues of inner-city crime.

However, despite the multiple problems encountered by inner city dwellers, Dodman's analysis of the assets utilized by an inner-city community in Kingston shows how residents can be resourceful and creative in seeking out and utilizing livelihood opportunities. Social capital is clearly important in these communities, as is a better understanding by decision-makers of peoples' perception of their own situation. Reconciling the sometimes-opposing perceptions of residents and governance is critical, as demonstrated eloquently by Jaffe.

Beyond Caribbean Vulnerability: Towards Resistance and Resilience

In chapter 17, the editors have drawn together the major themes of the volume and commented on their direction in terms of a research agenda. This synthesis is not in any sense comprehensive, but it reflects our individual authors' perspectives and our interpretation of them. The key issues discussed in moving forward with a better understanding of

Caribbean vulnerability include scales/dimensions of vulnerability, the political ecology of vulnerability and future directions for vulnerability research. Whereas a better understanding of vulnerability helps recognize the dimensions of the problem, a focus on resistance and resilience directs us towards appropriate ways to address vulnerability. In this chapter, we reflect on scales, political ecology, and future directions for research into resilience. Reflecting the plea by the Hon. Omar Davies (foreword), we conclude by looking at practical interventions for reducing vulnerability and increasing resilience, interventions which have been proposed by the chapters in this book, in order to bridge the gap between academic research and practical policy.

References

Allen, K. 2003. Vulnerability reduction and the community-based approach. In *Natural disasters and development in a globalising world*, ed. Mark Pelling, 170–84. London: Routledge.

Barker, D., and D.F.M. McGregor, eds. 1995. *Environment and development in the Caribbean: Geographical perspectives*. Kingston: The Press, University of the West Indies.

———. 2003. *Resources, planning and environmental management in a changing Caribbean*. Kingston: University of the West Indies Press.

Barker, D., and D.J. Miller. 1990. Hurricane Gilbert: Anthropomorphising a natural disaster. *Area* 22 (2): 107–16.

Bicknell, J., D. Dodman and D. Satterthwaite, eds. 2009. *Adapting cities to climate change: Understanding and addressing the development challenges*. London: Earthscan.

Blaikie P., T. Cannon, I. Davis and B. Wisner. 1994. *At risk: Natural hazards, people's vulnerability and disasters*. London: Routledge.

Boruff, B.J., and S.L. Cutter. 2007. The environmental vulnerability of Caribbean island nations. *Geographical Review* 97 (1): 24–45.

Brierley, J.S. 1988. A retrospective on West Indian small farming, with an update on Grenada. In *Small farming and peasant agriculture in the Caribbean: Manitoba Studies in Geography 10*, ed. J.S. Brierley and H. Rubenstein, 63–81. Winnipeg: Department of Geography, University of Manitoba.

Briguglio, L. 1995. Small island developing states and their economic vulnerabilities. *World Development* 23 (9): 1615–32.

Chambers, R. 1989. Editorial: Vulnerability, coping and policy. *IDS Bulletin* 20 (2): 1–7.

Conway, D, and P. Lorah. 1995. Environmental protection policies in Caribbean small islands: Some St Lucian examples. *Caribbean Geography* 6 (1): 16–27.

Day, M.J. 2007. The karstlands of Antigua, their land use and conservation. *Geographical Journal* 173 (2): 170–85.

———. Forthcoming. *Karst landscapes of the Caribbean: Their use and protection.* Coconut Creek, FL: Caribbean Studies Press.

Deep Ford, J.R., and G. Rawlins. 2007. Trade policy, trade and food security in the Caribbean. In *Agricultural trade policy and food security in the Caribbean: Structural issues, multilateral negotiations and competitiveness*, ed. J.R. Deep Ford, C. dell'Aquila and P. Conforti, 7–40. Rome: Food and Agriculture Organisation.

Dodman, D. 2004. Community perspectives on urban environmental problems in Kingston, Jamaica. *Social and Economic Studies* 53 (3): 31–59.

———. 2007. Post-independence optimism and the legacy of waterfront redevelopment in Kingston, Jamaica. *Cities* 24 (4): 275–86.

———. 2008. Developers in the public interest? The role of urban development corporations in the anglophone Caribbean. *Geographical Journal* 174 (1): 30–44.

———. 2009. Blaming cities for climate change? An analysis of urban greenhouse gas emissions. *Environment and Urbanization* 21 (2): forthcoming.

Dodman, D., and D. Satterthwaite. 2008. Institutional capacity, climate change adaptation and the urban poor. *IDS Bulletin* 39 (4): 67–74.

Emanuel, K. 2005. Increasing destructiveness of tropical cyclones over the past thirty years. *Nature*, no. 436: 686–88.

Huq, S., S. Kovats, H. Reid and D. Satterthwaite. 2007. Reducing risks to cities from disasters and climate change. *Environment and Urbanization* 19 (1): 3–16.

Intergovernmental Panel on Climate Change (IPCC). 2007. Summary for policymakers. In *IPCC Climate Change 2007: Synthesis Report*, 1–22. Cambridge: Cambridge University Press. Available at http://www.ipcc.ch/pdf/assessment-report/ar4/syr/ar4_syr_spm.pdf

Jaffe, R. 2006. Urban blight in the Caribbean: City, environment and culture in Curaçao and Jamaica. PhD diss., University of Leiden.

Jaffe, R., ed. 2008. *The Caribbean city*. Kingston: Ian Randle.

Krishnarayan, V., T. Geoghegan and Y. Renard. 2002. *Assessing capacity for participatory natural resource management.* Caribbean Natural Resources Institute Guidelines Series, no. 3. Laventille, Trinidad and Tobago: CANARI.

Lorah, P. 1995. An unsustainable path: Tourism's vulnerability to environmental decline in Antigua. *Caribbean Geography* 6 (1): 28–39.

McGranahan G, D. Balk and B. Anderson. 2007. The rising tide: Assessing the risks of climate change and human settlements in low elevation coastal zones. *Environment and Urbanization* 19 (1): 17–38.

McGregor, D.F.M., and D. Barker. 2003. Environment, resources and development: Some reflections on the Caribbean research agenda. In Barker and McGregor 2003, 1–10.

McGregor, D.F.M., D. Barker and S. Lloyd Evans, eds. 1998. *Resource sustainability and Caribbean development.* Kingston: The Press, University of the West Indies.

McGregor, D.F.M., and R.B. Potter. 1997. Environmental change and sustainability in the Caribbean: Terrestrial perspectives. In *Land, sea and human effort in the Caribbean Basin*, special volume of *Beitrage zur geographischen Regionalforschung in Latinamerika*, ed. B. Ratter and W. D. Sahr, 1–15. Hamburg: University of Hamburg.

Mimura, N., L. Nurse, R.F. McLean, J. Agard, L. Briguglio, P. Lefale, R. Payet and G. Sem 2007. Small islands. In Parry et al. 2007, 687–716. Available at http://www.ipcc-wg2.org/.

Momsen, J.H. 1998. Caribbean tourism and agriculture: New linkages in the global era? In *Globalization and neoliberalism: The Caribbean context*, ed. T. Klak, 115–33. Lanham, MD: Rowman and Littlefield.

Parry, M.L., O.F. Canziani, J.P. Palutikof, P.J. van der Linden and C.E. Hanson, eds. 2007. *Climate change 2007: Impacts, adaptation and vulnerability. Contribution of Working Group II to the Fourth Assessment Report of the Intergovernmental Panel on Climate Change.* Cambridge: Cambridge University Press. Available at http://www.ipcc-wg2.org/

Pelling, M. 2003a. Vulnerability, urbanization and environmental hazard in coastal Guyana. In Barker and McGregor 2003, 133–52.

———. 2003b. *The vulnerability of cities: Natural disasters and social resilience.* London: Earthscan.

Pelling, M., and J.I. Uitto. 2001. Small island developing states: Natural disaster vulnerability and global change. *Environmental Hazards* 3 (2): 49–62.

Potter, R.B. 1995. Urbanisation and development in the Caribbean. *Geography* 80: 334–41.

———. 2000. *The urban Caribbean in an era of global change.* Aldershot, UK: Ashgate.

Potter, R.B., D. Barker, D. Conway and T. Klak. 2004. *The contemporary Caribbean.* Harlow, UK: Pearson Education.

Roopnarine, L. 2006. Small-scale gold mining and environmental policy challenges in Guyana: Protection or pollution. *Canadian Journal of Latin American and Caribbean Studies* 31 (60). Available at http://findarticles.com/p/articles/mi_6971/is_/ai_n28355286.

Satterthwaite, D., S. Huq, H. Reid, M. Pelling and P. Romero Lankao. 2007. Adapting to climate change in urban areas: The possibilities and constraints in low- and middle-income nations. Discussion Paper Series, International Institute for Environment and Development (IIED), London.

Sletto, B. 1998. Fish, rice and the meaning of place: The political ecology of the Nariva swamp, Trinidad. *Caribbean Geography* 9 (1):14–29.

———. 2002. Producing space(s), representing landscapes: Maps and resource conflicts in Trinidad. *Cultural Geographies* 9 (4): 389–420.

Smith, K., and D.N. Petley. 2009. *Environmental hazards: Assessing risk and reducing disaster*. 5th ed. London: Routledge.

Thomas-Hope, E., and A. Jardine-Comrie. 2007. Caribbean agriculture in the new global environment. In *No island is an island: The impact of globalization on the Commonwealth Caribbean*, ed. G. Baker, 19–43. London: Chatham House.

Ulibarri, C.A., and K.F. Wellman. 1997. *Natural resource valuation: A primer on concepts and techniques*. Washington DC: United States Department of Energy.

United Nations Development Programme (UNDP). 2007. *Human Development Report 2007/8: Fighting climate change*. New York: UNDP.

Wilbanks, T., P. Romero Lankao, M. Bao, F. Berkhout, S. Cairncross, J-P. Ceron, M. Kapshe, R. Muir-Wood and R. Zapata-Marti. 2007. *Industry, settlement and society*. In Parry et al. 2007, 357–90.

CHAPTER 2

Caribbean Vulnerability

Development of an Appropriate Climatic Framework

DOUGLAS GAMBLE

Introduction

The issue of the vulnerability of Caribbean small island developing states (SIDS) is frequently discussed within a framework of global climate change. Many scholars link the outcomes of potential atmospheric warming and sea level rise to an increase in vulnerability across the Caribbean (for examples, see Granger 1997; Pelling and Uito 2001). Specific impacts of climate change on Caribbean SIDS can include land submergence, beach erosion, increased storm flooding, high water tables and reduced fresh water supply (Leatherman and Beller-Sims 1997). Such changes alone may pose a serious challenge to governments and planners, but when such environmental problems are combined with the SIDS characteristics of small size, limited range of natural resources, susceptibility to natural hazards, fragile ecosystems, isolation, extensive land/sea interface per unit area and susceptibility of economies to external shocks, the ability of many Caribbean SIDS to adapt to climate change is open to debate (Maul 1996; Nurse and Sem 2001).

One major obstacle in assessing the vulnerability of Caribbean SIDS to climate change is the difficulty of applying climate change projections to specific locations. Limitations of existing atmosphere-ocean coupled global circulation models (A-OGCMs) and the ever-evolving knowledge of climate change create an unfinished framework within which to set the issue of Caribbean vulnerability. Without such an appropriate climatic framework, assessment of this vulnerability is incomplete, and unable to offer realistic planning and mitigation strategies for climate change. The purpose of this chapter is to outline a starting point from which researchers and officials can move towards the development of an appropriate climatic framework that can be used to assess and address Caribbean SIDS vulnerability. The chapter will focus upon four key issues that must be addressed in order to construct an appropriate climatic framework: (1) a review of current literature concerning observed climate change in the Caribbean; (2) a review of current projections of climate change for the Caribbean; (3) a discussion of the limits of current climate change projections; and (4) a presentation of an example of how geographic research of climatic phenomena, in this case the Caribbean mid-summer drought, can be combined with existing climate change knowledge to create a more appropriate climatic framework in which to address and assess Caribbean vulnerability.

Observed Climate Trends in the Caribbean

Temperature

In terms of observed global temperature change, the IPCC's third report (TAR) (Houghton et al. 2001) indicates that the global average surface temperature has increased by about 0.6°C since 1900, and average night-time daily minimum air temperatures over land, between 1950 and 1993, have increased by about 0.2°C per decade. The recently released IPCC fourth report (FAR) (Solomon et al. 2007) indicates that eleven of the last twelve years (1995–2006) rank among the twelve warmest years in the instrumental record of global surface temperature since 1850. Further, the updated hundred-year linear trend (1906 to

2005) is 0.74°C, as compared to the TAR trend of 0.6°C, and the linear warming trend over the last fifty years is nearly twice that for the last one hundred years.

Observations for the Caribbean region tend to agree with these observed global trends in the twentieth century. Overall, the IPCC TAR indicates that Caribbean islands have experienced an increase in temperature exceeding 0.5°C in the twentieth century with temperatures increasing by as much as 0.1°C per decade (Nurse and Sem 2001). The more recent FAR indicates that new observations and reanalyses support this trend (Solomon et al. 2007). Singh (1997) indicates that the Caribbean temperature has been increasing at a somewhat higher rate, approaching 1°C for the twentieth century. In reviewing observations of temperature extremes for the Caribbean, Peterson et al. (2002) concluded that extreme intra-annual temperature range is decreasing, the number of very warm days and nights is increasing dramatically, and the number of very cool days and nights is decreasing.

Temperature trends have been identified for specific Caribbean locations and are listed in table 2.1. An increase in extreme warm temperatures, particularly in the winter, and little change in minimum temperature has been observed on New Providence and Long Island, Bahamas since 1905 (Martin and Weech 2001). Granger (1995) completed analysis of change in mean annual temperature for multiple locations across the Caribbean (San Juan, Puerto Rico; Kingston, Jamaica; Nassau, Bahamas; Raizet, Guadeloupe and Maracaibo, Venezuela) and found an increase at all locations except Maracaibo (table 2.1). Singh (1997) found mean annual temperature, since the 1950s, increasing at three locations in Trinidad (Piarco, Hollis Reservoir, Penal) (table 2.1). Aparicio (1993), based upon data records from 1951 to 1986, found temperature to be increasing by 0.1°C per decade along the north coast of Venezuela. Thus it is clear that the majority of the evidence indicates air temperature in the Caribbean has been increasing in the twentieth century, particularly since the 1950s. However, the rate at which such an increase is occurring is not consistent across the region, with some locations indicating a short term decrease in temperature.

Table 2.1 Temperature Trends Identified in Previous Literature for Specific Caribbean Locations

Station Location	Period of Analysis	Temperature Variable Studied	Temperature Change
New Providence, Bahamas[a]	1905–2000	January–January extreme maximum temperature	3.7°C/100 years
New Providence, Bahamas[a]	1905–2000	July–January extreme maximum temperature	3.3°C/100 years
Kingston, Jamaica[b]	1940–1990	Mean annual temperature	0.7°C
Maracaibo, Venezuela[b]	1950–1990	Mean annual temperature	–0.3°C
Nassau, Bahamas[b]	1940–1990	Mean annual temperature	0.4°C
Raizet, Guadeloupe[b]	1950–1990	Mean annual temperature	1.3°C
San Juan, Puerto Rico[b]	1940–1990	Mean annual temperature	2.3°C
Piarco, Trinidad[c]	1946–1995	Mean annual temperature	0.5°C
Hollis Reservoir, Trinidad[c]	1973–1993	Mean annual temperature	1.0°C
Penal, Trinidad[c]	1974–1986	Mean annual temperature	1.1°C
Coastal Venezuela[d]	1951–1986	Mean annual temperature	0.1°C/decade

Sources: [a]Martin and Weech 2001; [b]Granger 1995; [c]Singh 1997; [d]Aparicio 1993.

Precipitation

A global trend in precipitation over the past century is difficult to ascertain due to significant regional variability. However, more intense and longer droughts have been observed over large areas of the world since the 1970s, and the frequency of heavy precipitation events has increased over most land areas (Solomon et al. 2007). In terms of the Caribbean, identification of significant trends in precipitation can be much more difficult as compared to temperature, given precipitation's high natural variability within the region. Further, identification of trends becomes even more difficult since it appears that one result of climate change in the Caribbean is an increase in this already high precipitation variability. Nurse and Sem (2001) report that, since 1900, there has been a significant increase in rainfall variability with mean annual total rainfall declining by approximately 250 millimetres for Caribbean islands. Gray (1993) found a weak indication that rainfall is decreasing across the Caribbean. Taylor, Enfield and Chen (2002) also found a decrease in precipitation for the Caribbean with a marked negative trend starting in the 1960s. In an analysis of climate extremes in the Caribbean from 1958 to 1999, Peterson et al. (2002) found that the number of heavy rain events is increasing, and the number of consecutive dry days is decreasing.

In an analysis of eastern Caribbean (Lesser Antilles) precipitation, Walsh (1998) found large-scale changes in annual rainfall, dry season characteristics, drought and heavy rainfall magnitude-frequency, cyclone frequency and cyclone tracks over the past 130 to 150 years. Oscillations between dry and wet periods dominate the area with low rainfall between 1899 and 1928, higher rainfall between 1929 and 1958, and very low rainfall since 1959. This current dry period is characterized by a mean annual rainfall 10 to 20 per cent lower than between 1929 and 1958, with a longer dryer season, more frequent extreme droughts and fewer large rainstorms. One reason for this dry period is the sharp reduction in tropical cyclone activity. Is should be noted that Walsh's research was published before the recent increase in Caribbean tropical cyclone activity from 1995 to the present (Mimura et al. 2007).

Some trends in precipitation for specific locations are listed in table 2.2. Research has found that, for the Bahamas, the precipitation record is highly variable, primarily due to hurricanes (Martin and Weech 2001). After smoothing a precipitation time series with a ten year moving average, a decrease in rainfall was observed until 1963 and an increase from that time to present exists for New Providence (ibid.). Closer examination of data for the past forty years indicates that, opposite to the increase in rainfall on New Providence, rainfall has decreased on Inagua and Long Island (ibid.). Granger (1995) found precipitation to increase since the 1950s at Nassau, Bahamas and Maracaibo, Venezuela and precipitation to decrease at Kingston, Jamaica; San Juan,

Table 2.2 Precipitation Trends Identified in Previous Literature for Specific Caribbean Locations

Station Location	Period of Analysis	Precipitation Variable Studied	Precipitation Change
New Providence, Bahamas[a]	1960–2000	Total annual rainfall	554 mm/ 100 years
Inauga, Bahamas[a]	1960–2000	Total annual rainfall	–427 mm/ 100 years
Long Island, Bahamas[a]	1960–2000	Total annual rainfall	–264 mm/ 100 years
Kingston, Jamaica[b]	1940–1990	Total annual rainfall	–147.9 mm
Maracaibo, Venezuela[b]	1950–1990	Total annual rainfall	515.1 mm
Nassau, Bahamas[b]	1940–1990	Total annual rainfall	377.9 mm
Raizet, Guadeloupe[b]	1950–1990	Total annual rainfall	–246.8 mm
San Juan, Puerto Rico[b]	1940–1990	Total annual rainfall	–343.7 mm
Piarco, Trinidad[c]	1946–1995	Total annual rainfall	18.3 mm
Penal, Trinidad[c]	1974–1986	Total annual rainfall	–100 mm

Sources: [a]Martin and Weech 2001; [b]Granger 1995; [c]Singh 1997; [d]Aparicio 1993.

Puerto Rico and Raizet, Guadelope. For Trinidad, analysis of data from multiple stations found that in general, no significant trends exist in the data, but in some cases, declining dry season rainfalls, more or less wetter conditions in the wet season and increasingly wetter conditions in the transitional season exist at all stations (Singh 1997). Hanson and Maul (1993) found no significant change in precipitation in a 1886 to 1986 time series for Key West, Florida. Aparicio (1993) found that between 1951 and 1986 there was no trend in precipitation along the southern coastal boundary of the Caribbean Sea.

In summary, as indicated by the variety of trends in these analyses, it is difficult to make one broad statement about trends in Caribbean precipitation. However, multiple studies do indicate a decrease in precipitation for the central Caribbean and an increase in precipitation at peripheral locations such as Nassau and Maracaibo.

Tropical Cyclones

Changes in tropical cyclone frequency and magnitude due to climate change have significant implications for future Caribbean vulnerability. Almost all records indicate that high variability has dominated tropical cyclone occurrence in the Caribbean in recent history. One reason for this variability is the El Niño Southern Oscillation (ENSO), which can cause periods of high and low tropical cyclone frequency due to changes in upper level wind speed and position. The La Niña phase of ENSO has been found to be associated with increased probability of hurricane landfalls across the Caribbean region and the El Niño phase of ENSO is associated with a decrease in the probability of tropical cyclones in the east and west Caribbean (Trataglione, Smith and O'Brien 2003).

In terms of trends in Caribbean tropical cyclone data for the past four centuries, Reading (1990) found that high levels of cyclone activity are suggested for the whole or part of the Caribbean during the 1770–1780s, 1810s and 1930s–1950s with troughs in activity for the 1650s, 1740s, 1860s and during the early twentieth century. Further, a noticeable drift eastward in favoured tracks is reported from the mid twentieth century onwards. Despite these patterns, no clear increase or decrease in tropical cyclone frequency or intensity was found for the period 1770 to 1990.

However, a more recent study analysed data for the period 1970 to 2005 and found a large increase in the number and proportion of hurricanes reaching category four and five across the globe (Webster et al. 2005). The largest increases occurred in the North Pacific, Indian, and Southwest Pacific and the smallest increases occurred in the North Atlantic, the area containing the Caribbean. However, the authors of this study indicate that, in order to attribute these recent trends to global warming, a longer global record of tropical cyclones is required (ibid.). In accordance with this trend, the 2005 North Atlantic hurricane season was one of the most active on record and included the largest number of named storms (twenty-seven), the largest number of hurricanes (fourteen), and it was the only year on record with three category five storms, the most intense storm on record (Wilma: 882 hPa), and the most intense storm in the gulf of Mexico (Rita: 897 hPa) (Anthes et al. 2006). Given such recent trends and evidence, some scientists believe an increase in tropical cyclones, caused by an increase in anthropogenic greenhouse gas emissions, has begun, yet such a theory is not universally accepted in the scientific community (for example, see Anthes et al. 2006 versus Pielke et al. 2006).

Sea Level Change

Sea level attributed to global warming has been identified as the most serious threat to the future development of SIDS (Mimura et al. 2007). Global sea levels have been found to fluctuate by +100 metres throughout recent Earth history (Woodworth 1993). The global-averaged sea level rise computed for the period 1950–2000 is 1.8 + 0.3 mmyr^{1}, but decadal variability makes it impossible to detect a significant increase over this period (Church et al. 2004). The FAR (Solomon et al. 2007) indicates an average global sea level rise of 1.8 mmyr^{1} from 1961 to 2003 and a rate of 2.1 mmyr^{1} from 1993 to 2003. Previous estimates of relative sea level rise for the Caribbean include +2.37 mmyr^{1} (Aubrey, Emery and Uchupi 1988) and +3.0 mmyr^{1} (Pirazolli 1986; Hendry 1993; Hanson and Maul 1993). However, it should be noted that there is regional variation in these predictions, with Hanson and Maul reporting a range for the Caribbean of +1.0 cmyr^{1} for the northwest Gulf of Mexico and -3.0 mmyr^{1} for the southwest Gulf of Mexico and Lesser

Antilles. Hendry (1993) notes that regional and local forcing of sea-level change has become more important than glacio-eustatic forcing since the beginning of the late Holocene, when the overall rate of sea-level rise has decreased. The IPCC reports suggest for the Caribbean a sea level rise between 1 and 2 mmyr^1 over the past century (Nurse and Sem 2001; Solomon et al. 2007). In summary, sea level has been observed to increase for the majority of the Caribbean, but regional variations in this trend do exist.

Projected Climate Change for the Caribbean

Temperature

The majority of predictions for climate change under various scenarios of increased green house gas emissions are completed with A-OGCMs. Given the spatial resolutions of such models, projections are produced for regions as opposed to specific points locations, that is, the Caribbean Basin or portions of the Caribbean Basin (Singh 1997; Lal, Harasawa and Takahashi 2002). Based upon the most recent generation of A-OGCMs, the IPPC projects that globally averaged surface temperature will increase by 1.4°C to 5.8°C over the period from 1990 to 2100 (Houghton et al. 2001) and, more specifically, 0.15°C to 0.3°C per decade for 1990 to 2005 (Solomon et al. 2007). The IPCC (Houghton et al. 2001; Solomon et al. 2007) also indicates that, on a global scale, it is very likely that higher maximum and minimum temperatures, with more hot days, cold days and days with frost, will occur over land areas. The end result is a reduced diurnal temperature range and an increase in heat index over most land areas.

In the IPCC TAR, Nurse and Sem (2001) projected the Caribbean to experience moderate warming in the future, 2.0°C by 2050 and 3.1°C by 2080 with no significant difference in seasonal forecasts. Further, these projections include more pronounced increases in minimum temperature as compared to maximum temperature, causing a marginal decrease in diurnal temp range (0.3–0.7°C). Lal, Harasawa and Takahashi's (2002) and Singh's (1997) projections over the Atlantic Ocean and Caribbean Sea concur (2°C for the 2050s and 3°C for 2080) with Nurse and Sem, and these two studies also predict a more pronounced increase in minimum temperature as compared to maximum temperature. In terms of extreme temperature, Nurse and Sem project an increase in maximum summer temperatures that implies an increase in thermal stress conditions and less water availability. The latest projections included in the IPCC FAR (Solomon et al. 2007) indicate a 0.48 to 1.06 degree increase by 2039, a 0.79 to 2.45 increase by 2069, and a 0.94 to 4.18 increase by 2099.

Precipitation

The IPCC TAR (Houghton et al. 2001) projections regarding change in precipitation include an increase of 20 to 30 per cent over the tropical oceans. However, in regional analysis, a slight decrease in annual precipitation (–5.2 per cent + 11.9 per cent for 2050, –6.8 per cent + 15.8 per cent for 2080) is projected for the Caribbean, but this trend is not considered significant given the high natural variability of precipitation in the region and differences in model projections (Nurse and Sem 2001). Seasonal projections include a slight increase expected for winter (December to February) and a decrease in summer (June to August) (ibid.). Due to the variability in model results, the greatest confidence is placed in the summer projection. Projections of extreme precipitation events indicate that there will be a lower number of annual rain days but an increase in the daily intensity of precipitation (ibid.). The IPCC FAR (Solomon et al. 2007) precipitation predictions have a broad range for the Caribbean (that is, –14.2 to +13.7% by 2039) and also support the prediction of extreme precipitation events with a lower number of annual rain days but an increase in the daily intensity of precipitation.

For annual and seasonal projections, Lal, Harasawa and Takahashi (2002) concur with the IPCC TAR (Houghton et al. 2001). It is difficult to compare Singh's (1997) projections to Nurse and Sem (2001) and Lal, Harasawa and Takahashi (2002) due to a different seasonal definition. Singh (1997) created mean wet and dry seasonal rainfall projections for the Caribbean and found slight decreases (–1.0 to –1.5 mmday[1]) in the dry season (January to April) and a slight increase (0.1 to 2.0 mmday[1]) during the wet season (May to August). Further, analysis for the transitional season (September to December) indicates small changes in rainfall, with a tendency to a slight decrease in the northern Caribbean and a slight increase in the southern Caribbean.

Tropical Cyclones

Currently, a clear consensus on recent trends in tropical cyclone characteristics and projections under various warming scenarios does not exist in the scientific community (Nurse and Sem 2001; Walsh 2004). This lack of consensus was made even more clear after the active 2004 and 2005 Atlantic hurricane seasons, with some scientists suggesting the two seasons are evidence of a global warming induced increase in frequency and intensity of tropical cyclones (Emmanuel 2005; Anthes et al. 2006), while other scientists indicate that there is still not enough evidence of a direct link between the two factors (Pielke et al. 2006). However, despite this lack of consensus, projections have been made concerning tropical cyclones in a warmer world. The IPCC TAR and FAR (Houghton et al. 2001; Solomon et al. 2007) conclude that maximum tropical cyclone intensities are likely to increase 5 to 10 per cent in a doubled CO_2 climate. Some other individual studies have reported the likelihood of a possible increase of approximately 10 to 20 per cent in intensity of tropical cyclones under enhanced CO_2 conditions (Holland 1997; Tonkin et al. 1997).

Sea Level

The IPCC TAR and FAR (Houghton et al. 2001; Solomon et al. 2007) indicate that global mean sea level is projected to rise by 0.09 to 0.88 metres between 1990 and 2100. For areas with SIDS, it is projected that sea level may rise by as much as 5 $mmyr^1$ over the next one hundred years, although regional variation in this trend will exist (Nurse and Sem 2001).

Limitations to Application of Projected Climate Change

In order for Caribbean SIDS to develop effective climate change adaptation strategies, officials and planners must inevitably determine how climate change will impact their specific locations (Pelling and Uitto

2001). Thus the question becomes as follows: can it be assumed that projected climate change is correct for all locations in the Caribbean? Unfortunately the answer is no. A suite of reasons exist for the inability to directly extrapolate A-OGCM projections to specific locations in the Caribbean. This section will focus upon what the author, a geographer, believes to be the central challenges in predicting climate change impacts for a specific location: (1) the coarse resolution of A-OGCMs; and (2) the lack of data and knowledge concerning local-scale physical processes and intra-regional variability of physical processes.

The coarse scale of A-OGCM analysis and output is one of the greatest limitations in directly downscaling climate change projections to individual locations. A general mismatch exists between the regional spatial scale at which reliable model analysis and projections are available and the local scale at which data are required for planning purposes (Maul 1993; Walsh 1998). For example, Lal, Harasawa and Takahashi (2002) examined the response of the climate of SIDS to increases in greenhouse gases using seven different A-OGCMs. The finest resolution of these seven models is the 1° × 1° (latitude and longitude) cell for the National Center Atmospheric Research (NCAR) OGCM, while the remaining six models have a resolution ranging from 1.8° × 1.8° to 4.5° × 7.5° (latitude and longitude) grid cells. A grid cell from one of the finer resolution models, 1.8° × 1.8° latitude and longitude, can contain portions of Guadaloupe, Dominica, Martinique and St Lucia, indicating the difficulty in providing a climate change scenario for an individual SIDS. Consequently, Lal, Harasawa and Takahashi (2002) interpolate model results to a 1° × 1° (latitude and longitude) cell using a third order polynomial scheme. Such an interpolation is justified and represents an accepted methodology in climate modelling, but it does not ensure the ability of the model to accurately represent island-scale physical processes or intra-regional climatic variability. Further, Singh (1997) points out that most current GCMs characterize the wider Caribbean as entirely ocean-covered. Such parameterization is problematic, since regional precipitation is strongly modulated by the local orography, and such island scale processes considerably enhance the uncertainty of future regional precipitation projections by A-OGCMs (Wigley and Santer 1993; Lal, Harasawa and Takahashi 2002).

Adding to the difficulty of downscaling A-OGCM predictions to specific Caribbean locations is a lack of data and knowledge of local-scale physical processes on Caribbean islands and Caribbean intra-regional variability. The lack of knowledge creates a situation meaning that, even if an A-OGCM could provide projections on a finer scale, the projections would not be useful, since the island processes would not be accurately represented in the model or accurately linked to the model. Therefore, the conclusion that can be drawn from a review of the limitations of climate change is not that A-OGCM projections are unusable, rather as additional knowledge of intra-regional variability and island scale processes becomes available, it must be combined with A-OGCM projections in order to create a more accurate and appropriate climatic framework for assessment of Caribbean vulnerability.

Development of an Appropriate Climatic Framework to Assess Caribbean Vulnerability

Despite the limitations of A-OGCM predictions of climate change, it is possible to construct an appropriate climatic framework for assessment of Caribbean vulnerability by combining A-OGCM projections with analysis of island and intra-regional scale processes. Such a framework is very similar to the concept of a climate scenario, a plausible, internally consistent description of a climatic future (Lamb 1987). Such a framework or scenario should not be viewed as a prediction, rather it is a description of conditions that could potentially occur in the future. Climate scenarios can be constructed by couching current global climate change projections within assessment of past regional environmental conditions to develop a future analogue. One example of a climate change scenario for the Caribbean is Gallegos et al.'s (1993) assessment of changes in Caribbean ocean circulation in a future warming environment.

To illustrate such an appropriate climatic framework or scenario with current climate change projections, a recent spatial analysis of the Caribbean mid-summer drought (MSD) from 1960 to 1995 (Gamble, Parnell and Curtis 2008) will be discussed and then combined with A-OGCM projections to develop an appropriate climatic framework for assessment of Caribbean vulnerability. One reason for choosing the MSD for the illustration is its applicability across the region. The MSD is pervasive across the Caribbean region (as shown in figure 2.1) with all locations experiencing some form of drought during the summer. The other reason is that it has been argued that the impacts of drought are "more insidious and persistent" and "understated" within the region as compared to the other pervasive meteorological hazard, the hurricane (Watts 1995, 6). Thus, this climatic framework offers insight into an often overlooked meteorological hazard in the Caribbean.

The Caribbean MSD is a mid-summer rainfall minimum (July to August) between the two rainfall maxima (May to June and September to October) in the annual bimodal distribution of rainfall across the region (Magaña, Amador and Medina 1999; Chen and Taylor 2002; Ashby, Taylor and Chen 2005; Gamble, Parnell and Curtis 2008). It is important to note that this "drought" is not characterized by an absence of rainfall during this period, rather the MSD is a decrease in rainfall over the summer months as compared to the spring and late summer and fall. Thus, in order to identify the MSD, consecutive months must be compared and a decrease in rainfall from month to month indicates "drought conditions". The most widely accepted theory as to the cause of the Caribbean MSD is the intensification and expansion of the North Atlantic Subtropical High Pressure Cell (NAHP) into the region in July (Hastenrath 1976, 1978, 1984; Granger 1985; Knaff 1997; Giannini, Kushnir and Cane 2000; Curtis and Gamble 2008). According to this theory, the intensification and expansion of the NAHP translates into stronger trade winds, cooler sea surface temperatures (SSTs), increased subsidence and diminished Caribbean rainfall.

Review of current climate change projections suggests that, in a warming environment, the Caribbean MSD will become more severe. The first reason is a projected increase in temperature, particularly maximum summer temperatures, that will lead to a decrease in water availability concurrent with the Caribbean MSD (Nurse and Sem 2001). Further, at least two studies (ibid.; Lal, Harasawa and Takahashi. 2002) call for a decrease in summer rainfall (June to August). Accordingly, through a combination of the projected decrease in water availability and decrease in summer rainfall, drought conditions are expected to be more severe in the future. However, such a projection assumes that the Caribbean MSD is consistent, temporally and spatially, across the region.

Such an assumption is not appropriate since multiple authors have noted spatial variability in the Caribbean MSD. Granger (1985) indicates that the MSD is more discernable in the more westerly islands of

Table 2.3 List of Caribbean Weather Stations with 85 Per Cent of Monthly Precipitation Records, 1960–1995, and Used in Analysis

Station Name	Latitude (North)	Longitude (West)
Raizet, Guadeloupe	16.00	–61.43
Belize/Int.Airport, Belize	17.50	–88.30
Cabo San Antonio, Cuba	21.87	–84.95
Camaguey, Cuba	21.57	–77.85
Casablanca, Cuba	23.17	–82.35
Coro, Venezuela	11.40	–69.70
Grantley/Seawell, Barabados	13.10	–59.50
Guantanamo Bay, Cuba	19.90	–75.15
Juliana Aero, Dutch Antilles	18.05	–63.12
Key West Wso Ap, USA	24.55	–81.75
Kingston/Manley I, Jamaica	17.90	–76.80
Maracaibo–La Chin, Venezuela	10.70	–71.60
Miami Flo, USA	25.80	–80.30
Nassau Int.Airp, Bahamas	25.10	–77.40
Punta Maisi, Cuba	20.30	–74.20
Roosevelt Roads, Puerto Rico	18.25	–65.63
San Juan/Int.Ai, Puerto Rico	18.50	–66.00
Santo Domingo, Dominican Republic	18.40	–69.90
Sesquicentena, Colombia	12.40	–81.40
Simon Bólivar, Colombia	11.10	–74.10

the Caribbean. Giannini, Kushnir and Cane (2000) support Granger's claim by finding that the bimodal pattern of annual rainfall, centred around a MSD, accounts for 66 per cent of the total variance of the annual cycle of Caribbean rainfall and is most pronounced in the western Caribbean. However, despite these observations, a rigorous analysis of the intra-regional variability of the Caribbean MSD has not been available. Such an analysis is essential in developing an appropriate framework for assessment of the impacts of future climate change on the Caribbean MSD.

In order to determine the spatial variability in the onset and duration of the Caribbean MSD, a mapping analysis of loadings from a Principal Components Analysis (PCA) of monthly Caribbean rainfall data was recently completed (Gamble, Parnell and Curtis 2008). Data included in this analysis were chosen from Caribbean weather stations in the Global Historical Climate Network Version 2 (as available from the NOAA NCDC website). Twenty locations were chosen for analysis, each with 85 per cent of monthly precipitation records from 1960 to 1995, a bimodal distribution of annual precipitation and location in an expanded definition of the Caribbean Basin (5°–25°N, 50°–90°W) (table 2.3).

Raw data were manipulated to focus on the MSD season and standardized in regard to magnitude. Specifically, for each station, only monthly rainfall from March through October was retained for the PCA. The reason for use of March through October is that it allows for a conservative temporal definition of the MSD. The definition of the MSD season was extended two months earlier due to observations during initial review for the bimodal distribution that the first maximum in rainfall sometimes occurs in April. March was included due to the need for an extra month to calculate monthly differences in rainfall.

To standardize for magnitude across the region, rainfall was expressed as a proportional difference. This proportional difference was calculated by subtracting the current month (m) from current month plus one (m+1) and dividing the difference by the current month (m). For example, the proportional difference for March through April was determined by subtracting March precipitation from April precipitation and dividing the difference by total precipitation for March. A positive difference represents an increase from one month to the next and a negative difference represents a decrease from one month to the next.

The monthly proportional differences in precipitation for the twenty stations were then used as a raw data matrix in a PCA. The resultant PCA factor pattern was rotated orthogonally through a varimax rotation. The loading of each weather station on the rotated components was mapped to determine regions of the Caribbean MSD. Regions of similar MSD were defined as areas that encompass all weather stations with a 0.5 or greater loading on a specific component (Henderson and Vega 1996).

Table 2.4 Factor Loadings for Each of the Twenty Weather Stations included in the Caribbean Monthly Difference in Precipitation Principle Component Analysis

Station Name	Factor 1	Factor 2	Factor 3	Factor 4	Factor 5
Raizet, Guadeloupe	–0.054	**0.937**	0.125	–0.050	0.262
Belize/Int.Airport, Belize	**0.834**	–0.347	–0.058	0.053	0.097
Cabo San Antonio, Cuba	**0.834**	0.312	–0.035	0.048	0.385
Camaguey, Cuba*	**0.655**	0.262	0.086	0.008	**0.701**
Casablanca, Cuba	**0.919**	–0.212	0.217	0.236	0.056
Coro, Venezuela	–0.123	0.173	0.077	**0.962**	0.150
Grantley/Seawell, Barabados	0.032	–0.230	**–0.940**	0.016	–0.244
Guantanamo Bay, Cuba	0.157	–0.051	0.290	0.092	**0.933**
Juliana Aero, Dutch Antilles	–0.134	**0.935**	0.246	–0.209	0.052
Key West Wso Ap, USA	**0.946**	0.178	0.056	–0.229	0.106
Kingston/Manley I, Jamaica*	**0.536**	0.316	0.249	–0.280	**0.577**
Maracaibo–La Chin, Venezuela	0.032	**0.804**	–0.092	0.373	–0.239
Miami Flo, USA	**0.931**	0.161	0.024	–0.145	0.220
Nassau Int.Airp, Bahamas	**0.882**	–0.031	–0.216	–0.097	0.392
Punta Maisi, Cuba*	0.056	–0.126	**0.630**	0.322	**0.614**
Roosevelt Roads, Puerto Rico	0.356	0.295	0.253	–0.039	**0.847**
San Juan/Int.Ai, Puerto Rico	0.181	**0.850**	0.024	0.266	0.416
Santo Domingo, Dominican Republic	0.397	0.259	–0.045	0.157	**0.843**
Sesquicentena, Colombia*	**0.768**	–0.006	–0.164	0.287	**0.541**
Simon Bólivar, Colombia	0.370	**0.761**	–0.113	0.484	0.063
Per cent variance explained by component	33.0	22.2	8.7	9.3	22.8

Note: The per cent variance explained by each factor is listed in the last row. Shaded cells represent multi-station regions, numbers in bold represent loadings over 0.5, and location with asterisk represents transition stations.

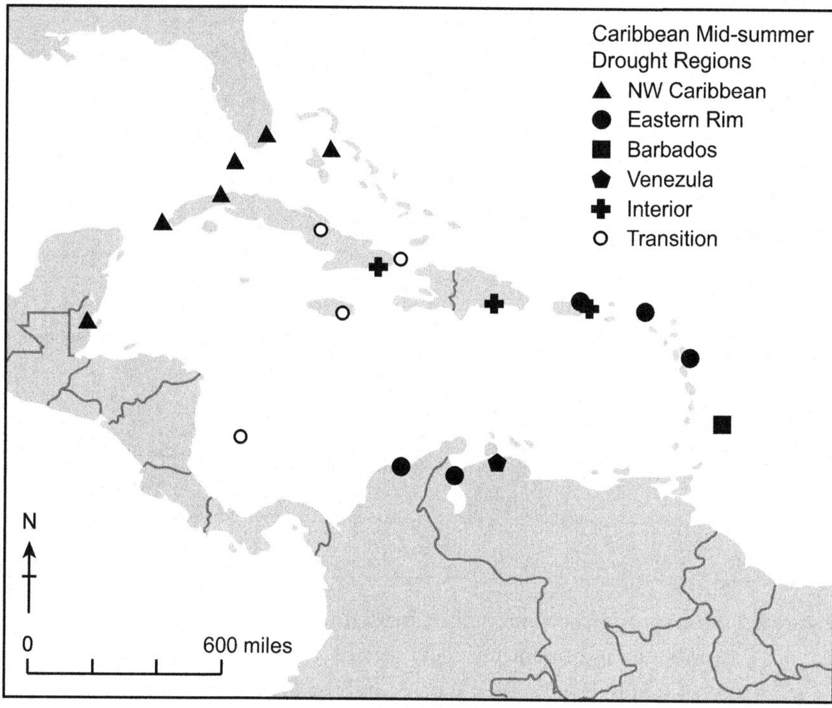

Figure 2.1 Map of Caribbean mid-summer drought regions as determined by principal component analysis.

Upon examination of the PCA scree plot and map of component loadings, a total of five MSD regions were identified in the Caribbean that explained 96 per cent of the total variance in proportional difference of monthly MSD season precipitation for the twenty weather stations (table 2.4). All weather station locations had component loadings over 0.5 and were therefore placed within a MSD region (table 2.4). The five factors identified three MSD regions with multiple stations: the northwestern Caribbean, the interior Caribbean, and the eastern rim Caribbean; and two regions with single stations, Coro, Venezuela and Grantley, Barbados (table 2.4, figure 2.1). The PCA also identified one transition zone between MSD regions. The transition zone has stations with loadings of greater than 0.5 on multiple factors and thus cannot be assigned to an individual region. The transition zone is located in the central portion of the study area and has stations with high loadings on northwest, interior and eastern rim region components (table 2.4).

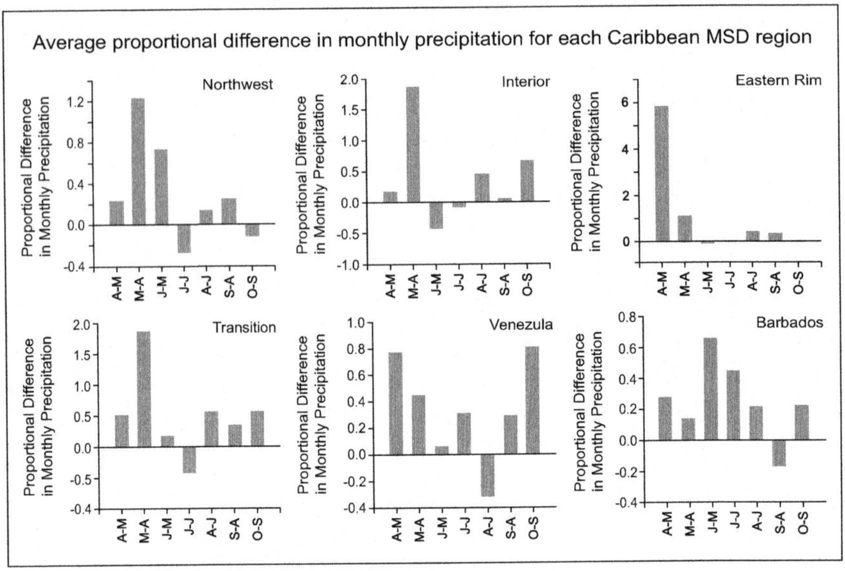

Figure 2.2 Mean proportional monthly rainfall difference (1960–1995) for Caribbean mid-summer drought regions. (A–M = April–March, M–A = May–April, J–M = June–May, J–J = July–June, A–J = August–June, S–A = September–August, O–S = October–September).

Review of the mean proportional differences in monthly precipitation for the MSD season indicates that each of the regions has a different month of MSD occurrence or a different MSD duration (figure 2.2). For the northwestern Caribbean region, the MSD separating the two precipitation maxima occurs in June–July. For the transition zone, the MSD occurs June–July but has a greater magnitude than the Northwest region MSD. The Interior region MSD occurs May–June, has the greatest magnitude of all regions, and extends into June–July. The Eastern Rim region MSD also occurs in May–June and has the smallest magnitude of all the regions. For the two single station MSD regions, the MSD occurs much later in the year, July–August for Venezuela and August–September for Barbados. The magnitude of the MSD in these two single station regions is of moderate magnitude as compared to the multiple station regions.

A combination of these PCA results with A-OGCM projections for the region provides an assessment of climate change impact on the

Caribbean MSD that includes intra-regional spatial variability. The combination of the projections and analysis indicates that for the northwest MSD region, the reduction in precipitation may cause the June to July decrease in precipitation to become greater. Since there is a marginal increase in precipitation from July to August, the decrease in precipitation may continue into this month, the end result is an increase in the magnitude and duration of the MSD in this region (figure 2.2). For the interior region, since the increase in precipitation before and after the MSD is a greater magnitude, it is most likely that the warming and decrease in precipitation would only increase the magnitude of the MSD and not affect the duration of the drought (figure 2.2). For the eastern rim region, the decrease in precipitation will cause the initial June to July decrease in precipitation to increase in magnitude, and given the marginal increase in precipitation July to August and August to September, a decrease in precipitation may cause a negative monthly difference in precipitation, extending the length of the MSD (figure 2.2). For Coro, Venezuela, there is potential for the magnitude of the MSD to be more severe, since the MSD in these locations occurs from July to August. However, since the increase in precipitation is very small in Coro from May to June, the MSD may also begin to occur earlier in the year (figure 2.2). Grantley, Barbados may potentially experience a minimal impact from climate change on the MSD, since the MSD occurs August to September, a month later than the seasonal forecast for a decrease in precipitation (figure 2.2).

Conclusion

Review of existing literature indicates air temperature in the Caribbean has been increasing in the twentieth century, particularly since the 1950s, yet this trend is not consistent across the region. In terms of precipitation, a variety of observed trends are reported in the literature, making it difficult to make one broad statement about trends in Caribbean precipitation. Despite these differences, multiple studies do indicate a decrease in precipitation for the central Caribbean. Climate change projections indicate that the Caribbean is expected to experience moderate warming in the future with an increase in maximum summer temperatures. Projections for future Caribbean precipitation suggest a slight decrease in both summer (June to August) and annual precipitation, but these trends are not considered significant, given the high natural variability of precipitation in the region and difference in model projections. The greatest limitation in the use of the A-OGCM projections is the coarse resolution of the models.

In order to develop an appropriate climatic framework that diminishes these limitations, a coupling of intra-Caribbean spatial patterns in precipitation and A-OGCM projections has been developed here to assess the Caribbean mid-summer drought in a future warm climate. This analysis indicates that subtle differences can be expected in future Caribbean MSD occurrence, increased mid-summer drought magnitude and duration in the northwest Caribbean, and increased magnitude in the interior Caribbean. The analysis also indicates an increase in magnitude and duration for the eastern rim Caribbean, an increase in magnitude and earlier occurrence in Coro, Venezuela, and potentially minimal impact in Barbados. Such a technique of combining knowledge of island and intra-regional scale processes with A-OGCM projections offers one option in overcoming the limits of existing A-OCGM projections and moving toward an appropriate climatic framework to assess Caribbean vulnerability.

References

Ashby, S.A., M.A. Taylor, and A.A. Chen. 2005. Statistical models for predicting rainfall in the Caribbean. *Theoretical and Applied Climatology* 82: 65–80.

Anthes, R.A., R.W. Corell, G. Holland, J.W. Hurrel, M.C. MacCracken and K.E. Trenberth. 2006. Hurricanes and global warming: Potential links and consequences. *Bulletin of the American Meteorological Society* 87 (5): 623–28.

Aparicio, R. 1993. Meteorological and oceanographic conditions along the southern coastal boundary of the Caribbean Sea, 1951–1986. In *Climatic change in the Intra-Americas Sea: Implications for future climate on the ecosystems and socio-economic structure in the marine and coastal regions of the Caribbean Sea, Gulf of Mexico, Bahamas, and the northeast coast of South America*, ed. G.A. Maul, 100–114. London: Edward Arnold.

Aubrey, D.G., K.O. Emery and E. Uchupi. 1988. Changing coastal levels of South America and the Caribbean from tide-gauge records. *Tectonophsyics* 154: 269–84.

Chen, A.A., and M.A. Taylor. 2002. Investigating the link between early season Caribbean rainfall and the El Niño + 1 year. *International Journal of Climatology* 22: 87–106.

Church, J.A., N.J. White, R. Coleman, K. Lambeck and J.X. Mitrovica. 2004. Estimates of the regional distribution of sea level rise over the 1950–2000 period. *Journal of Climate* 17: 2609–25.

Curtis, S., and D.W. Gamble. 2008. Regional variations of the Caribbean mid-summer drought. *Theoretical and Applied Climatology* 94 (1–2): 25–34.

Emmanuel, K.A. 2005. Increasing destructiveness of tropical cyclones over the past thirty years. *Nature* 436: 686–88.

Gallegos, A., S. Czitrom, J. Zavala and A. Fernandez. 1993. Scenario modelling of climate changes on the ocean circulation of the Intra-Americas sea. In Maul 1993, 55–76.

Gamble, D.W., D.B. Parnell, and S. Curtis. 2008. Spatial variability of the Caribbean mid-summer drought and relation to the North Atlantic High. *International Journal of Climatology* 28 (3): 343–50.

Giannini, A., Y. Kushnir and M.A. Cane. 2000. Interannual variability of Caribbean rainfall, ENSO, and the Atlantic Ocean. *Journal of Climate* 13: 297–311.

Granger, O.E. 1985. Caribbean climates. *Progress in Physical Geography* 9 (1): 16–43.

———. 1995. Climate change, environmental degradation and eco–refugees: Crises on nonsustainable development in the Greater Caribbean region. In *Climate Institute 1995 Area studies: Environmental degradation and eco–refugees*. Washington, DC: Climate Institute.

———. 1997. Caribbean island states: Perils and prospects in a changing global environment. Special issue, *Journal of Coastal Research* 24: 71–93.

Gray, C. 1993. Regional meteorology and hurricanes. In Maul 1993, 87–99.

Hanson, J., and G.A. Maul. 1993. Analysis of temperature, precipitation and sea level variability with concentration on Key West, Florida for evidence of trace-gas-induced climate change. In Maul 1993, 193–214.

Hastenrath, S. 1976. Variations in low-latitude circulation and extreme climatic events in the tropical Americas. *Journal of Atmospheric Science* 33: 202–15.

———. 1978. On the modes of tropical circulation and climate anomalies. *Journal of Atmospheric Science* 35: 2220–31.

———. 1984. Interannual variability and annual cycle: Mechanisms of circulation and climate in the tropical Atlantic sector. *Monthly Weather Review* 112: 1097–107.

Henderson, K.G., and A.J. Vega. 1996. Regional precipitation variability in the southern United States. *Physical Geography* 17 (2): 93–112.

Hendry, M. 1993. Sea-level movements and shoreline change. In Maul 1993, 152–61.

Holland, G.T. 1997. The maximum potential intensity of tropical cyclones. *Journal of Atmospheric Science* 54: 2519–41.

Houghton, J.T., Y. Ding, D.J. Griggs, M. Noguer, P.J. van der Linden, and D. Xiaosu. 2001. *Climate change 2001: The scientific basis. Contribution of working group I to the third assessment report of the Intergovernmental Panel on Climate Change (IPCC).* Cambridge: Cambridge University Press.

Knaff, J.A. 1997. Implications of summertime sea level pressure anomalies in the tropical Atlantic region. *Journal of Climate* 10: 789–804.

Lal, M., H. Harasawa and K. Takahashi. 2002. Future climate change and its impacts over small island states. *Climate Research* 19: 179–92.

Lamb, P.J. 1987. On the development of regional climatic scenarios for policy-oriented climatic-impact assessment. *Bulletin of American Meteorological Society* 68 (9): 1116–23.

Leatherman, S.P., and N. Beller-Simms. 1997. Sea-level rise and small island states: An overview. Special issue, *Journal of Coastal Research* (24): 1–16.

Magaña, V., J.A. Amador and S. Medina. 1999. The midsummer drought over Mexico and Central America. *Journal of Climate* 12: 1577–88.

Martin, H.C., and P.S. Weech. 2001. Climate change in the Bahamas? Evidence in the meteorological records. *Bahamas Journal of Science* 5: 22–32.

Maul, G.A. 1993. Implications of the future climate on the ecosystems and socio-economic structure in the marine and coastal regions of the intra-Americas Sea. In Maul 1993, 3–28.

———, ed. 1996. *Coastal and estuarine studies fifty-one small islands: Marine science and sustainable development.* Washington, DC: American Geophysical Union.

Mimura, N., L. Nurse, R.F. McLean, J. Agard, L. Brigulio, P. Lefale, R. Payet and G. Sem. 2007. Small islands. In *Climate change 2007: Impacts, adaptation and vulnerability. Contribution of the working group II to the fourth assessment report of the Intergovernmental Panel on Climate Change*, ed. M.L. Parry, O.F. Canziani, J.P. Palutikof, P.J. van der Linden, and C.E. Hanson, 687–716. Cambridge: Cambridge University Press.

Nurse, L.A., and G. Sem. 2001. Small island states. In *Climate change 2001: The scientific basis. Contribution of working group I to the third assessment report of the Intergovernmental Panel on Climate Change (IPCC)*, ed. J.T. Houghton, Y. Ding, D.J. Griggs, M. Noguer, P.J. van der Linden and D. Xiaosu, 843–75. Cambridge: Cambridge University Press. Available at http://www.ipcc-wg2.org/

Pelling, M., and J.I. Uitto. 2001. Small island developing states: Natural disaster vulnerability and global change. *Environmental Hazards* 3: 49–62.

Peterson, T.C., M.A. Taylor, R. Demeritte, D.L. Duncombe, S. Burton, F. Thompson, A. Porter, M. Mercedes, E. Villegas, S.R. Fils, A. Klein Tank, A. Martis, R. Warner, A. Joyette, W. Mills, L. Alexander and B. Gleason. 2002. Recent changes in climate extremes in the Caribbean region. *Journal of Geophysical Research* 107 (D21), 4601: 16–1 to 16–9.

Pielke, R.A. Jr, C. Landsea, M. Mayfield, J. Laver and R. Pasch. 2006. Reply to "Hurricanes and global warming – potential linkages and consequences". *Bulletin of American Meteorological Society* 87: 628–31.

Pirazolli, P. 1986. Secular trends of relative-sea level changes indicated by tide-gauge records. Special issue, *Journal of Coastal Research* (1): 1–26.

Reading, A.J. 1990. Caribbean tropical storm activity over the past four centuries. *International Journal of Climatology* 10: 365–76.

Singh, B. 1997. Climate changes in the greater and southern Caribbean. *International Journal of Climate* 17: 1093–114.

Solomon, S., D. Qin, M. Manning, Z. Chen, M. Marquis, K.B. Averyt, M. Tignor and H.L. Miller. 2007. Summary for policymakers. In *Climate change 2007: The physical science basis. Contributions of working group 1 to the fourth assessment report of the Intergovernmental Panel on Climate Change (IPCC)*. Cambridge: Cambridge University Press. Available at http://www.ipcc-wg2.org/

Taylor, M.A., D.B. Enfield and A.A. Chen. 2002. The influence of the tropical Atlantic vs. the tropical Pacific on Caribbean rainfall. *Journal of Geophysical Research* 107 (C9): 3127.

Tonkin, H., C. Landsea, G.J. Holland and S. Li. 1997. Tropical cyclones and climate change: A preliminary assessment. In *Assessing climate change results from the Model Consortium for Climate Assessment*, ed. W. Howe and A. Henderson-Sellers, 327–60. Sydney: Gordon and Breach.

Trataglione, C.A., S.R. Smith and J.J. O'Brien. 2003. ENSO impact on hurricane landfall probabilities for the Caribbean. *Journal of Climate* 16: 2925–31.

Walsh, K. 2004. Tropical cyclones and climate change: Unresolved issues. *Climate Research* 27: 77–83.

Walsh, R.P.D. 1998. Climatic changes in the eastern Caribbean over the past 150 years and some implications in planning sustainable development. In *Resource sustainability and Caribbean development*, ed. D.F.M. McGregor, D. Barker and S. Lloyd Evans, 51–68. Kingston: The Press, University of the West Indies.

Watts. D. 1995. Environmental degradation, the water resource and sustainable development in the eastern Caribbean. *Caribbean Geography* 6 (1): 2–15.

Webster, P.J., G.J. Holland, J.A. Curry and H.R. Chang. 2005. Changes in tropical cyclone number, duration, and intensity in a warming environment. *Science* 209: 1844–46.

Wigley, T.M.L., and B.D. Santer. 1993. Future climate of the Gulf/Caribbean Basin from the global circulation models. In Maul 1993, 31–44.

Woodworth, P.L. 1993. Sea level changes. In *Climate and sea level change: Observations, projections and implications*, ed. R.A. Warrick, E.M. Barrow and T.M.L. Wigley, 379–91. Cambridge: Cambridge University Press.

PART 2

Managing Vulnerable Environments

CHAPTER 3

Hurricanes or Tsunami?

Comparative Analysis of Extensive Boulder Arrays Along the Southwest and North Coasts of Jamaica: Lessons for Coastal Management

DEBORAH-ANN ROWE, SHAKIRA KHAN AND EDWARD ROBINSON

Introduction

Over the past four years the Marine Geology Unit at the University of the West Indies, Mona, has been evaluating natural hazards of the Jamaican coastline in selected coastal communities. Although the evaluation has concentrated on the effect of tropical storms and hurricanes on beach changes, particularly beach erosion, attention has also been paid to those coasts without beaches, where the shoreline is formed by cliffs cut in Quaternary limestone terraces. Common features of such rocky coasts are the presence of more or less large quantities of sedimentary debris deposited on the terrace surfaces, including boulders exceeding 80 tonnes in mass (Robinson, Rowe and Khan 2005, 2006) and the existence of a shore-parallel debris ridge at some distance from the coastline. Such deposits have been noted in areas where housing and hotel development is taking place along the shoreline. The presence of such debris on many Caribbean islands is indicative of past giant wave events, whether from storms/hurricanes (Scheffers 2005; Khan et al. 2008), tsunami (Camfield 1994, especially his figure 7; Scheffers 2002; Scheffers and Kelletat 2003) or, perhaps, even "rogue" or "freak" waves that can develop through interaction among ocean wave trains and have been observed off the east coast of Jamaica (Cdr. John McFarlane, Jamaica Coastguard, pers. comm.; Bruun 1994).

The practical ramifications of the ongoing controversy between those who invoke tsunami events as the generators of most boulder-sized coastal debris and those who suggest that most, if not all, coastal boulders result from hurricane events are important. Whereas the return periods for damaging wave events associated with Caribbean tropical storms and hurricanes are measurable in decades, the return periods for tsunamis, resulting from earthquakes and associated submarine landslides, are much longer, presently unknown for Jamaica and may not be measurable. Meaningful return periods for significant tsunami occurrences have also not yet been established for the Caribbean despite the efforts of researchers such as Scheffers (2002, 2005) in dating ancient boulder and debris assemblages. In assessing risk to coastal structures; for example, for insurance purposes, it is most desirable to be able to distinguish between the two kinds of event. If the observed boulders were indeed all emplaced by hurricanes, then the risks for those zones with boulders are much higher than would be the case for tsunami emplacement.

Our two main objectives in assessing these deposits are as follows: first, to determine to what extent they result from storms and hurricanes, rather than tsunami or other comparatively rare events, and second, to emphasize the value of using the deposits as a guide to determining setbacks in coastal construction and management.

Here we evaluate the size range of what we term boulders (Blair and McPherson 1999), elsewhere referred to as megaclasts (Noormets, Crook and Felton 2004), which appear to have been emplaced and/or moved mainly during two recent severe hurricanes, Allen (1980) and Ivan (2004) (see figure 3.1). The two hurricanes generated significant storm surges along the north and south coasts of Jamaica (figure 3.2). The last well-documented tsunami event coincided with the 1907 earthquake and affected the Jamaican coastline between Port Antonio and St Ann's Bay (end map in Cornish 1908) (figure 3.2).

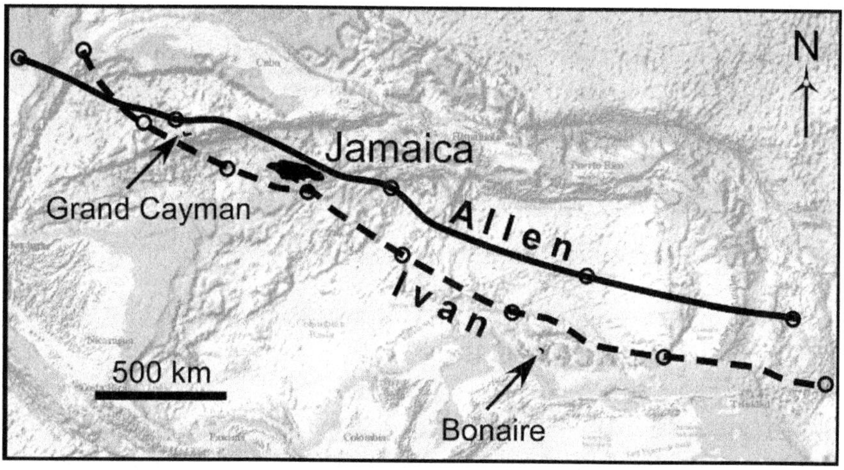

Figure 3.1 Tracks of hurricanes Allen (1980) and Ivan (2004) across the Caribbean.

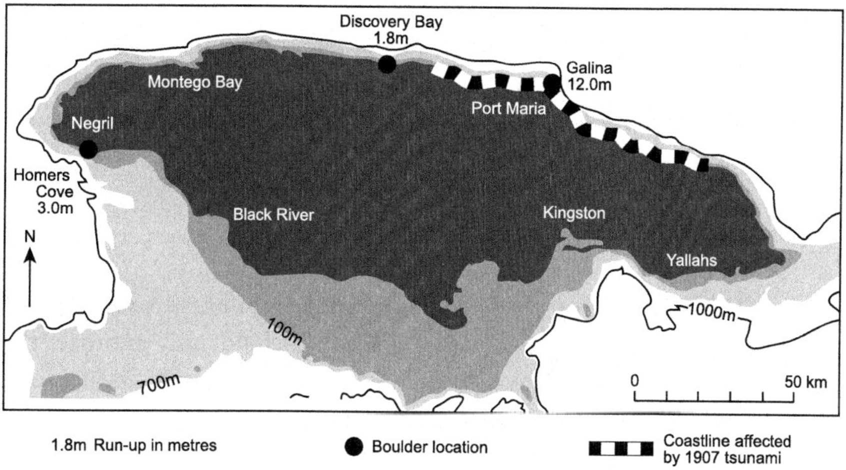

Figure 3.2 Jamaica, showing general features of bathymetry, recorded storm surge heights for Hurricane Allen (1980) for Discovery Bay and Galina (Wilmott-Simpson 1980), and surge plus run-up estimates measured at Homers Cove for Hurricane Ivan (2004). Extent of the 1907 tsunami is indicated.

Mapping of surface, as well as buried, wave-emplaced coastal deposits serves to provide concrete historical evidence of the extent and significance of wave incursion onto the coastline. It therefore provides an objective basis for guiding both present and future development and management of the coastal zone.

In the following account the term "debris" is used to include all stranded material ranging from sand to boulder size.

Localities

This investigation focused on three sites; two on the north coast and one on the south coast of Jamaica (figure 3.2).

North Coast

On the north coast one of the two localities is at Galina, in the parish of St Mary, where there is a well-documented record of damage from Hurricane Allen (Wilmot-Simpson 1980) and where the Marine Geology Unit carried out detailed mapping of boulder strews and debris ridges (Robinson, Rowe and Khan 2006). The other extends from the harbour and bauxite loading terminal of Discovery Bay eastwards for about 6 kilometres (figure 3.2).

South Coast

The south coast locality chosen for this study is situated south east of the Negril Lighthouse in the Parish of Westmoreland, and embraces the community of Homers Cove, also called Brighton (figure 3.2). Along this coastline there are extensive arrays of boulders, and the passage of Hurricanes Ivan (2004), Dennis and Wilma (2005), and Dean (2007) all resulted in debris emplacement and movement on shore.

Methodology

The methodology used was modified from that of previous investigations by the unit (Robinson, Rowe and Khan 2006). The a, b and c axes of boulders were measured and a volume estimate applied by multiplying the a × b × c volume by a correction factor of 0.5. The correction factor is based on the mean of several measurements of irregular blocks of which the volumes were obtained by immersion in a graduated tank of water, and by analysis of oriented photographs of *in situ* boulders (Khan et al. 2008). A relative density of 2.5 was used to estimate the mass, following Scheffers (2002), although the tank experiments suggest that a somewhat lower value be used (Khan et al. 2008). The position of each boulder was determined using a hand-held, WAAS-enabled global positioning system (GPS). The results were analysed for size distribution, distance from shore, and whether or not they formed part of a boulder or debris ridge. The occurrence of accumulations of coral debris, frequently imbricated, was also noted. Anecdotal evidence was acquired through interviews with residents to determine emplacement and/or movement of boulders by recent storms. The current positions of these transported boulders were marked using a hand held GPS unit. In some cases, estimates of the distances moved were calculated based on information provided by local residents. The Marine Geology Unit undertook bathymetric surveys of the island shelf adjoining Homers Cove and Galina in support of the field investigations. These were carried out using a hand-held GPS and hand-held depth sounder, the measurements being conducted from rented canoes.

Results

Galina

The limestone platform at Galina (figure 3.3) is part of a more or less continuous system of late Pleistocene terraces extending along the north coast of Jamaica (Cant 1972), dated from 120,000 to 140,000 years before present (BP) (Robinson, Rowe and Khan 2006). As well, for the most part, the platform rocks probably result from deposition during high sea levels attained in oxygen isotope stage 5e. The platform forms a low cliff ranging from 1 metre to 10 metres above sea level. The terrace is cut by joints and minor faults, of which the dominant trend is 060°, clearly seen on aerial photographs of the region. The surface of the platform is rugged, consisting of jagged, decimetric-scale phytokarst, barren of vegetation near the coastline,

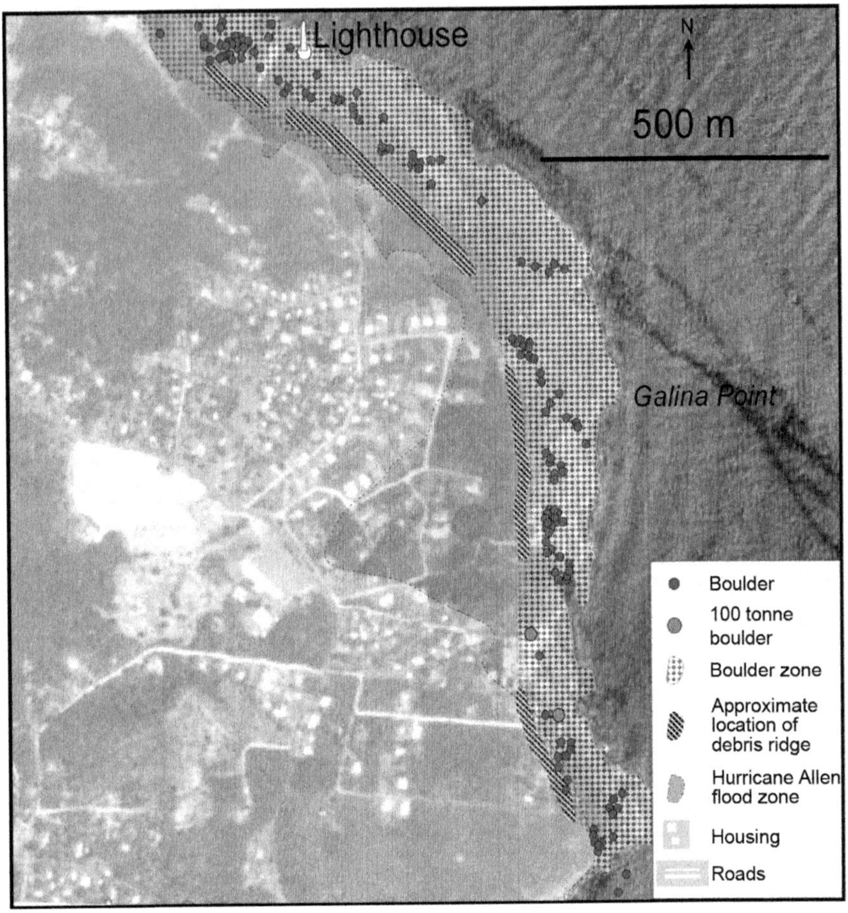

Figure 3.3 Location of boulders and debris ridge at Galina, St Mary.

grading landwards into more subdued karst. Remnants of a red, lithified palaeosol occupy depressions and joint planes in the karst surface. This is interpreted here to have been formed as a *terra rossa* soil during the last glacial period and low sea level stand, followed subsequently by lithification, and partial to complete removal while sea level was still below present levels. The depressions are partly filled by unconsolidated sandy carbonate sediment, evidently deposited onshore during recent storms.

The size distribution of the 180 mapped boulders is plotted against distance from the shoreline on figure 3.4, with boulders ranging in mass

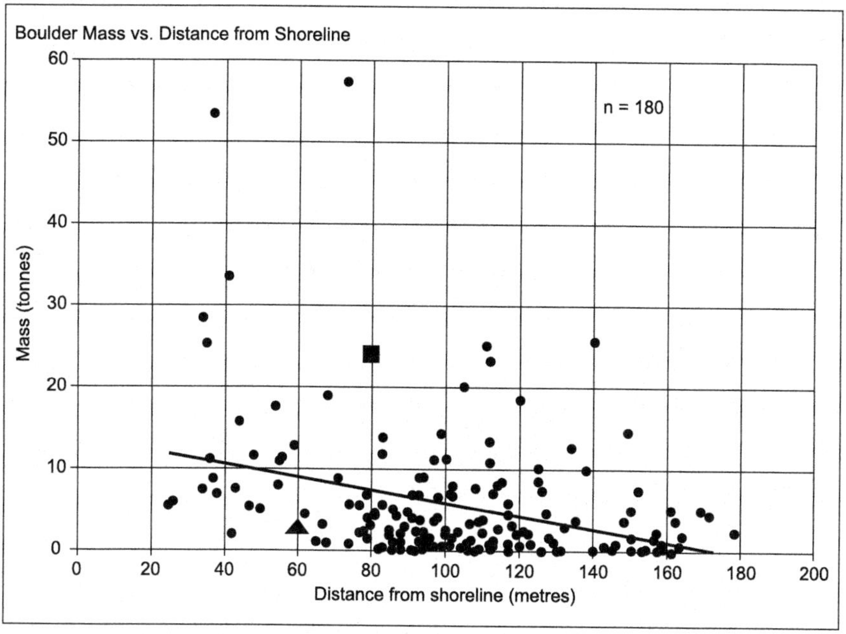

Figure 3.4 Mass vs. distance from shoreline for boulders at Galina. Square, boulder emplacement attributed to the 1907 tsunami; triangle, boulder emplaced during Hurricane Allen, 1980. Trend line indicates reduction in mean mass with increasing distance from shore.

from tens of kilograms to as much as 60 tonnes. Many of the larger boulders have vegetation growing on their upper surfaces both here and at the other sites examined (figure 3.5). A debris ridge, overlying the red paleosol, occurs to the rear of the boulder strew and effectively separates the bare limestone platform from the vegetated areas to landward. A more detailed description of the Galina area is included in Robinson, Rowe and Khan (2006).

Discovery Bay

The surveys at Discovery Bay (figure 3.6) are of a preliminary nature and only a handful of boulder measurements have been made. The four main morphological elements typical of rocky shorelines impacted by giant waves are well seen (figure 3.7). A barren zone at the shoreline is frequently swept by storm waves, inhibiting soil formation and

Figure 3.5 Large boulder with vegetation growing on it, Discovery Bay, St Ann.

vegetation growth. Storms frequently emplace imbricated coral rubble arrays on the rock platform as perched beaches (*sensu* Hopke, Gmirkin and Richmond 2005). Less frequently, intense storms, and perhaps tsunami, generate boulders torn from the rocky cliff face and transported ashore. At the rear of the platform, a prominent debris ridge is typically formed. It is composed of a mixture of sand and pebbles; small, mostly rounded cobbles of coral; and boulders of Pleistocene limestone derived from the platform bedrock. This material has evidently accumulated over many years, since sea level has risen to, more or less, its present position (for example, see the sea level curve of Toscano and McIntyre 2003).

At Discovery Bay, boulders are situated at positions ranging from the shoreline to some 50 metres back from the coast on the upper Pleistocene limestone terrace (figure 3.6). Of the seven larger boulders measured, the smallest was 10 tonnes and the largest 50 tonnes. Some are much smaller, and others appear to have been moved to clear space for development. The most distinctive feature at this locality

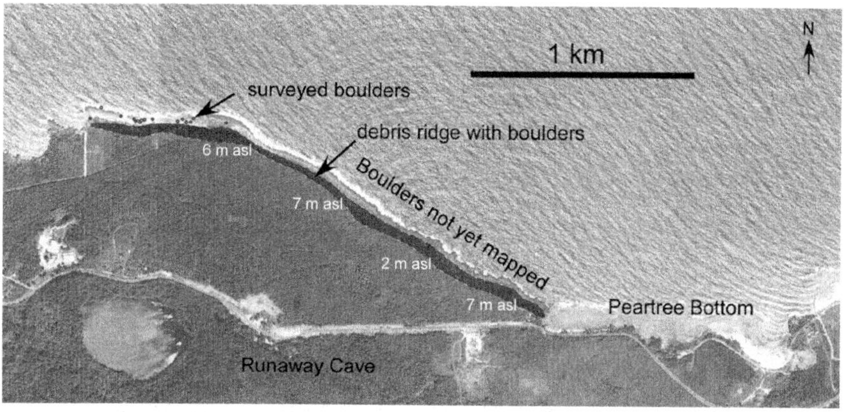

Figure 3.6 Map of boulder ridge and other isolated boulders at Discovery Bay, St Ann.

Figure 3.7 Debris ridge (right) at Discovery Bay. The four major morphological elements of rocky shorelines impacted by giant waves are all seen on this figure (annotated arrows).

is the height of the debris ridge, marking the contact between the vegetation-free platform and the densely vegetated zone behind the platform. In some places it reaches, or exceeds, 6 metres above sea level, some 4 to 5 metres above platform height (figure 3.7). Ridges this high have not yet been observed elsewhere on this stretch of coast.

Homers Cove

Homers Cove is a small but distinctive embayment in the terrace that extends from Negril lighthouse in the west, to Southwest Point (Salmon Point) in the east (figure 3.8). The settlement of Brighton has grown up around the cove. The cove is flanked by a rocky terrace, forming cliffs at the coastline, ranging from 1 to 4 metres in height. The terrace consists of coral and molluscan limestone, closely resembling the late Pleistocene limestone of the terrace along the north coast of the island, as at Galina and Discovery Bay (Cant 1972), but no dating has been undertaken. A number of houses were destroyed or badly damaged in the area by the passage of Hurricane Ivan, some from high wind and others from storm waves. The island shelf here is relatively narrow compared with other south coast locations. Our bathymetric survey indicates a width of about 300 metres (figure 3.8).Water depth at the shelf edge is 16 to 18 metres and the island slope is steep down to 300 metres.

Shoreline boulders are extremely numerous in this region (figure 3.9) and only a short stretch of the coast was surveyed, in the area immediately encompassing Homers Cove (figure 3.8). Two hundred and fifty-eight boulders were measured along the cliffs, this being a sample of many more hundreds that were noted. Boulder masses ranged from less than 1 tonne to more than 50 tonnes. Most of the boulders are within 50 metres of the cliff edge. In addition to the boulders, there are a number of low, shore-parallel, perched beaches composed of debris from imbricated pebble to cobble-sized coral, some of which were emplaced by hurricanes Emily and Wilma in 2005 (figure 3.10). Most of the coral rubble forming these beaches is *Acropora palmata*. The bare limestone platform is backed by a prominent boulder or debris ridge that, in areas without housing development, lies at or just within the edge of the forested part of the platform (figure 3.10). At Homers Cove

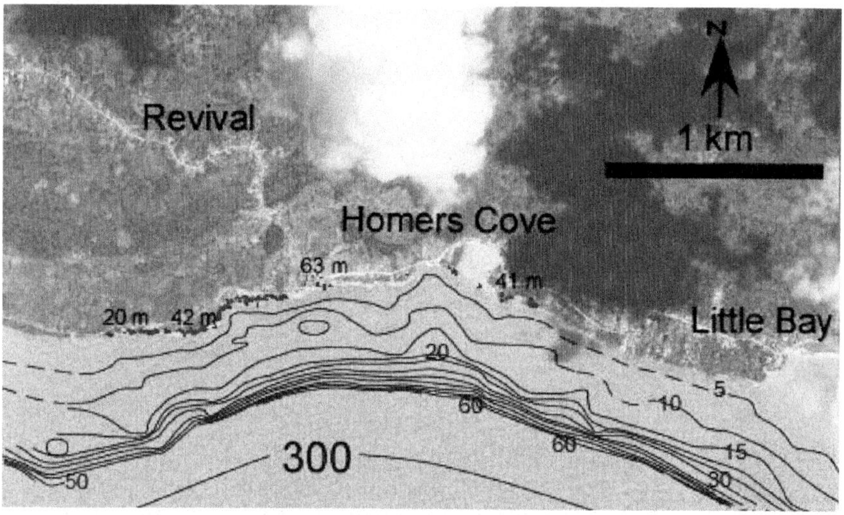

Figure 3.8 Boulder positions (small triangles) and off-shore bathymetry (metres) at Homers Cove. Figures at points along the coast indicate distances of main groups of boulders from the shoreline.

Figure 3.9 Numerous boulders emplaced by Hurricane Ivan blocking pathway to housing at Homers Cove.

Figure 3.10 Imbricate clasts in coral ridge (foreground) and boulder/debris ridge (background) at Homers Cove.

there is first-hand evidence of boulders that were freshly emplaced onshore by Hurricane Ivan and, most recently, by Hurricane Dean (August 2007).

Discussion

Movement of Shore Platform Boulders by Storm Surge and Waves

Movement of boulders by hurricane events are documented for Homers Cove and Galina, and compared with similar observations made on the Caribbean islands of Bonaire and Grand Cayman. No anecdotal reports of movement of boulders are available from Discovery Bay, although a single radiocarbon date from the coral *Diploria*, attached to a 10-tonne boulder about 1 metre above sea level, and 3 to 4 metres from the shoreline, indicates a post-1950 age of emplacement.

At Homers Cove, residents living along the shoreline reported boulder movement and deposition of debris. Roads were blocked and houses destroyed by debris and/or boulders transported across the platform during hurricanes Ivan and Wilma. Although no records of wave heights were measured for these events, residents reported storm surge overtopping the cliffs, indicative of surge heights (including breaking wave runup) of at least 3 to 4 metres in some places. Inland inundation greater than 90 metres was reported in some areas. During a preliminary visit to Homers Cove one boulder of 33 tonnes was observed to be about 8 metres from the shore and displayed evidence of fresh breakage from bedrock (Robinson, Rowe and Khan 2006). A local resident indicated that this boulder had been dragged and tossed ashore from a point in the sea some distance from the shoreline. He cited a trail of scrape marks and broken coral on the seafloor as evidence. Anecdotal evidence from a second visit included eyewitness reports of horizontal movement of 8.5 metres for a 50-tonne boulder during Hurricane Ivan. Inspection of the terrace surface between the initial and final locations of this boulder revealed some chipped pavement but no scrape marks. Assuming the observations to be correct, movement by flotation is indicated. Several other boulders were reported to have moved in the size range from less than 1 tonne to 30 tonnes. A few boulders in the range of 5 to 30 tonnes were pointed out as having been emplaced during a 1932 and/or 1933 hurricane event, and some of those (up to 11 tonnes) were subsequently moved by Ivan. One boulder of 12 tonnes was reported to have been emplaced by Hurricane Gilbert in 1988 but had evidently not moved recently as vegetation was growing, seemingly undisturbed, on its upper surface. Similar observations and accounts by residents have resulted from the passage of Hurricane Dean (19 August 2007), and these are being written up elsewhere.

At Galina, Hurricane Allen produced 12-metre-high breaking waves at the platform edge (Wilmot-Simpson 1980). Anecdotal reports indicate movement of several boulders. Eyewitnesses identified three boulders that allegedly were moved considerable distances during the passage of Allen. These were estimated to weigh 0.5, 7 and 10 tonnes respectively, moved over distances of up to 50 metres during that storm. A radiocarbon date on encrusting barnacles and vermetids on a boulder of 5 tonnes, also indicated removal from the surf zone at the cliff face and emplacement and movement inland of 60 metres during Hurricane Allen (or perhaps Hurricane Gilbert, as one person insisted), over a platform at 6 metres above sea level (Robinson, Rowe and Khan 2006). One boulder damaged a dwelling near the shore, killing a person inside.

Comparison with the Netherlands Antilles and Cayman Islands

Scheffers (2002) provided a table of published observations on boulders and other objects moved by storms in the Netherlands Antilles, concluding that boulders weighing more than 2 or 3 tonnes would not normally be moved even by the largest hurricane waves. In later publications, Scheffers (2005) and Scheffers and Scheffers (2006) provided more detailed observations for Bonaire following the passage of Hurricane Ivan (figure 3.1). This category four storm created exceptional, 12-metre-high waves off the northeast coast of Bonaire as it passed 140 kilometres to the north. Water fountains rose as much as 30 metres high from breaking waves. The run-up/inundation over the 6-metre-high limestone terrace was 1.5 metres deep and several hundred metres inland. Fragments up to 50 centimetres were commonplace on the terrace, pushed against trees and into the vegetation. In a few cases only was there evidence for transport of boulders from the cliff edge inland. A 6-tonne boulder was moved 30 metres inland on a platform 6 metres above sea level. Although Ivan produced some of the largest waves ever recorded for Bonaire, actual movement of large boulders was limited. One ancient boulder of 50 tonnes, probably emplaced by a tsunami, and 110 metres from the shoreline, was pushed about 10 metres by Ivan's surge (Scheffers 2005, table 2). On Bonaire, hurricane debris is reported to be superimposed on tsunami debris of an age of about 1300 years BP or older (Scheffers 2002). An important difference between the situations on Bonaire and Jamaica is the lack of dense forest extending up to the coastline in Bonaire. This may account for the generally wider spreads of debris we observed on Bonaire compared with Jamaica.

On Grand Cayman (figure 3.1) the evidence of boulder movement is provided by "before and after Hurricane Ivan" surveys of boulder positions (Jones and Hunter 1992; Robinson, Rowe and Khan 2006). Ivan passed much closer to the island than for Bonaire (figure 3.1).

Some changes in location of previously emplaced boulders on the shore platform surveyed by Jones and Hunter (1992) were noted during the re-survey carried out by one of us (DAR) in June 2005 (Robinson, Rowe and Khan 2006). If not due to locational errors, these must have happened during the passage of Ivan in 2004, as no other damaging storms occurred between the dates of the two surveys. Limited movement of boulders up to 40 tonnes and for distances up to 15 metres appeared to have taken place. These movements are similar in scope to those reported by Scheffers (2005). In addition, two large boulders were mapped that do not appear on the Jones and Hunter map, suggesting emplacement by Hurricane Ivan of clasts as massive as 50 tonnes up to distances of over 50 metres. Anecdotal reports (B. Carbym pers. comm.) indicate generation of two boulders from the platform surface itself at Blowholes, and emplacement of many boulders during passage of a severe hurricane in 1932.

Debris Ridges

Although the main debris ridges at the three sites differ in character, they resemble one another in that they effectively separate the foreshore area, which is more or less lacking in vegetation, from the backshore and interior regions, which are more or less extensively forested. At Homers Cove, a debris ridge, composed of boulders armouring polymodal debris, occurs just within the edge of the forested area, and the delineation between forest and platform is sharp where there has been no human disturbance. The boulders appear to have been inhibited in their landward movement by the forest and, over time, have piled up, one upon the other. An example of this kind of inhibition, albeit on a small scale, was reported by Keating et al. (2005) following the 2004 Indian Ocean tsunami. Shoreward of this ridge, other smaller discontinuous perched beaches are evidently single-event emplacements at Homers Cove, and anecdotal reports indicate emplacement by hurricanes Ivan (September 2004), Emily (July 2005) and Wilma (October 2005).

At Galina, the single debris ridge includes fewer boulders and more large cobbles, in a matrix of sand and gravel-sized clasts, but the well-vegetated zone is found mainly behind the ridge. The whole ridge is set much further back, up to 200 metres from the shoreline, and a shallow brown soil has developed over much of its surface. Some growth of bushes and trees has also occurred over and in front of the ridge, and the transition from wave-swept, barren platform to vegetated interior is more gradual. It seems likely that severe, debris-transporting wave events reached into the ridge region less frequently than elsewhere.

The ridge at Discovery Bay is similar to that at Homers Cove in being relatively close to the shoreline, except that it is much higher. There is a sharp separation of shore platform from forested interior. As at Homers Cove, there are perched sand and imbricated cobble beaches in front of the main ridge. These appear to change their characteristics on an annual time-scale, probably through the activity of northers.

Hurricanes or Tsunami?

Based on the information discussed above, it can be surmized that boulders up to about 10 tonnes have been moved considerable distances (more than 10 to 15 metres) by severe hurricane waves and surge. Some boulders up to about 35 tonnes, and perhaps as much as 50 tonnes in weight, have been moved shorter distances. At a locality on the east coast of Jamaica, an 80-tonne boulder (the largest measured in Jamaica so far), over 50 metres back from an 8-metre-high sea-cliff, was moved at least 2 metres by Hurricane Dean (August 2007), and a 12-tonne boulder was wrenched from the cliff face at Homers Cove, also by Hurricane Dean (Khan et al. 2008).

Our observations therefore suggest that boulder formation and movement of large (50- to 80-tonne) boulders during hurricane events may be relatively common, in contrast with previously developed models. Nott (2003) developed models for wave heights required to overturn boulders at sea level. Most boulders of more than 20 tonnes would require storm wave heights which would be, for Nott, considered exceptional along the world's coastlines. Noormets, Crook and Felton (2004), in a study of Hawaiian boulders, concluded that the emplacement of large boulders onto the shore platform requires extreme sea waves with periods longer than those Pacific storms or swells normally generate.

Whether or not boulders in this size range can be emplaced or moved by storms probably depends on their shape (Nott 2003), along with the nature of joint sets in the cliff face and platform surface. Other than the observations reported here for Jamaica and Grand Cayman, noted above, and the single 50-tonne example from Bonaire (Scheffers 2005), there appear to be no Caribbean records of substantial movement of boulders bigger than about 50 tonnes by storm activity.

On the other hand, there is evidence that clasts exceeding 200 tonnes have been moved by tsunami (Whelan and Keating 2004; Nott and Bryant 2003). Until further evidence becomes available, we suggest that boulders up to 80 tonnes mass, already emplaced on platforms, can be moved laterally by hurricane surge and run-up (although perhaps initially emplaced by tsunamis).

Anecdotal reports of the Jamaican tsunami of 1907, from coastal settlements extending from St Ann's Bay to Port Antonio (Taber 1920; Tomblin and Robson 1977, Lander, Whiteside and Lockridge 2002), indicate wave heights ranging from 2.5 metres over low-lying ground to as much as 9 metres against rising ground, indicating an intensity of III to V on the scale of Soloviev (1978, in Scheffers 2002). Based on the formulae used by Scheffers (2002), boulders of up to some 40 tonnes could have been emplaced by such an event, depending on shape. The emplacement and movement of a radiometrically dated 30-tonne boulder at Galina (Robinson, Rowe and Khan 2006), from the shoreline 80 metres inland, may have occurred at this time, as this locality was within the region affected by the 1907 event (figure 3.2; Cornish 1908).

Wave Inundation Distances

As the map of Galina shows (figure 3.3), the storm surge and wave inundation distances of a moderately powerful storm like Hurricane Allen significantly exceeded the distance of the conceivably 4000-year old debris ridge from the coastline. This was also the case at Homers Cove, where inundation distances from Ivan were as much as 100 metres or double the distance of the ridge from the shoreline. In mapping debris distributions on rocky shorelines for hazard purposes, a factor to allow for debris-free inundation estimates needs to be added, based on parameters such as slope and topographic roughness of the areas behind the ridges. Work is still in progress to address this problem.

Implications for Setbacks and Coastal Management

The evidence of boulder strews along cliffed shorelines indicates just how destructive severe hurricanes can be when passing by. A hurricane does not have to hit directly to be dangerous. At the coastline, it is the waves, breaking against and over-topping the cliff face, that are hazardous, and they do not necessarily have to be accompanied by a barometric surge to be destructive (figure 3.9). In October 2005 at West End, Hoggs Haven and Homers Cove, huge swell waves generated by a distant, category four to five Hurricane Wilma, caused considerable damage to buildings situated on top of cliffs. Emplacement of small boulders also occurred (Carey Wallace, personal observation and pers. comm.; Wallace is of Negril Escape Resort and Spa, West End, Negril).

On those coastal platforms, where boulder strews are evident, one can use the relative sizes and locations of boulders as a guide for estimating the distance that permanent structures need to be set back to prevent damage from moving debris generated by hurricane surge and waves.

The positions of the debris ridges are the most reliable indicators of the maximum distances from the shore, over which damage from storm-entrained debris is likely to occur. As boulders are pushed further and further from the shore by successive events, the ability of subsequent inundations to move them must diminish (although the effect of rising sea level in the future should be considered), and the presence of forest vegetation provides an additional inhibitor. We suggest that the boulder-rich debris ridges have grown over time in an equilibrium location, beyond which hurricanes and even tsunami events have diminishing ability to move debris. The ridges themselves would also provide increasingly effective natural barriers against movement over time. The total time period over which the ridges have grown must depend, at least in part, on the time since post-glacial sea level rise reached more or less its present position to put the present coastline within range of potentially destructive forces. Reconstructed sea level curves, such as those for Jamaica (Digerfeldt and Hendry 1987), the Caribbean (Toscano and McIntyre 2003) and globally (Solomon et al. 2007), suggest that conditions suitable for the growth of coastal debris ridges have probably existed for the past four thousand to five thousand years. Published isotopic dates on material transported onto rocky shorelines (for example, see Scheffers 2002; Nott and Bryant 2003) also suggest there may be an upper limit of some 5000 years BP for giant wave impacts on today's coastal platforms.

At Galina, community members are well aware of the danger posed by hurricane surge and waves breaking over the platform, and buildings are generally set well back from the cliffs. At Homers Cove, on the other hand, many buildings close to the shore (and indeed, in front of the debris ridge) were destroyed during the passage of Hurricane Ivan (figure 3.9). At Discovery Bay, coastal development has yet to occur in the area covered by boulder arrays, although there is evidence that it may happen soon.

Conclusions

As we previously noted (Robinson, Rowe and Khan 2006), boulders and other deposits are particularly common along raised rocky shorelines, fronted by a narrow island shelf with a steep island slope into deep water beyond.

Our investigations show that boulders up to 50 tonnes and, rarely, more than 80 tonnes have been moved on coastal platforms by severe storms, such as Hurricane Ivan. Fifty-tonne boulders have also been torn from the cliff face and emplaced by hurricanes or tsunami, and could have been emplaced by tsunami of a magnitude similar to that of the 1907 event, affecting the northeast coast of Jamaica. These observations on Jamaica and Grand Cayman suggest that the modelling approaches for the size of wave-emplaced boulders, used by Nott (2003) and Noormets, Crook and Felton (2004) may be too conservative.

There is a close relationship between the position of the main debris ridge, whether composed of boulders or other debris, and the front edge of the forest vegetation in those regions where the forest environment has not been destroyed. Although the passage of Hurricane Ivan destroyed and damaged trees on the shore platform near Homers Cove (Robinson, Rowe and Khan 2006), there was little loss of trees behind the main ridge.

There appears to be only one major ridge at all the places examined (although the density of the forest in some places makes it difficult to see details) suggesting that there is a long term balance in which the forest may inhibit penetration of debris behind the ridge. The ridge is probably composed of debris of varying ages, perhaps extending back over the past four thousand to five thousand years, when sea level had reached more or less its present position and giant waves had the opportunity to encroach on cliffed shorelines. The presence and growth of the ridge, in turn, has provided an increasing degree of protection over time for the back coast region. As such, the position of the ridge may be interpreted to indicate the limit of storm surge inundation involving megaclasts for the "worst case scenario giant wave event" over the past four to five millenia.

Destruction of the debris ridge and the natural forest behind the ridge during housing and hotel development upsets the equilibrium achieved between forest protection and boulder incursion. This opens the way for greater penetration onshore of debris during future extreme events, and it will increase the vulnerability of coastal communities to the consequences of increasing frequency of giant wave activity that will accompany future sea-level rise through climate change (Solomon et al. 2007).

The calculations of setbacks for permanent structures on cliff coasts need to be site-specific, and they should take into account the evidence presented in this chapter. A generalized, simple height-above-sea-level formula, such as has been suggested by some authorities, does not take into account local variations in on-shore topography and off-shore island shelf morphology. Both variations significantly influence the way in which waves and surges impact on the coast. The assessment of boulder size range and encroachment inland, as well as the location of the debris ridge, where these features occur, are the best guides to use to assist in assigning site-specific setbacks for shoreline construction.

On the southwest coast, residents have a fresh memory of the effects of hurricanes Ivan, Emily, Wilma and, most recently, Dean. However, on the coast between Montego Bay and Galina, the memory of Allen is fading, and the 1907 tsunami has been long forgotten. Meanwhile, the north coast has seen tourist-related coastal construction taking place on a massive, accelerating scale since the last decade of the twentieth century.

Historical records suggest that tsunami generation in the Jamaican region results mainly from submarine slides generated by local earthquakes (Ahmad, Robinson and Rowe 2005) and, perhaps, extreme meteorological events (Robinson and Rowe 2005). Should a tsunami-generating, magnitude nine, seismic event occur off the south coast of Cuba, at a distance similar to the lesser event of 4 February 2007, which originated 100 kilometres west-north-west of Montego Bay, and was widely felt in Jamaica (earthquake unit website: www.mona.uwi.edu/earthquake/), the first tsunami waves would arrive at the Jamaican coast less than ten minutes later. This time is significantly less than the minimum time which can be provided by any tsunami warning system at present.

In summary, our investigations suggest that boulder-generating impacts of giant waves have been occurring around Jamaica for many years into the past. These phenomena have been detailed here for Jamaica, and evidence of similar phenomena in the Netherlands Antilles and Grand Cayman was discussed earlier in the chapter. Similar boulder strews and ridges have been recognized throughout the insular Caribbean, for example, in the Bahamas, Guadeloupe and Isla de Mona (Hearty 1997; Kelletat, Scheffers and Scheffers 2004; Morton et al. 2008; Scheffers, Scheffers and Kelletat 2005).

The implications for coastal zone management are significant. However, as yet there is relatively little recognition of the threat that these possibly tsunami and/ or storm-related phenomena represent. Although storm surge modelling is now an accepted tool in terms of vulnerability to hurricanes, as part of the environmental impact assessment exercise in the Caribbean region and tsunami hazard mapping it is becoming an integral part of the coastal development process (for example, Rowe et al. 2008), the official adoption of revised or upgraded guidelines for setbacks for buildings and other infrastructure still leaves much room for improvement.

With accelerated sea-level rise and increased incidence of severe hurricanes being projected for at least the next century, such impacts are very likely to increase in number and magnitude. There is still too little provision for increasing the effectiveness of coastal management programmes through detailed, island-wide, site-specific coastal vulnerability analysis, both in developed areas and pristine areas that may be the targets for development in the future. Meanwhile, with the prospect of continuing rapid urbanization of the coastline, the critical geological evidence for making sound, scientifically based management decisions is being bulldozed away or otherwise modified and concealed by buildings. This evidence needs to be mapped before it is destroyed. Coastal development and management projects must include consideration of the geological parameters indicative of giant wave events in their terms of reference when environmental impact assessments are carried out and developments in the coastal zone are undertaken.

Acknowledgements

The Environmental Foundation of Jamaica provided funding and vehicular support throughout this study. Radiocarbon dating of encrusting coral on a boulder at Discovery Bay was funded through a Special New Initiative Grant from the University of the West Indies and carried out by Beta Analytic, Miami, Florida. We benefited from many instructive discussions with Bruce Jaffe, Robert Morton and Bruce Richmond of the US Geological Survey during an extended field excursion in Jamaica in January 2007. Deborah-Ann Rowe is supported by a Gladstone Fellowship from the University of Chester, United Kingdom.

References

Ahmad, R., E. Robinson and D-A. C. Rowe. 2005. Storm surge and tsunami coastal flooding processes in Jamaica. Proceedings of the seventh conference, Faculty of Pure and Applied Sciences, 16–19 May 2005, University of the West Indies, Mona, Jamaica.

Blair, T.C., and J.G. McPherson. 1999. Grain size and textural classification of coarse sedimentary particles. *Journal of Sedimentary Research* 69: 6–19.

Bruun, P. 1994. Freak waves in the ocean and along shores, including impacts on fixed and floating structures. In *Coastal Hazards: Perception, Susceptibility and Mitigation*, ed. C.W. Finkle, 163–75. *Journal of Coastal Research*, special issue no. 12.

Camfield, F.E. 1994. Tsunami effects on coastal structures. In *Coastal Hazards: Perception, Susceptibility and Mitigation*, ed. C.W. Finkle, 177–87. *Journal of Coastal Research*, special issue no. 12.

Cant, R.V. 1972. Jamaica's Pleistocene reef terraces. *Journal of the Geological Society of Jamaica* 12: 13–17.

Cornish, V. 1908. The Jamaica earthquake (1907). *Geographical Journal* 31 (3): 245–71

Digerfeldt, G., and M.D. Hendry. 1987. An eight-thousand-year Holocene sea-level record from Jamaica: Implications for interpretation of Caribbean reef and coastal history. *Coral Reefs* 5: 165–69.

Hearty, P.J. 1997. Boulder deposits from large waves during the last Interglaciation on North Eleuthera Island, Bahamas. *Quaternary Research* 48 (3): 326–38.

Hopke, C.J., R. Gmirkin and B.M. Richmond. 2005. *Coastal change rates and patterns: Kaloko-Honakohau NHP, Kona Coast, Hawaii*. United States Geological Survey, Open-File Report 1069 of 2005.

Jones, B., and G. Hunter. 1992. Very large boulders on the coast of Grand Cayman: The effects of giant waves on rocky shorelines. *Journal of Coastal Research* 8 (4): 763–74.

Keating, B.H., C. Helsley, Z. Waheed and D. Dominey-Howes. 2005. 2004 Indian Ocean tsunami on the Maldive Islands: Initial observations. *Science of Tsunami Hazards* 23 (2): 19–70.

Kelletat, D., A. Scheffers and S. Scheffers. 2004. Holocene tsunami deposits on the Bahaman islands of Long Island and Eleuthera. *Zeitschrift fur Geomorphologie, N.F.* 48 (4): 519–40.

Khan, S.A., E. Robinson, D. Rowe and R. Coutou. 2008. Size of shoreline boulders moved and emplaced by recent hurricanes, Jamaica. Second international tsunami field symposium, Ostuni (Puglia), Italy and Ionian Islands, Greece, 22–28 September 2008 (IGCP Project 497): 135–38.

Lander, J.F., L.S. Whiteside and P.A. Lockridge. 2002. A brief history of tsunamis in the Caribbean Sea. *Science of Tsunami Hazards* 20 (2): 57–94.

Morton, R.A., B.M. Richmond, B.E. Jaffe and G. Gelfenbaum. 2008. Coarse-clastic ridge complexes of the Caribbean: A preliminary basis for distinguishing tsunami and storm-wave events. *Journal of Sedimentary Research* 78 (9): 624–37.

Noormets, R., K.A.W. Crook and E.A. Felton. 2004. Sedimentology of rocky coastlines 3: Hydrodynamics of megaclast emplacement and transport on a shore platform, Oahu, Hawaii. *Sedimentary Geology* 172: 41–65.

Nott, J. 2003. Tsunami or storm waves? Determining the origin of a spectacular field of wave emplaced boulders using numerical storm surge and wave models and hydrodynamic transport equations. *Journal of Coastal Research* 19 (2): 348–56.

Nott, J., and E. Bryant. 2003. Extreme marine inundations (tsunamis?) of coastal Western Australia. *Journal of Geology* 111: 691–706.

Robinson, E., and D. Rowe. 2005. The great Sav submarine slide. *Jamaica Gleaner*, 27 September 2005, A5.

Robinson, E., D. Rowe and S.A. Khan. 2005. Mystery boulders at Galina Point. *Jamaica Gleaner*, 13 October 2005, C10.

———. 2006. Wave-emplaced boulders on Jamaica's rocky shores. *Zeitschrift für Geomorphologie*, Supplement-Band 146: 39–57.

Rowe, D.A., M. Degg, D. France and S. Miller. 2008. Tsunami hazard mapping, SW Jamaica. Second international tsunami field symposium, Ostuni (Puglia), Italy and Ionian Islands, Greece, 22–28 September 2008 (IGCP Project 497): 147–48.

Scheffers, A. 2002. Paleotsunami in the Caribbean: Field evidences and datings from Aruba, Curaçao and Bonaire. *Essener Geographische Arbeiten* 33: 185.

———. 2005. Coastal response to extreme wave events: Hurricanes and tsunami on Bonaire. *Essener Geographische Arbeiten* 37: 100.

Scheffers, A., and D. Kelletat. 2003. Sedimentologic and geomorphologic tsunami imprints worldwide: A review. *Earth-Science Reviews* 63: 83–92.

Scheffers, A., and S. Scheffers. 2006. Documentation of the impact of Hurricane Ivan on the coastline of Bonaire (Netherlands Antilles). *Journal of Coastal Research* 22 (6): 1437–50.

Scheffers, A., S. Scheffers and D. Kelletat. 2005. Paleo-tsunami relics on the southern and central Antillean Island Arc (Grenada, St Lucia and Guadeloupe). *Journal of Coastal Research* 21 (2): 263–73.

Solomon, S., D. Qin, M. Manning, Z. Chen, M. Marquis, K.B. Averyt, M. Tignor and H.L. Miller. 2007. *Climate change 2007: The physical science basis. Contribution of Working Group I to the Fourth Assessment Report of the Intergovernmental Panel on Climate Change*. Cambridge: Cambridge University Press. Available at http://www.ipcc-wg2.org/

Taber, S. 1920. Jamaica earthquakes and the Bartlett Trough. *Bulletin of the Seismological Society of America* 10 (2): 55–89.

Tomblin, J.M., and G.R. Robson. 1977. A catalogue of felt earthquakes for Jamaica, with references to other islands in the Greater Antilles, 1564–1971.

Mines and Geology Division Special Publication no. 2. Kingston: Ministry of Mining and Natural Resources, 243.

Toscano, M.A., and I.G. McIntyre. 2003. Corrected western Atlantic sea-level curve for the last eleven thousand years based on calibrated 14C dates from *Acropora palmata* framework and intertidal mangrove peat. *Coral Reefs* 22: 257–70.

Whelan, F., and B. Keating. 2004. Tsunami deposits on the island of Oahu, Hawai'i. In *Geographie der Meer und Küsten; Coastline reports 1 (2004)*, ed. G. Schernewski and T. Dolch, 77–82. Warnemünde: EUCC – Die Küsten Union Deutschland e.V.

Wilmot-Simpson, C. 1980. Effects of Hurricane Allen along the north coast of Jamaica: Geotechnical Report 42, in *Geotechnical Reports 4*. Kingston: Geological Survey Division, Ministry of Mining and Natural Resources.

CHAPTER 4

An Investigation into the Causes and Consequences of Coastal Flooding in Guyana

A Case Study of Six Villages on the East Coast of Demerara

PATRICK WILLIAMS AND LINDA JOHNSON-BHOLA

Introduction

Freeman (2006) noted that direct losses from natural disasters in the developing world average about US$35 billion annually. Floods have been identified as one of three categories of disasters that account for 90 per cent of the world's direct losses. However, it is impossible to eliminate floods as one of nature's most devastating powers, despite the fact that it is possible to minimize their effects through programmes, projects, and activities aimed at reducing the vulnerability of economic and social infrastructure to their onslaught.

Despite continuous and steady efforts worldwide to reduce flood disasters, statistics reveal that there is an upward tendency, contrary to the expectations of many engineers, scientists and development planners. Deaths from flooding around the world average approximately ten thousand per annum (Rodda and Rodda 1999), and there is a possibility that, with the spectre of global warming, the number will rise, contributing to increasing severity and frequency of coastal and inland flooding. Dasgupta et al. (2007) predict that a one-metre rise in sea level is likely to turn at least 56 million people in the developing world into refugees.

Types of flood disasters range from the inundation over large delta areas, such as the Mississippi Delta, Mekong Delta and Bangladesh, to urban area inundation which is sometimes accompanied by flooding into underground space. Among them, flash floods and landslides in upper and mid reaches are small-scale events, but often with a high mortality rate. For instance, a 2007 Reuters report observed that a number of Asian countries were hit by flash floods and landslides due to torrential monsoon rains. The death toll included 700, 400 and 150 in India, China and Bangladesh respectively. Additionally, a total of approximately 3,527,000 persons were displaced in the three countries, compounding the calamity faced by the population.

In Central America, data acquired by the Unit for Sustainable Development and Environment indicate that floods constitute the most prevalent natural disaster in the region. For instance, in 2002, a Natural Hazards Project report from the Organization of American States (OAS) revealed that of the 850 natural disasters registered in Central America between 1960 and 1995, 68 per cent were caused by floods. In general, increasing flood disasters are mainly attributed to the proliferation of urbanization into areas vulnerable to flooding. The promotion of urbanization has been attributed to economic growth and accompanying development pressures brought on by an increase in the inflow of population into urban areas caused by unemployment and poverty problems in rural areas. In areas where the legal and institutional framework is weak, the ability to streamline development activities away from flood-prone areas becomes almost impossible.

The Guyanese Situation

In Guyana, problems with tides and floods have posed a continuous threat to the population even before the twentieth century, when the colonial powers occupied the coastal areas and the banks of the lower reaches of the main rivers for the production of sugar and other export crops. In that early period, the waterlogged coast and near-inland wetlands were drained and are now characterized by drainage and irrigation canals, sluice gates and elaborate sea defence infrastructures. However, as observed by Westmaas (2005) and Kandasammy (2006), several historians (including Rodney, Young, Adamson, Rodway and Webber) have noted that, despite several engineering interventions and heavy capital outlay for nearly a century, flooding along Guyana's coast has persisted.

Measures such as drainage and irrigation canals, sluice gates and elaborate sea defence infrastructure did not avert the perennial problem of flooding. As Westmaas (2005) observed, in 1921, a heavy rainstorm lasting over twenty-four hours flooded some city streets in Georgetown and swept away bridges. Flooding and sea breaches in the country have occurred at various scales every decade since the 1920s. In 1934, there was a major flooding disaster that affected Georgetown and the entire East Coast, and this has recurred at intervals since.

In early 2005, Guyana was confronted with one of the worst floods in its history after experiencing prolonged and unusually heavy rainfall for almost three weeks. Almost the entire populated coastal area of the country became inundated or was threatened. For several weeks the response institutions were overwhelmed by the scale of the devastation, contributing to loss of lives and large-scale population displacement as people migrated from their homes that were covered almost to the roofs in water. The plight of the displaced people was aggravated by the disruption of social services and economic activities. Estimates of the magnitude of the suffering varied among the different relief agencies that responded to the crisis. For example, a Catholic Relief Service report (CRS 2005) put the figures at 66,000 displaced, and 128,000 affected by the destruction of drainage, sewerage and drinking water systems; while a leading UN Organization, the United Nations Children's Fund (UNICEF), reported in February of 2005 that over 300,000 people were affected, and about 92,000 had water in their homes at the time the report was being compiled.

This study investigates the 2005 flooding in Guyana, focusing specifically on six selected villages on the east coast of Demerara, roughly 30 kilometres east of Georgetown. This area is one of the worst affected. It is persistently threatened by floodwaters from both the Atlantic Ocean and rivers that flow northward from the forested upland regions through the coastal zone (and associated wetlands).

The chapter first gives a broad contextual framework on the issues of flooding, including its impact on the human environment and on the global, regional and local scenarios. It then deals specifically with the flood situation in the selected villages. It gives a brief outline of the study area, discusses the methodology used in collecting data for the case study and analyses the consequences of flooding for communities in the study area, including analysis of the response mechanisms of various agencies to the plight of the flood victims. Recommendations for the avoidance of such a disaster in Guyana are made.

Coastal Flooding in Guyana

Flooding along Guyana's coastal areas has been such a regular feature for several decades that it has become an almost accepted part of coastal living. However, in January 2005, some exceptional weather conditions prevailed over the coast, and the country experienced one of the most devastating floods in its recorded history. Despite several decades of heavy capital investment in infrastructural development, this occurrence has shown that the level of vulnerability of the coast, particularly along the east coast of Demerara, which is the most populated area outside of urban areas, continues to be high.

According to a UNICEF Report (March 2006) the flood affected around 290,000 persons, about 39 per cent of the national population, of which more than 50 per cent were females and approximately 33 per cent children. The report also observed that almost 192,000 people experienced flooding in their homes, and the most affected were of the low socio-economic strata of the society. Much of the discussion to date has attributed the causes of this major flood in Guyana to a combination of natural and anthropogenic causes (UNICEF Guyana 2006).

Natural Causes of Flooding

According to Daniel (1991), there are three much-reported natural causes of flooding along the coastal areas of Guyana: unusual and fluctuating weather conditions, the geologic and geomorphologic nature of the coast, and marine and tidal influences.

1. *Flooding and weather conditions*

Figure 4.1 provides monthly rainfall data for 2005 when Guyana experienced one of the worst floods for nearly a century. Records of monthly rainfall for the period 1990–2005 reveal that January 2005 received rainfall far in excess of the amount obtained for any other month during the period. The steep decline in the graph in February 2005 indicates that the situation normalized by the end of January.

2. *Flooding and coastal geology and geomorphology*

The coastal plain of Guyana is divided into two geomorphological regions; the young and the old coastal plain areas (figure 4.2). The young coastal plain is covered by clays of the Demerara formation of the Holocene period and the old coastal plain by mottled and leached clays of the Coropina formation of the Pleistocene period (Daniel 1984). The young coastal plain does not rise more than 2.4 metres above mean sea level, and, in areas where the villages are located, it is 1.5 metres below mean sea level.

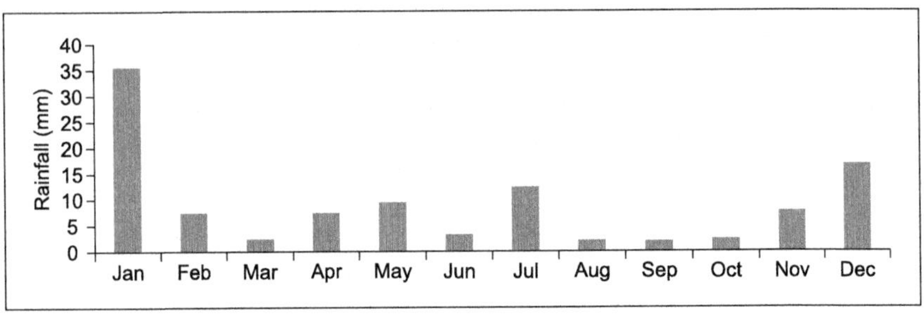

Figure 4.1 Monthly rainfall for 2005.
Source: Guyana Hydrometeorological Office.

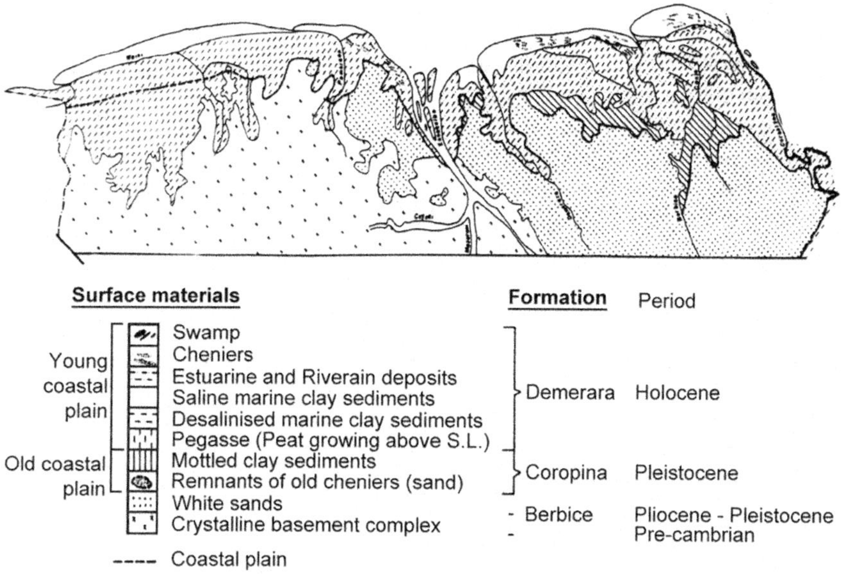

Figure 4.2 Geomorphology of the Guyana coast (Daniel 1984).

There are three ways in which the geological and geomorphological characteristics of the coast contribute to flooding. First, the area below mean sea level becomes submerged during periods of heavy rainfall or when there are sea defence breaches during high tide. Second, clay soil is poorly drained. This allows water to accumulate much quicker and remain on the land much longer, thereby giving rise to flooding. Third, the gradient of the land slopes downward from the more elevated old coastal plain in the south to the Atlantic Ocean, where it meets the built-up sea-defence structures. This creates a slight depression that allows floodwaters from the backlands to move northwards and accumulate in the housing areas on the coast.

3. *Flooding and marine and tidal effects*

Wave energy is considerably reduced on the coast because of the continental shelf that stretches for an average distance of 140 kilometres. According to Daniel (1991), the tide is semi-diurnal and averages about 2 metres near Georgetown. The coast experiences wave height above 3.5 metres only about 2 per cent of the time; only 2 metres from December

Figure 4.3 One of the several drainage pumps installed along the coast. *Photo by Gary Clarke and Patrick Williams.*

to June and 1 to 1.5 metres from July to November. However, between the period of November and February when higher than average spring tides occur, the tidal effects are felt as far as 80 kilometres to 100 kilometres inland in the major rivers. The combined effects of low wave energy and inland tidal effects serve to impact on coastal flooding by increasing siltation along the coast and in estuaries. This has the effect of reducing both the capacity and rate of flow of the drainage systems, especially during periods of heavy rainfall, giving rise to floodwaters escaping over levees and into inhabited areas. In an attempt to offset this problem, very expensive, high-powered drainage pumps are installed to dispel floodwaters impounded by sea defence structures (figure 4.3).

Anthropogenic Causes of Flooding

There are four anthropogenic factors that have been attributed to flooding in Guyana over the years. These have been identified as poor maintenance of drainage infrastructure, lack of proper land use planning, weak legal and institutional framework, and limited knowledge of flood-related issues.

1. *Poor maintenance of drainage infrastructure*

Over decades, Guyana has had to face the tremendous technical and financial burden of ensuring that the drainage and irrigation systems along the length of Guyana's coast between the Pomeroon and Corentyne Rivers are properly maintained and functioning. This includes the clearing of canals and outfalls to the Atlantic Ocean, and maintenance of protective embankments and of drainage pumps.

Key agencies with responsibility for the management and maintenance of the drainage and irrigation infrastructure include the Guyana Sugar Corporation (GUYSUCO), the Drainage and Irrigation Board (D and I Board) and the Regional Democratic Council (Region 4) in the case of East Coast Demerara.

2. *Lack of land use planning*

Land use planning is a very complex and controversial issue in Guyana, partly because of the political sensitivity surrounding the issue of land allocation, and partly due to the multiplicity of stakeholders involved in the use of, and demand for, land. The country has never had a comprehensive land use plan to guide development; consequently, numerous land use conflicts have emerged. The government, through the Central Housing and Planning Authority, embarked in 1995 on a massive housing development programme aimed at upgrading and regularizing squatter settlements as well as developing new housing areas. The result is that large tracts of land, some formerly occupied by sugar plantations, were brought into development.

This housing development programme, instituted in the absence of an approved land use plan, created two main situations that exacerbated the problem of flooding in the coastal areas. First, many housing developments took place in low-lying, flood prone areas without proper drainage and infrastructure. Second, buildings were constructed in a haphazard manner in some areas. This blocked existing waterways or prevented the use of mechanical equipment for cleaning and rehabilitating of some of the primary waterways. In periods of heavy rainfall, therefore, many of the housing areas quickly become waterlogged and pose a threat to the health of residents.

In the 2005 flood, the issue of land use planning and human health emerged as a major discussion point. Health personnel were concerned with the many waterlogged cemeteries and garbage disposal sites that were in close proximity, or adjacent, to housing areas. Their major focus centred on the possibility of an outbreak of cholera and other water borne diseases. However, while there were no reported cases of cholera, a number of persons were hospitalized with leptospirosis infection, and there were a few reported cases of death. As reported by the International Federation of Red Cross and Red Crescent Societies (IFRC) on 27 January 2005, a meeting was convened between the Ministry of Health and various NGOs and international organizations to discuss the distribution of antibiotics to combat the spread of the disease. At that meeting it was also reported that up to that point there had been nine deaths from leptospirosis. Similar concerns were also raised on 7 February 2005 by the Pan American Health Organization (PAHO), which warned of a possible outbreak of leptospirosis. On that same day, the United States Agency for International Development (USAID) Guyana had reported 112 leptospirosis cases admitted to the Georgetown Public Hospital Corporation (GPHC) of which 12 were estimated to have died of the disease.

3. *Weak legal and institutional framework*

The problem of land use planning and flooding is exacerbated by weak legal and institutional framework that regulates the use of land and its resources. The weaknesses lie partly in the lack of sufficiently skilled personnel to undertake the tasks of planning and, then enforcing, the planning regulations, together with the datedness and fragmented nature of the regulations themselves. In addition, there seemed to have been a lack of political will to enforce the existing regulations, limited as they were, as the authorities seemed either unprepared or unwilling to face the political fallout that might have resulted from such action.

The result is that development is unimpeded in areas that are vulnerable to flooding. In some areas along the coast, for example, there is illegal occupation of lands adjacent to sea defence structures and on the embankment of critical waterways, undermining their durability and exposing areas to flooding. The ability of authorities to remove such developments has been hamstrung by weak legislation and staff constraints. Fines have also often been too lenient to serve as a deterrent to violators.

4. *Community's limited knowledge of flood issues*

The manner in which communities treat the environment is often a reflection of their perception of their surroundings, as well as in some cases the inability to understand the consequences of their actions. In many of the communities, flooding occurs due to dumping of solid waste in waterways. This is a major source of bother to the authorities, as it proves very costly to maintain the drains and canals. The limited knowledge and, to an extent, lack of awareness of many communities of the consequences of illegal waste dumping is a major factor that contributes to the persistence of the flood problems in many areas.

Flooding in East Coast Villages

The main objective of this case study is to identify the characteristics of victims of the flood, the extent of their losses, the effectiveness of support they received, who or what they felt was responsible for the flood and finally what they felt should be done to avert the problem. The study area is roughly 30 kilometres east of Georgetown. It extends over one of the most densely populated areas on the coast of Guyana outside of the major urban centres. In most of the settlements, the population extends for a distance of approximately 5 kilometres inland from the ocean.

This area forms part of the larger coastal plain that is dissected by large rivers, creeks and canals. The soils of the area are developed from different types of parent materials but are predominantly clays on the frontlands and pegasse in the backlands. Pegasse soil is a type of tropical peat which occurs behind the coastal clays and along the river estuaries. The spongy nature of the soil allows for a high moisture-retention capacity and, as such, contributes to flooding in areas where it exists. Natural vegetation is virtually non-existent, except for patches along the shoreline and in swamp areas. Most of the frontlands have been utilized for agriculture, in particular sugar cane cultivation, pasture, and housing development.

As in most of the rural coastal areas of Guyana, the population in the villages is segregated according to the two main ethnic groups; people of East Indian descent (referred to here as Indo-Guyanese) and African descent (referred to here as Afro-Guyanese). About 80 per cent of the households within the study area belong to the low socio-economic group, earning their livelihood primarily from the agricultural sector either as farmers or as hired labourers on the two sugar plantations at Enmore and La Bonne Intention (LBI).

The economic status of the residents in the study area is also characterized by their relatively low-income earnings, the part-time nature of their plantation work as cane cutters, the small size of their agricultural plots, widespread squatting and their widespread occupation of cheap rented dwellings. Survey data indicate that the average earnings of the most severely affected households fall between G$20,000 (US$100) and G$30,000 (US$150) per month, which is significantly less than the average monthly income of G$100,000 (US$500) for those households in the area that were less impacted upon by the flood and live largely in privately owned dwellings.

Research Methodology

Purpose of the Study

The purpose of the study was threefold. First, the study was designed to identify those households or socio-economic groups that were impacted upon by flooding. Second, it set out to determine the severity of the floods among the selected communities and to identify the major factors that respondents consider are responsible for flood occurrence. Third, efforts were also made to consolidate responses in order to highlight the various factors that contributed to the floods, and to investigate the response mechanisms and the steps required to confront the flood problems on the coastal areas of Guyana. Questionnaire surveys were developed, with series of questions focused around these themes. The responses to specific individual questions are highlighted in the discussion below.

Selection of the Study Area

The field study focuses on six villages on the East Coast of Demerara, east of the capital city, Georgetown, and spanning a distance of approximately 24 kilometres. The villages were Cummings Lodge, Industry,

Plaisance, Bachelors Adventure, Paradise and Enmore (figure 4.4). The area reflects many of the aspects of physical and human vulnerability described above. Three criteria were used in selecting the study area and the villages. The criteria used were as follows: the vulnerability of the area, the severity of the flooding during the period under consideration and the desire to ensure that the two dominant ethnic groups affected by the flooded are proportionately represented in a manner that reflected the population size and composition of the area.

The villages of Cummings Lodge, Industry and Enmore are East Indian dominated while Plaisance, Bachelors Adventure and Paradise are villages where inhabitants are of African descent. The other ethnic groups in the area are quite insignificant in number and therefore not given consideration in the selection of the study area.

The shaded area in figure 4.4 shows the location and extent of the flooded area on the East Coast Demerara and the water conservancy adjacent to it. The conservancy provides water for domestic, agricultural and industrial use along the East Coast of Demerara as well as Georgetown.

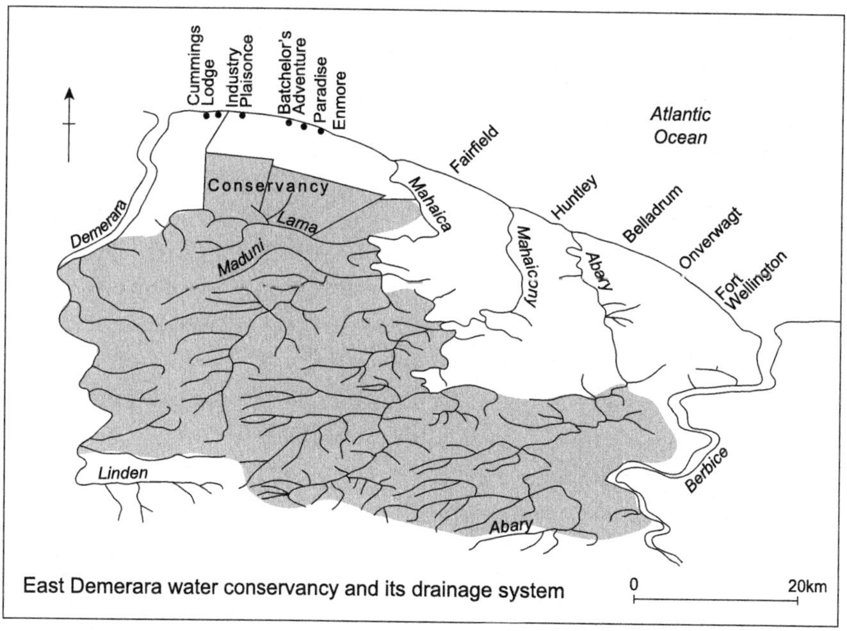

Figure 4.4 The study area.
Courtesy of the Cartographic Unit, University of Guyana.

Data Collection and Processing Procedures

Data for the research paper were collected over a period of two days from a total of 165 households randomly selected from the six villages identified above. A team of six assistant researchers carried out the questionnaire survey. The target group comprised males and females between nineteen and seventy years old. The household survey data were analysed using the Statistical Package for Social Sciences (SPSS).

Analysis of Flooding in the Villages

The primary focus of the discussion in this section of the study is analysis of the perceptions on behalf of the flood victims on the East Coast of Demerara with respect to the causes and consequences of the flood that affected their communities in early 2005. To undertake this task, four aspects are examined. These are the profile of the victims, their perceptions of the causes of the flooding, the impact of the flood on their households and communities, and their views on the response mechanisms that were put in place to address the flooding.

Demographic and Socio-economic Characteristics of Flood Victims

Studies such as Pelling (1996, 1997) which focused on the environmental and economic causes of coastal flooding in Guyana, and a case study on the determinants of vulnerability to flooding in Georgetown, have shown that there is a strong positive correlation between the levels of vulnerability, coping strategies and the impacts of flooding on communities on the one hand and the demographic and socio-economic characteristics of those affected communities on the other.

The majority of the respondents (74.5 per cent) are below fifty years of age with the largest proportion (27.3 per cent) being between ages of thirty and thirty-nine years. Only a small percentage of the respondents, about 9.7 per cent, are in the age groups that could be classified as old, namely fifty-nine years and over. Most of the respondents who were affected were below fifty years old. These are considered the working population of the study area.

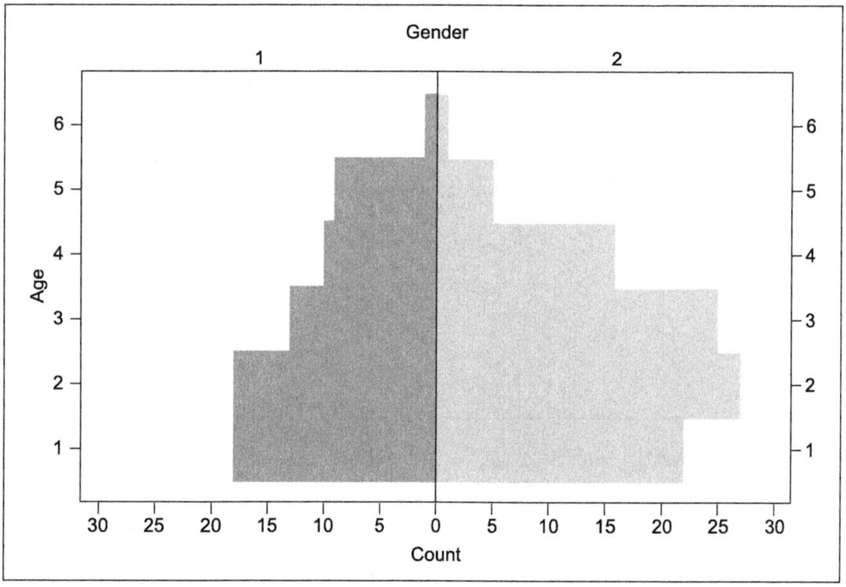

Figure 4.5 Age-sex distribution of the respondents.
Note: Age group: 1 = 20–29; 2 = 30–39; 3 = 40–49; 4 = 50–59; 5 = 59–69; 6 = 70+; Gender: 1 = Male; 2 = Female.

The age-sex pyramid (figure 4.5) shows that more females were interviewed than males except in the fifty to sixty-nine age groups. This is due mainly to the fact that the interviews took place on weekdays and during working hours when most of the males were away at work. In these rural communities, the males are the often the ones who work outside of the home.

The largest proportion of the interviewees was Indo-Guyanese, with Afro-Guyanese the second largest ethnic group and other groups represented in much smaller proportions. As seen in table 4.1, of the 165 households interviewed, 98 (59.2 per cent) were Indo-Guyanese, 50 (30.5 per cent) Afro-Guyanese and the remaining 17 (10.3 per cent) comprising people of Portuguese descent, Amerindians and Mixed groups. This distribution compares favourably with the ethnic distribution of the population in the study area (Bureau of Statistics 2002).

Table 4.2 attempts to determine which of the two major ethnic groups, Indo-Guyanese and Afro-Guyanese, within the study area which together comprised 89.7 per cent of the households interviewed,

Table 4.1 Ethnic Distribution of Flood Respondents

Ethnic Group	Frequency	Per cent
East Indian	98	59.2
African	50	30.5
Portuguese	1	0.6
Amerindian	4	2.4
Mixed	12	7.3
Total	165	100.0

Source: Questionnaire survey.

was most affected by the floods. The survey revealed that 75.5 per cent of the 98 Indo-Guyanese households interviewed were affected a great deal by the flood compared with marginally fewer (68 per cent) of the 50 Afro-Guyanese households.

With respect to education, the majority of the interviewees attained primary and secondary education. Of the 165 households interviewed, about 49 per cent obtained primary education and 35 per cent secondary

Table 4.2 Effects of Flood on Households by Ethnicity (Per Cent)

Flood Effects	Ethnic Groups				
	East Indian	African	Portuguese	Amerindian	Mixed
Not at all	4.1	4.0	0	25.0	0
A little	11.2	14.0	1.0	25.0	25.0
Not very much	8.2	14.0	0	0	0
A great deal	75.5	68.0	0	50.0	75.0
Don't know	1.0	0	0	0	0
Total	100.0	100.0	100.0	100.0	100.0

Source: Questionnaire data.

education. Collectively those two categories account for 84 per cent of educational status of the respondents. Although most of the interviewees attained education, they were not fully aware of the long term consequences that flooding could have on their livelihoods. For example, regardless of educational status, most people disposed of domestic effluent in waterways.

Data collected on the marital status of the flood victims reveal that there were family units within the households to provide support for the distressing situation that confronted many families. Forty-seven per cent of the respondents were married and 10 per cent had a common law relationship and lived as a family unit. However, some of the interviewees were divorced (7 per cent), some widowed (8 per cent), and others single, having never married (21 per cent). For those interviewees without partners, it was noted that they were particularly hard hit by the flood as they faced the pressures alone. The flood was particularly difficult for single parents.

Many of the individuals (66 per cent) owned their own homes while 23.6 per cent lived in rented buildings. Also, many of the respondents could not provide estimates of the cost of the damaged building, due in part to the significant proportion of renters. Due to the low economic status of a significant proportion of the flood victims, it now seems hardly likely that a number of them would be in a position to either rebuild or procure their own homes due to the extensive damage caused by the flood.

Estimates of the value of the homes affected by the flood were between G$1.0 million (US$5,000) and G$6.0 million (US$30,000) (table 4.3). This, by local standards, represents the homes of lower-income groups in the community. Visits to these homes also revealed that many of them were still in a poor condition six months after the flood and the owners described the lack of available funds to effect repairs. Also, most of the respondents did not participate in community-based work, a vital element during times of disasters like this in order to improve their social and economic status. Pelling's (1996) research on four contrasting neighbourhoods in Georgetown, for example, concluded that community-based activities are more common in high-income areas, where some economic resources are available. They are also common in squatter areas, where there is a strong tradition of community

Table 4.3 Home Value of Flood Respondents (G$)

Home value ($)	Percentage of Respondents
<1,000,000	11
1,000,000–1,999,000	7
2,000,000–2,999,000	12
3,000,000–3,999,000	15
4,000,000–4,999,000	11
5,000,000–5,999,000	10
> 6,000,000	34

Note: Exchange rate US$1 = G$200.
Source: Questionnaire survey.

participation and self-help. In a similar study, he also attributed the poor socio-economic conditions in Guyana as being a major determinant of the degraded state of the flood protection infrastructure on the coast of the country (Pelling 1997).

The plight of the flood victims is clearly revealed when their income status is examined. The data showed that 57 per cent of the respondents earned a monthly income of under G$40,000 (US$200) while 29 per cent received between G$40,000 and G$69,000 per month. The salaries for the households indicated that a large proportion of them fell within the low-income grouping and this explains why they have been unable, post-flood, to refurbish their dwellings or replace many of the damaged items.

For a number of the victims, the problem is compounded by the fact that they are part-time, self-employed and unemployed persons (table 4.4). These three categories accounted for 42 per cent of the respondents. During the flood and for a considerable period after, self-employed persons, many of whom were construction workers, auto mechanics, farmers and shopkeepers, could not work. These individuals therefore faced immense financial difficulties similar to those of the unemployed persons.

The information collected suggests that flooding is a major problem in the study area. While only 10 per cent of the households reported

Table 4.4 Employment Status of Flood Respondents

Employment Status	Percentage of Respondents
Full-time	30
Part-time	7
Self employed	26
Homemaker	17
Unemployed	8
Student	5
Retired	6
Other	1

Source: Questionnaire data.

that they never experienced flooding, 61 per cent experienced flooding at least once, 29 per cent of the households were flooded more than once and 9 per cent more than three times. This cycle of flooding could be seen as contributing to the impoverishment of the already poor communities.

The burden of facing floods in their communities has also forced some of the households to contemplate the option of relocation from the area. Of the 165 residents, 12 per cent revealed that if given the option they would seek another area in which to live. However, for most of them, the cost of relocation is too exorbitant to consider, and 70 per cent of the respondents stated that they will not consider relocation as an option.

An examination of the extent to which respondents felt their homes were damaged shows that 76 per cent of the households indicated that their homes were either not damaged at all or were only very minimally or slightly damaged. On the other hand, 17 per cent observed that their homes were either moderately or very severely damaged by the flood. The limited structural damage to a number of the homes from flooding may have been due in part to the types of the material used in their construction as well as the duration of the flooding. In Guyana, a greater number of persons are now utilizing concrete materials in housing construction, making houses more resistant to flooding. Also, in many of the seriously affected areas, the floodwaters receded within a period of ten days.

While the data have shown that damage to homes was not staggering, when consideration is give to damage to other property, such as cars, furniture, crops and livestock, and appliances, where respondents could identify any significant damage, a different picture is presented (table 4.5).

The overall picture therefore is much bleaker than one may imagine. A number of the household items (furniture and appliances in particular) would have been obtained through hire purchase, and full payment might not have been effected. Given that 79 per cent of the flood victims did not procure insurance on their property or items contained within their dwellings, it can be assumed that it may take several years for some of the respondents to replace articles damaged or destroyed by the flood. The poor economic circumstances of a number of the households would have prevented them from acquiring insurance policies in the first place, and replacement for most of them would therefore be almost impossible.

Table 4.5 Flood Damage to Property

Types of Property	Nos. Affected	Percentage
Car	10	4.8
Livestock	34	16.5
Crops	29	14.1
Appliances	48	23.4
Household furniture	59	28.7
Other	19	9.2
No damage	3	1.4
None of the above	3	1.4
Total	205	100.0

Source: Questionnaire data.

Often in disaster situations there is a tendency for victims to exaggerate their losses. However, when one considers the monthly incomes of the households, and the fact that very few had alternative means of livelihood, the estimated damages from the flood suggested by them is also quite revealing. Table 4.6 indicates that almost half (48 per cent) of the respondents suggested that their losses were between G$16,000 and G$26,000.

The flood also impacted on people's lives by depriving them of their means of livelihood. Fourteen per cent of the respondents reported losing their employment as a result of the flood. This was particularly the case with those persons who worked in private businesses that had to close their doors due to destruction of stocks. However, the fact that 84 per cent of those interviewed indicated that their jobs were not affected is to be seen as a positive indicator for the communities.

Finally, one of the most severe impacts of the floods on communities was the general effect on the health of the population. While there were no reported cases of death from the floods among those interviewed, there were at least two dozen cases of flood-related deaths (from drowning and diseases such as leptospirosis) in other communities. However, the survey did reveal that 27 per cent of those interviewed attributed health-related problems to the floods.

Table 4.6 Estimated Value of Loss from Flooding

Value of loss ($)	Percentage of Respondents
< 1,000	6
1,000–5,999	11
6,000–10,999	5
11,000–15,999	8
16,000–20,999	10
21,000–25,999	37
Don't know	23

Source: Questionnaire data.

Effectiveness of the Response to the Floods

As was stated earlier, the depressed socio-economic status of the households in the study area suggested that flood relief during and after the flood was an urgent matter that needed to be given priority by the government and relief agencies. Results of the study revealed that 73 per cent of the flood victims were able to get emergency relief in the form of food, water and clothing. Shelter was also provided to many of the victims by most of the relief organizations, especially children and the elderly, whose homes were made inaccessible.

One of the main problems that confronted the flood victims was the difficulty of getting sufficient quantities of food and water. As all the food stores were closed for a number of weeks, many households ran short of essential food items, and they therefore depended on supplies provided by the relief bodies. Of the 165 households interviewed, 30 per cent indicated that they experienced difficulties getting food and drink during the flood, while 66 per cent revealed that they had no problems.

In the study, respondents were requested to identify from a list of agencies those that provided support to the flood victims. The survey showed that almost 56 per cent of the respondents were given relief of one kind or another by the government while private individuals contributed to 10 per cent of the households (table 4.7). One aspect of the data that needs to be given attention is the contributions of NGOs, Red Cross and other agencies. Because they worked in collaboration with the government, and other bodies affiliated with government such as the Civil Defence Commission, it was widely and mistakenly viewed that the relief effort was almost entirely government executed. In fact, the financial, logistical and technical support of such bodies was under-estimated by the flood victims. One such organization is UNICEF. This UN organization almost single-handedly addressed the situation faced in the education sector by taking on the responsibility of housing displaced children and later overseeing the clean up and sanitization of the schools.

An attempt was made to discover who the flood victims felt was responsible for the flooding in their communities, whether it was the government or nature or the communities themselves. As shown in table 4.8, the single largest proportion of the respondents, 31 per cent, expressed the view that the government was to be blamed for the

Table 4.7 Main Support Agencies That Provided Flood Relief

Agencies cited	Frequency	Per cent
Government	92	55.8
Private individual	17	10.3
Private company	3	1.8
NGOs	6	3.6
Local Council	6	3.6
Red Cross	9	5.5
Other	32	19.4
Total	165	100.0

Source: Questionnaire survey.

Table 4.8 Respondents' Views on Liability for Floods

	Government	Community
Strongly Disagree	10.0	34
Disagree	20.0	9
Undecided	15.0	9.1
Agree	31.0	26.0
Strongly Agree	24.0	14.0
Total	100.0	100.0

Source: Questionnaire survey.

magnitude of flooding. Fifty-five per cent of the respondents reported that the government was liable, as it did not provide the necessary financial and technical resources to avert the flood and protect the population. On the other hand, 29 per cent of the interviewees rejected this position and felt that the government did the best it could at the time with the resources at its disposal.

On the question of whether the villages were liable for the flood, 43 per cent of the respondents felt that they were not liable for the flood while 40 per cent were of the opinion that the communities should be blamed for the disaster. They suggested that by squatting on critical waterways and adjacent to sea defence structures, as well as dumping solid waste in drainage canals, the villages contributed to the flooding.

Finally, on the issue of whether nature played a part in the flooding, 43 per cent of the households suggested that the flood was an "act of God" while 45 per cent did not support this proposition.

Suggested Mitigation Measures to Avert Flooding

The survey concluded with a qualitative and semi-structured discussion of the views of the households on what they thought was necessary to avert flooding in their communities in future. A wide range of suggestions were put forward during these less formal discussions. The most important ones included the following:

- Upgrade the drainage canals and the capacity of the pumps
- Rectify the design and other technical problems associated with the conservancy dam
- Undertake better monitoring of the weather patterns
- Relocate some of the communities to higher areas
- Improve the solid waste disposal system
- Take steps to prevent the dumping of solid waste in drainage canals
- Install warning systems and prepare an emergency evacuation plan
- Undertake flood vulnerability studies before identifying areas for housing construction
- Introduce building codes and regulations to take account of floods

These, in effect, form a list of recommendations to be evaluated and taken forward by government. It is suggested that the authorities place greater emphasis on the upkeep of the coastal drainage systems through improved technology and proper maintenance. Further, reference was made to the need to improve weather forecasting. This suggestion is considered vital, as it will allow the authorities to be more proactive and take precautionary actions that will minimize large-scale impacts.

Some respondents suggested that consideration should be given to relocating some coastal settlements. At present, this suggestion seems impractical because of costs, logistics and other difficulties. However, there is at least one primary lesson that can be learned: in future, human settlement development in Guyana should make land use planning an essential component, and this should include vulnerability assessment studies, as well as social and environmental impact assessment studies. In addition, there is a need to update the country's building codes, ensuring that they are rigidly enforced. Many of the affected households in this study are located in swampy areas with poor drainage and on the banks of canals which often serve as reservoirs for domestic wastes. Proper site planning should avoid some of these occurrences.

The research has pointed to the fact that some residents utilize the drainage systems to dispose of their solid waste, and these serve to block the flow of water which on occasions depends on gravity. An appropriate and enforced system of penalties would assist in addressing this issue, but this requires government to provide reliable means of collecting waste.

Conclusion

Flooding along Guyana's Atlantic coast and riverine areas is not an unusual occurrence. Rivers often burst their banks after periods of intermittent rainfall or unusually high tides cause breaches along sections of the sea defence. However, the flood along Guyana's Atlantic coast and near inland areas in January 2005 was an exceptional event which severely impacted the livelihoods of a large number of households. Homes and household items were damaged or destroyed, crops and livestock lost and public health impaired as a number of persons died of diseases and others hospitalized. The flood was described as one of the worst to hit Guyana for centuries. The magnitude of its impact was felt mainly because the communities affected belonged to the low socio-economic group. These communities possess limited coping strategies and restoring them will be a long term and painful experience for many of the victims. Some may never recover.

The survey revealed that people's vulnerability to coastal flooding in Guyana is linked to a number of clearly identifiable factors. These included their economic status, including job security, income generation capacity and the general physical conditions of the areas in which they built their homes. While the information gathered indicated that not all the respondents suffered equally, it was seen that the poorest segments of the population bore the brunt of the flood effects. This was due, in part, to the fact that they did not possess flood insurance. Many of them also lost the jobs, their only source of income, as well as crops and livestock upon which they and their families depended for sustenance. In addition, the population's susceptibility to flooding was linked to the fact that the areas in which they settled were reclaimed swamp lands. These areas were those that the authorities found difficult to protect because of the need for expansive and expensive investment in infrastructure and sophisticated technology. Decades of critical infrastructural works neglect had put large numbers of people at great risk.

Overall, the study has identified three main areas of concern among the surveyed respondents, for which actions should be taken to address the problem of coastal flooding in Guyana. The first is attributed to the current status of the technology used to control and monitor floods. The view is widely held that the existing infrastructure is inadequate to deal with periods of heavy and prolonged rainfall and that the drainage systems are outdated and poorly maintained. Second, the questions of settlement relocation, and more rigid enforcement of planning and building codes are issues to be urgently considered. Third, efforts should be made to improve solid waste disposal in the relatively densely populated coastal areas of Guyana.

References

Bureau of Statistics. 2002. *Population and housing census* (NDC level data). Georgetown, Guyana: Bureau of Statistics.

Catholic Relief Service (CRS). 2005. *Report on Guyana flood.* Georgetown, Guyana: CRS.

Daniel, J.R.K. 1984. *Geomorphology of Guyana: An integrated study of natural environment*. Occasional Paper no. 6. University of Guyana, Georgetown.

———. 1991. Potential impacts of sea level rise on the Guiana coast. International Sea Level Rise Studies Project, Division of Coastal and Environmental Studies, Rutgers University and the State University of New Jersey.

Dasgupta, S., B. Laplante, C. Meisner and D.Wheeler. 2007. Impact of sea level rise on developing countries: A comparative analysis. World Bank Policy Research Paper no. 4136, Washington, DC.

Freeman, P.K. 2006. Estimating chronic risks from natural disasters in developing countries: A case study of Honduras. Paper prepared for the Annual Bank Conference on Development Economics: Europe Development Thinking at the Millennium, September 2006.

International Federation of Red Cross and Red Crescent Societies (IFRC). 2005. *Report on Guyana flood*. Georgetown, Guyana: IFRC.

Kandasammy, L. 2006. A brief history of floods in Guyana. *Stabroek News*, 16 February 2006.

Organization of American States (OAS). 2002. *Natural Hazards Project report*. Washington, DC: Office for Sustainable Development and Environment, OAS.

Pelling, M. 1996. Coastal flood hazard in Guyana: Environmental and economic causes. *Caribbean Geography* 7 (1): 3–22.

———. 1997. What determines vulnerability to floods: A case study in Georgetown, Guyana. *Environment and Urbanization* 9 (1): 203–26.

Rodda, J.C., and H.E. Rodda. 1999. *Hydrological forecasting in dealing with natural disasters*. London: Royal Society, Science Advice Section.

UNICEF Guyana. 2006. UN flash flood appeal. Consolidated donor report for CIDA, Governments of Italy and New Zealand. Georgetown, Guyana.

Westmaas, N. 2005. *Flooding and getting dry*. Posted by E. Kwayana. Commentary: http://www.guyanacaribbeanpolitics.com/kwayana/kwayana_021605.html.

CHAPTER 5

Potential Impacts of Anthropogenic Environmental Change on the Caribbean Karst

MICK DAY AND SEAN CHENOWETH

Introduction

The Caribbean, defined broadly here as all the islands of the Greater and Lesser Antilles, the Bahamas, Trinidad, Tobago, and the islands of the Netherlands Antilles, corresponding to the West Indies (Watts 1987), is one of the premier karst regions in the world, with a limestone area of nearly 130,000 square kilometres, more than half the total land area of the region (Day 1993a). Over 50 per cent of the region's population depends upon the karstlands for water supply, agricultural land or other purposes, and the karst provides a critical backdrop for many environmental, agricultural and economic issues.

The karst is challenging to human habitation, since it possesses a broad array of natural hazards, but it is also inherently fragile and vulnerable to environmental change (ibid.). In particular, karst landscapes are vulnerable because of their unique hydrology, with sporadic surface drainage, limited surface water storage and rapid subterranean flow. Soils within the karst are often patchy and thin, and karst landscape ecology is adapted to these hydrologic and edaphic constraints, the parameters of which can be altered radically by anthropogenic environmental change. This dual vulnerability to natural and human-induced hazards renders the karst a particularly important regional barometer of environmental wellbeing.

In 1997, the International Union for the Conservation of Nature (IUCN) World Commission on Protected Areas (WCPA) recognized karst landscapes, including those in the Caribbean, as being at risk of degradation and warranting protection (Watson et al. 1997). In response to the call for information about protection and conservation efforts in karst landscapes (Gillieson 1996), an ongoing assessment of protected karst areas in the Caribbean region has been initiated (Kueny and Day 1998, 2002), and a series of country-wide studies of karst land use and conservation is being undertaken (Day 1993b, 1996, 2003b, 2006, 2007a, 2007b; Mujica-Ortiz and Day 2001; Day and Chenoweth 2004).

Caribbean Karst

The limestones of the Caribbean range from Holocene to Jurassic in age (0–200 million years BP). Dissolution of these carbonate rocks has produced a range of karst landscapes including dry valleys, dolines (sinkholes), cockpits, residual towers and extensive cave systems. Karst landscapes in the Caribbean have been and are still influenced by tectonic, eustatic and climatic changes (Day 1993a).

The largest karst areas in the Caribbean are on the islands of the Greater Antilles (Cuba, Hispaniola, Jamaica and Puerto Rico) with a total karst area of approximately 115,000 square kilometres. The islands of the Bahamas, the Lesser Antilles, Trinidad, Tobago and the Netherlands Antilles contain an additional 13,000 square kilometres of karst (Day 1993a; figure 5.1). There is much topographic variability, although three distinct karst terrain styles are recognized: doline, cockpit and tower styles (ibid.). Doline karst consists of enclosed depressions and subdued inter-depression divides, and occurs throughout the Caribbean. Cockpit karst, in which enclosed depressions and residual hills attain roughly equal prominence, with the residual hills forming ridges that act as interfluves between the polygonal or linear depressions, occurs

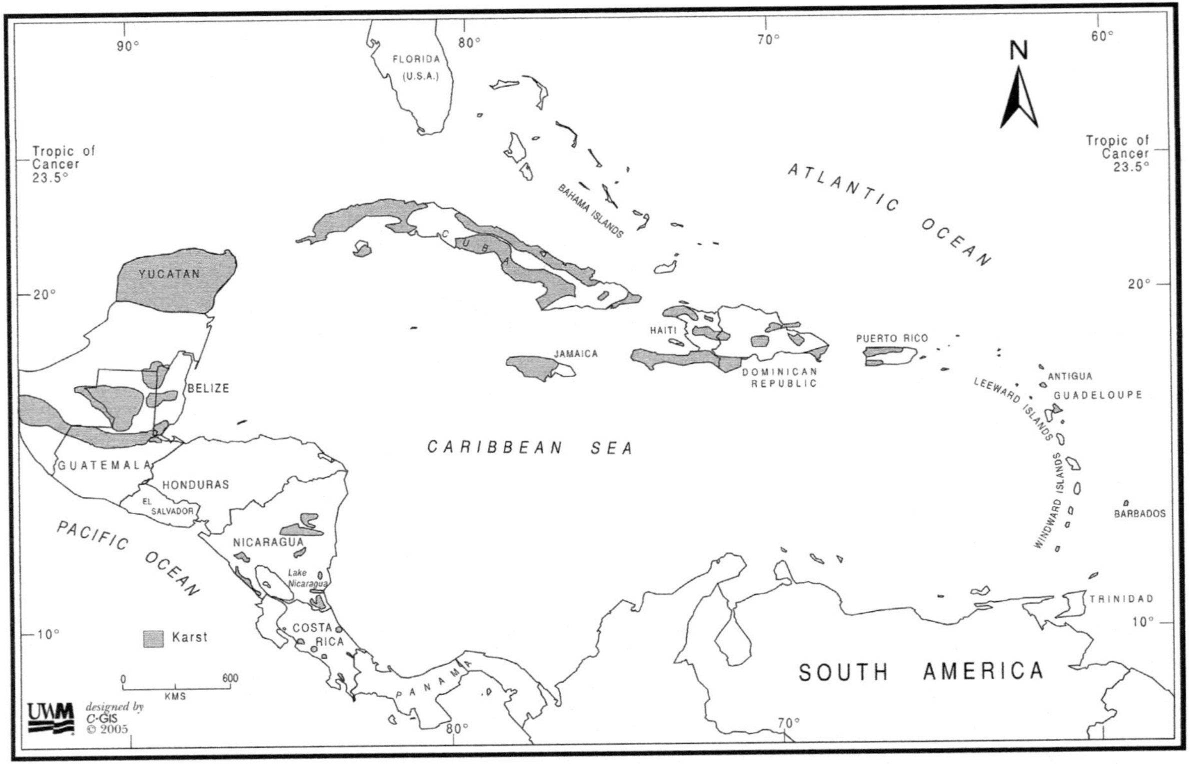

Figure 5.1 The Caribbean karst.

particularly in the Greater Antilles (Cuba, Hispaniola, Puerto Rico and Jamaica). Tower karst, in which isolated residual hills are separated by a near planar surface, is essentially restricted to the Greater Antilles. Caves, formed by the underground dissolution of the carbonate rock, are abundant.

The Caribbean karstlands are far from homogeneous with respect to geologic and geomorphic factors. Moreover, climate, soils and biota are also highly variable, leading to a wide range of specific karst environments. The karst rocks themselves range from pure, dense, hard, fractured, crystalline limestones, some much altered from their original state, to impure, powdery, soft, porous, amorphous carbonates. Some are covered by volcanic ash, and others have been folded and faulted by tectonic forces. Karst landscape elevations range from below sea level up to 3,000 metres; some are mountainous, others planar. Some are hydrologically isolated, while others receive surface drainage from higher, adjacent non-karst terrains.

Climate varies, too, with mean annual precipitation ranging from less than 1,000 millimetres to over 3,000 millimetres. Rainfall generally increases with elevation, and leeward locations experience higher temperatures and lower precipitation than karst areas to the windward. There are distinct, winter dry periods of differing onset, intensity and duration, and there are also brief midsummer droughts (Gamble, chapter 2). Summer convection storms result in spatially uneven rainfall distribution. Late-summer hurricanes and tropical depressions can cause severe flooding in normally dry karst areas, particularly in valleys and depressions.

Karst soils are also extremely variable, but generally tend to be clay-rich, heavily leached, patchy and thin. Steep slopes tend to be devoid of soils, which are thicker in depression and valley bases, where they are often associated with bauxitic infills.

Vegetation varies from xerophytic scrub to wet tropical broadleaf forest, with many species endemic to specific islands. Caribbean forests include both deciduous and coniferous trees, although much of the original forest has been cleared, with only fragments remaining in remote karst areas. The limestone forests are often floristically diverse with high rates of endemism (for example, see Proctor 1986). Wildlife is diverse, although relatively impoverished, and also with a high degree of endemism (for example, see Vogel 1998). Specifics of many regional karstland flora and fauna warrant additional studies.

Contemporary Conditions in the Karst

Beyond the overall roughness of much of the karst terrain, which makes access and construction challenging in general, the major problem is drought and water supply, particularly in rural areas where groundwater utilization is limited by finances and logistics, and there is still considerable dependence on rainwater and springs. The perennial problem of the paucity of surface drainage assumes even greater seasonal dimensions because rainfall itself is sparse and/or temporally unreliable. Short term dry season drought is a recurrent problem, but may also extend over longer periods and become severe, leading to crop and livestock losses, bush fires and emergency distribution of water supplies (Lashley and Bandara 2001).

Although considerable progress has been made throughout the karst in the provision of reliable urban water supplies via wells and pumps, much of the rural water supply still relies in large part upon collection of rainwater in tanks from roofs and gutters. These collection systems, although simple and environmentally friendly, have their limitations. Storage capacity is limited and supply is unreliable, being characterized by periods of deficit or surplus. Storage may also be further compromised by evaporation and leakage, or by accidental contamination, and the water may be deficient in minerals such as calcium that would otherwise be dissolved during percolation.

Springs remain the other important source of rural water supply, particularly around the karst peripheries. Major perennial springs are the most reliable, but some communities also utilize seasonal and ephemeral springs when discharge is adequate. Drought conditions often require authorities or individuals to bring in water by trucks from non-karst areas or from remaining sources within the karst itself.

Conversely, and perversely, flooding actually poses a greater hazard than drought and has more serious short term consequences, including human death, injury and displacement, and damage to homes and other structures (Lashley and Bandara 2001). Flooding of karst landscapes is an integral component of the natural karst hydrology and to be expected on a seasonal or intermittent basis, although it does not occur everywhere and remains problematic to predict. Flooding within the karst occurs through a number of distinct, but often complementary, mechanisms and affects only certain parts of the karst landscape. Heavy and/or prolonged rainfall, such as that associated with tropical storms, may produce significant overland flow when surface and epikarstic infiltration capacities are exceeded (Day 1979, 2007a). Surface flow is most common where the surface is mantled by bauxite or other regolith, and it poses hazards. These hazards occur particularly where natural drainage paths are provided by elongated, compound depressions or normally dry valleys.

Flooding in the karst may occur where rivers carrying increased allogenic drainage from adjacent non-karst terrain exceed the intake ability of the karst drainage system and overflow their channels. This is particularly likely in poljes, where such occurrences are sufficiently frequent as to be predictable on a broadly seasonal basis, and where flooding may also occur via dispersed estavelles (locations of alternating groundwater recharge and discharge which may function both as sinks and springs).

Flooding may also result from filling and overtopping of the epikarstic reservoir, which lies close to the surface within the upper portion of the vadose zone. Such flooding is unpredictable and uncommon, but is usually localized and temporary. Also, flooding on a broader scale may result from elevation of groundwater levels and upwelling of groundwater via estavelles. This may occur on a more-or-less predictable seasonal basis in larger, low-lying depressions, and it is this regional groundwater upwelling that represents the most serious flooding hazard, particularly in more densely populated, low-lying areas. Increased groundwater levels and volumes may also lead to temporary increases in spring discharge, which may in turn lead to stream flooding on the downstream sides of the karst.

Rapid surface failure is not common within the Caribbean karst, but occasional collapse and subsidence does occur, although it has yet to cause more than minor localized damage and inconvenience. Subsurface dissolution, cavity formation, surface subsidence and collapse represent major hazards in many karstlands (Waltham, Bell and Culshaw 2005), but the Caribbean has not yet suffered their effects as catastrophically as elsewhere. Moreover, certain sites, such as lower slopes and depression bases, are especially prone to collapse, and only about 10 to 15 per cent of sites show any evidence of collapse or subsidence, suggesting that collapse probabilities are generally low (Day 1984, 2003a). Despite these relatively low probabilities, both ground surface collapse and subsidence represent an increasing threat to developing infrastructure, such as highways and public service facilities, plus a minor hazard to rural dwellings and livestock.

There also exists the potential for karst-related deterioration or catastrophic karstic failure of industrial facilities such as dams and retention ponds, such as the storage facilities associated with bauxite mining and processing. Detailed, site-specific, subsurface studies provide the most acceptable way to detect imminent cavity collapse and sinkhole formation, but probabilistic studies allied with site assessment are valuable tools where such investigations are not warranted or not possible for other reasons.

Slope failure also poses a minor to moderate hazard to buildings, roads and other structures, although one that is rarely recognized. Although karst slopes are often considered to be stable, localized failure does occur, particularly along weathered planes of weakness, such as joints, and through shifting of talus. Karst bedrock failure may occur as slab failure, rock collapse or rockfall, the precise mechanism depending essentially on the configuration of the planes of weakness and the internal structure of the rock itself (Day 1978). Slab failure is uncommon, but does occur in more crystalline formations where near vertical joints are undercut, for example, by fluvial erosion of tower bases. Rock collapse and rockfall are potential failure mechanisms, and talus is potentially unstable, particularly when disturbed, representing a potential threat not only to individuals traversing talus slopes but also to construction projects.

Environmental Change and the Caribbean Karst

As examined by Gamble in chapter 2, predictions for the regional Caribbean in the twenty-first century are that anthropogenic climatic change will lead to higher atmospheric carbon dioxide contents, increasing air and water temperatures, rising sea levels, and changing weather patterns, including decreasing precipitation totals, and a potentially increasing frequency of extreme events such as droughts and hurricanes (Nurse and Sem 2001; Pelling and Uitto 2001; Peterson et al. 2002; Church et al. 2004; Webster et al. 2005; Simms and Reid 2006). The most significant and immediate changes that will affect the Caribbean karst will be related to rising sea levels and changes in rainfall regimes, soil moisture budgets, prevailing wind speeds and directions, and patterns of wave action (Nurse and Sem 2001). Coastal locations, where socioeconomic activities, infrastructure and population are concentrated, will be particularly vulnerable (Urich and Day 2005), and the effects of all these changes will be magnified in the coastal karst.

Atmospheric carbon dioxide levels are predicted to reach 550 ppm by the mid twenty-first century, up 55 per cent since 1990 (Hulme and Viner 1998). Heath et al. (2005) suggest that although these increased levels of atmospheric CO_2 may lead to increased tree growth, associated increases in microbial respiration will lead to a decreasing quantity of CO_2 being sequestered through the roots of trees into the forest soil. As CO_2 levels increase in the atmosphere, forests will not be able to perform their role as carbon sinks as well as they do under lower concentrations, in turn increasing the level of CO_2 that will stay in the atmosphere and exacerbate global warming. It is unclear what impact these changes might have upon karst processes, but changes in soil CO_2 levels have clear implications for dissolution rates, as the soil is the primary source of CO_2 dissolved by percolating rainwater.

In the karst, the most damaging results of climate change will be those arising from changes in the karst hydrology as a result of more frequent dry season droughts and wet season floods. Overall, the impact will be to exacerbate and magnify the existing situation, with increasing frequency of high magnitude events and greater extremes of drought and flooding. Climate change may alter the balance of chemical and mechanical processes, and will lead to increases in surface runoff, decreases in surface and underground water storage, decreased spring discharge, and increased sedimentation within karst catchments and caves. It may also have profound effects on future carbonate deposition rates, since increasing CO_2 levels will lead to sea-water acidification, which will have implications for future karst development and global carbon cycling. Climate change may also have a severe negative impact on coral reefs and upon the abundance and distribution of reef-based marine species (Kleypas et al. 1999; Gardner et al. 2003; Bellwood et al. 2004).

Water resources are already limiting, and disruption of the karst hydrological cycle may lead to increasing aridity and desertification, with concomitant impacts on geomorphic processes, ecology and potential land use. This would be similar to that experienced already in southern China (Yuan 1997). Increasing drought will further limit human access to rainwater and will also result in a diminution of recharge to the groundwater, placing an increased burden upon water resources in general (daCunha 1989). In addition, floodwaters may contribute relatively little to groundwater resources, as well as potentially increasing contaminant loads. The availability of water resources will become increasingly critical in the karst, where water is already in short supply and is reliant upon rainwater from small catchments or limited freshwater lenses. Rising sea levels will also increase the risks of saline water intrusion into the restricted freshwater lenses, especially in smaller islands such as Antigua and Barbados, where increasing groundwater salinity is already problematic.

An example of the existing and increasing pressures on Caribbean karst water resources is provided by Antigua (Day 2007b), where there is limited fresh water supply and a risk of saltwater intrusion throughout the karst aquifer, and increasing need for comprehensive water resource and landscape management to meet rapidly increasing water demands (Williams 1994; Cooper and Bowen 2001). Vulnerability to drought throughout the karst is rated as high or very high (OAS 2001) and mitigated only slightly by the current extraction of about 0.6 million cubic metres of groundwater each year (UNCCD 2005). Saltwater intrusion is a problem throughout the karst area, and high salinity levels pose a problem with much of the groundwater, particularly following drought or excessive pumping (Sandberg and Barnes 2004). High levels of total dissolved solids (TDS) also generally limit the direct use of ground water for human consumption, although brackish water in the fractured bedrock aquifers is potentially of considerable importance to the tourism and agricultural sectors (Bisson and Maharaj 2003). There are two desalinization plants in Antigua's karst; during drought these provide up to 83 per cent of all freshwater (UN 2002).

Over-extraction of groundwater and increasing salinity are also increasingly problematic in the karst of the larger islands. For example, excessive groundwater pumping from the northern karst belt of Puerto Rico resulted in an artesian head decline near the coast of nearly 50 metres and a decline of 23 metres in the inland unconfined aquifer between 1970 and 1989 (Lugo et al. 2001). As a consequence, landward advance of the saline/freshwater interface has rendered many public water supplies unpotable.

Related to water supply issues, arable land for crop agriculture is limited throughout the Caribbean karst, and the prospect of land loss and increased aridity as a consequence of climate change and sea-level rise will threaten the sustainability of both subsistence and commercial agriculture. Because water resources and agriculture are so climate-sensitive, it is expected that these sectors will be adversely affected by future climate and sea-level change (daCunha 1989; Nurse and Sem 2001; Fischer, Shah and van Velthuizen 2002; Parry et al. 2004).

The other major impact of projected climate change will be an increase in flooding within the karstlands. Floods will increasingly produce more serious short term consequences, including human death, injury and displacement, and damage to buildings. Flooding may still be expected on a similar seasonal or intermittent basis, but it will remain problematic to predict its geographical distribution accurately. Flooding will continue via distinct, but often complementary, mechanisms, and will affect increased areas of the karst landscape. Heavy and/or prolonged rains, associated with more intense tropical storms, will produce increased overland flow when surface and epikarstic infiltration capacities are exceeded, particularly in elongated, compound depressions as well as dry valleys.

There will also be increased flooding by allogenic drainage from non-karst terrains, particularly in poljes, which may be inundated for longer periods seasonally, if not permanently. Such flooding may affect the function of estavelles and other discharge/recharge features. Flooding as a result of filling and overtopping of the epikarstic reservoir will also become more common and extensive but will remain unpredictable and temporary. Widespread flooding may result from elevation of groundwater levels and upwelling of groundwater via estavelles. This may still occur on a more-or-less predictable seasonal basis in larger, low-lying depressions, and it will remain the most serious flooding hazard, particularly in densely populated, low-lying areas. Increased groundwater levels and volumes may also lead to temporary increases in spring discharge, which may in turn lead to stream flooding on the downstream sides of the karst. Coastal flooding will also increase as sea level rises (Urich and Day 2005).

Changes in the karst hydrologic regime may also promote and exacerbate surface collapse and subsidence, particularly where the karst is mantled by surficial deposits (Waltham, Bell and Culshaw 2005). Lower slopes and depression bases may become increasingly prone to collapse, and collapse probabilities may increase generally, posing increasing threats to developing infrastructure, dwellings, businesses and livestock. Similarly, there will be an increased potential for karst-related deterioration or catastrophic karstic failure of industrial facilities, necessitating detailed, site-specific monitoring.

Hydrologic changes may also increase the risk of slope failure along weathered planes of weakness, such as joints, and through redistribution of talus. Rock collapse and rockfall may become increasingly common failure mechanisms, representing potential threats to individuals, roads, houses and construction projects.

Human Pressure on the Caribbean Karst

Human impacts on Caribbean karst landscapes have been long term and severe (Day 1993a), in particular through clearing of remaining natural vegetation, species extinction or introduction, expanding agriculture, increasing utilization and contamination of water resources, tourism, urbanization, and industrial activities, including quarrying and mining. Important archeological sites are a feature of karstlands throughout the Caribbean, indicating that human disturbance extends back to pre-colonial times.

With a total land area of 233,927 square kilometres (Watts 1987) and a population approaching 40 million people (UNEP 2005; Blouet 2006), pressures on the Caribbean's karstlands are already severe. Increasing population and further economic development will exacerbate human impacts, including those mentioned above.

Potential loss of biodiversity is an important element of environmental change in the Caribbean karst. The 2006 Global Biodiversity Outlook (GBO) identifies four primary "drivers" of biodiversity loss as (1) habitat destruction, largely through the expansion of agriculture; (2) the accumulation of nutrients via fertilizers, sewage and pollution; (3) over-exploitation of species; and (4) the introduction of alien species (Secretariat of the Convention on Biological Diversity 2006).

Much, perhaps 70 per cent, of the Caribbean karst was originally covered by dry tropical forest, with the remainder classified as moist tropical forest or scrub. About 14 per cent of the worldwide total, or 147,000 square kilometres, of tropical dry forest remains in North and Central America, but this remaining area is at particular risk from human induced threats, particularly climate change (Miles et al. 2006). Moreover, little of the tropical dry forest in the region is within some form of protected area, and these protected areas are becoming increasingly isolated (DeFries et al. 2005). Taking their high biodiversity value into consideration, the remaining tropical dry forests should be accorded high conservation priority, and expansion of the protected area network in Mesoamerica should be given urgent consideration (Miles et al. 2006).

Human impact on the karst has been long term and is increasing, particularly, as mentioned earlier, via mining, quarrying, increased groundwater abstraction, urbanization, agricultural development and tourism (Day 1993a). Human impacts on the karst may be schematized by reference to a broad impact-process-consequence model involving human activity, geomorphic process change and landscape response, and problems for human occupancy and activity. This can be illustrated by reference to the karst hydrology. Through time, human activities have increasingly disturbed this system through forest clearance, soil utilization, groundwater abstraction and modification of surface drainage courses. This disturbance has resulted in process changes such as increased runoff, decreased infiltration, increased surface sediment transport, and decreased spring discharge. In turn, this has impacted the human population through increased flood susceptibility, desiccation of springs, accelerated soil erosion and surface subsidence (Day 1993a).

Karst hydrology, in particular, responds rapidly to climate change, especially precipitation change, and anthropogenic activities such as groundwater withdrawal (Ma, Wang and Guo 2004). The impact of contemporary climate change has yet to be documented convincingly in the Caribbean karst, in part because seasonal and short term annual variability is superimposed upon and "masks" longer-term trends. By contrast, anthropogenic impacts via changing land use patterns and groundwater extraction are relatively well documented, although there is a critical lack of long term hydrological and environmental monitoring (Day 1993a).

Recent trends in the karst hydrology of Jamaica illustrate these points, suggesting that climate change and more direct anthropogenic impacts may have a cumulative convergent effect (Day 2007a). Seasonal drought is a longstanding problem in the Jamaican karst, but it is of increasing concern as the karst becomes more populated. Drought has affected the karst increasingly recently, with serious problems in 1996–98, 2000, 2004 and 2005 leading to crop and livestock losses, bush fires, and emergency distribution of water supplies. Much of the rural water supply still relies in large part upon collection of rainwater in concrete catchments and in tanks from roofs and gutters. These collection systems, although simple and environmentally friendly, have their limitations. Storage capacity is limited, and supply is unreliable, being characterized by periods of deficit or surplus. Storage may be further compromised by evaporation and leakage, or by accidental contamination. Springs remain the other important source of rural water supply, particularly around the periphery of the karstlands. Major perennial springs are the most reliable, but communities also utilize seasonal and ephemeral springs when discharge is adequate. Anecdotal evidence suggests that the latter springs are increasingly undependable, but independent long term data are not available to substantiate this phenomenon.

Conversely, and perversely, flooding poses a greater hazard than drought and has more serious short term consequences, including human death, injury and displacement, and damage to homes and other structures. Historically, flooding caused little disruption in sparsely settled areas, but the hazard is now accentuated by hydrologic trends and by increasing settlement within the karst. In recent decades, serious floods in the Jamaican karst occurred in 1963, 1974, 1977, 1979, 1986, 1988, 1993, 1995, 1998, 2001 and 2002. The 1977 floods claimed eleven victims, with damage estimated at J$6 million

(US$2.8 million at that time). Severe flooding in the Newmarket to Hopewell area of St Elizabeth, St James, Hanover and Westmoreland parishes, in June 1979, killed over forty people and caused damage estimated at J$70 million (US$33 million). Rainfall on 12 June 1979 totalled up to 865 millimetres, following three weeks during which precipitation was almost three times the normally recorded amount. Intense surface runoff began extremely rapidly and substantial surface water collection appeared almost immediately, flooding valleys, glades and depressions. The depression containing the Shafston Blue Hole, a permanent vauclusian (confined or artesian) spring, was inundated, and it overflowed, with water cascading 86 metres into a second depression, which filled with 55,000 cubic metres of water. This in turn was overtopped, with water descending another 22 metres to fill a third depression with 113,000 cubic metres of water. This itself overflowed, devastating the village of Brighton. Elsewhere, the floodwaters eroded a channel up to 6 metres deep and 4 metres wide over a distance exceeding 1 kilometre. The resultant floods formed ninety individual lakes, covering more than 10 square kilometres, some of which took longer than a year to recede (O'Hara 1990). More recently, the karst island-wide was flooded during heavy rains during May and June 2002. At this time, Manchester parish was further disrupted by flooding associated with tropical storms Isidore and Lili in September 2002, causing displacement of more than forty families.

Forest clearance and agriculture also have had profound effects on the Caribbean karst landscape. Much of the karst is utilized for some form of agriculture, and these activities are increasing steadily. Sediment eroded from agricultural lands frequently has blocked sinkholes and other karst drainage features. Flooding in the karst may be exacerbated by human activities, such as infilling of sinks and depressions

Limestone quarrying has had pronounced impacts on the Caribbean karstlands. Limestone quarrying for roadbed and general construction has had an impact both locally and nationally, but the most significant quarrying has been for cement production. Regionally, the annual production of limestone in 2004 was about 16 million metric tons, and cement production was nearly 8.5 million metric tons (USGS 2006). By contrast, annual production of limestone in 1990 was about 9 million metric tons, and cement production was about 7 million metric tons (Day 1993a). Lime production in 1990 was about 300,000 metric tons

(Day 1993a), rising to over 465,000 metric tons by 2004 (USGS 2006). There is also a limited amount of marble, gypsum and aragonite mining. Over 10,000 cubic metres of marble were mined in the Dominican Republic in 2004, and 1.2 million metric tons of aragonite were dredged in the Bahamas in each of the years 2000–2002, although only 20,000 metric tons were extracted in 2004 (USGS 2006). Additionally, small-scale mining of phosphate rock and of bat guano from caves still takes place, although less than in previous centuries. Petroleum reserves are associated with limestone in Cuba, Trinidad, Aruba and Barbados.

Commercial bauxite production began in the Caribbean in the 1950s. Jamaica is the world's third largest bauxite producer, with mining occurring mostly in the parishes of Clarendon, St Elizabeth, St Ann, Trelawny and, to a lesser extent, Manchester and St Catherine. Some 100,000 hectares of northern Jamaican karst, particularly in the Dry Harbour Mountains, have been exploited for bauxite and alumina by surface mining, with similar areas affected in the south-central karst. Annual production in 1988 was about 8 million metric tons of dry bauxite and about 2 million metric tons of alumina (Day 1993a). By 2004, bauxite and alumina production amounted to 13.3 million metric tons and 4.08 million metric tons respectively (USGS 2006). Beyond the physical devastation of the surface karst landscape by bauxite mining, the operations have caused deforestation, ecological damage, air pollution and the displacement of thousands of local residents, particularly from St Ann parish. Bauxite mining also reduces soil water retention capabilities. Although topsoil is removed prior to mining and then replaced, the increase in surface area resulting from the mining, and the removal of considerable thicknesses of underlying bauxite, leaves the soil less capable of retaining water. This leads to locally increased surface runoff, and more generally to rapid infiltration and percolation to underground aquifers.

The bauxite mining and alumina production facilities have also been linked to local increases in respiratory illness and degradation of coral reefs. In Jamaica, the bauxite/alumina industries produce 60 per cent of SO_2 emissions, 32 per cent of NOx emissions and 55 per cent of total particulate matter emissions (Claude Davis and Associates 2002). In Discovery Bay, Jamaica, Perry and Taylor (2004) recorded high levels of trace metal contamination (iron, 13,701 ppm; manganese, 237 ppm and zinc, 74 ppm) associated with wind-blown inputs from the bauxite industry.

This contamination also reduces sediment calcium carbonate levels by up to 75 to 80 per cent, having a diluting effect on the carbonate substrate. Similar problems afflict bauxite mining areas in the Dominican Republic.

Ground and surface water contamination is also a serious hazard, particularly where industrial effluents and urban runoff enter underground drainage systems via point recharge. Groundwater contamination has been documented throughout the Caribbean karst where urban and industrial effluents enter the underground flow system. One particular problem has been the contamination of surface waters and groundwater by "red mud", the caustic waste from alumina production (Bell 1986).

Potential Responses to Environmental Change and Population Pressure

Environmental change and that brought about more directly by human land use and land cover modifications within the Caribbean karst must be understood in tandem (Dale 1997), and responses must be integrated to address the entire panoply of issues (UNEP 2005; Simms and Reid 2006). Climate change will have environmental impacts at a variety of spatial and temporal scales, but more direct human impacts will override these in many instances. For example, climate change may produce changes in soil CO_2 levels, but these are also affected by human activities, such as burning (Day 1999). Climate change and other human impacts on the Caribbean karst will result in already at-risk and vulnerable ecosystems as well as human communities experiencing even greater threats. Combinations of vulnerabilities relating to factors such as resource availability and poverty will render certain communities and groups particularly at risk, necessitating integration of climate change parameters and the adoption of appropriate risk management measures (Challenger 2002).

Initially, there is an urgent need for comprehensive regional, national and local surveys of existing conditions and hazards within the Caribbean karst. Recognition of the inherent characteristics of the karst needs to be at the forefront of such surveys, and the region's karstlands need to be clearly identified and delimited. There is a particularly urgent need for increased awareness of the karst area drought hazard and, even more so, of the flooding hazard, especially in known hazard locations. Although such surveys will probably need to be coordinated by government agencies, it is important that other stakeholders, such as local governments, NGOs, local communities and individuals be consulted and involved. Although there are clear roles for official bodies, including national and international NGOs, inclusion of local and indigenous knowledge about conditions and hazards is paramount. Scale issues also warrant attention, since impacts and responses will occur at a variety of spatial and temporal scales. Proactive steps are required from regional bodies and national governments, but local community actions may ultimately prove more important within the karst.

Once the contemporary situation is understood, then strategies for mitigating future environmental change can be developed rationally and with specific reference to the karst, although they may also apply in non-karst areas (UNEP 2005; Simms and Reid 2006). Strategies directed towards human wellbeing within the karst may include the improvement and protection of infrastructure, the development and implementation of comprehensive disaster management programmes, changes in land use policies, and the adoption and enforcement of appropriate building codes. Changing public health risks may be addressed through improved sanitation and health care infrastructure, emergency plans, and public education about the health risks posed by floods and droughts.

Provision of adequate water resources within the karstlands will involve measures such as improved rainfall retention, improved storage and distribution systems, development of alternative water sources, better management of supply and infrastructure, increased conservation, and application of improved technology, including desalinization. Agricultural strategies may include growing more drought-resistant crops, and increasing attention to the maintenance of soil and water resources. The potential effects of climate change and anthropogenic pressures also need to be integrated into broader community planning and into tourism development.

In a broad sense, the severity of both climatic and other anthropogenic impacts within the karst can best be reduced by appropriate land management and sustainable land use planning, including the expansion and maintenance of protected areas (Kueny and Day 1998; Day and Kueny 2006). All elements of natural and human-modified ecosystems within the karst need to be taken into account. Increased monitoring is required (Day and Koenig 2002), and priority should be allocated to the maintenance and protection of natural "buffers" within the karst, such as natural vegetation, surface watersheds, caves and groundwater aquifers.

Although most Caribbean nations now recognize the importance of resource protection for environmental, economic and social reasons, the ramifications of climate change and other human impacts increasingly require this to be a priority, particularly within the region's karstlands, where the potential risks to nature and human wellbeing are magnified and accentuated. Changing environmental conditions within the Caribbean karst may well be a portent for the overall environmental health of the region, and the karst thus represents a potential barometer of human ability to respond to the very real threats to environmental sustainability.

Acknowledgements

Research on Caribbean karst has been supported by grants from UWM's Center for Latin American and Caribbean Studies, and the Mary Jo Read bequest. Figure 5.1 was prepared by UWM's GIS and Cartographic Services.

References

Bell, J. 1986. Caustic waste menaces Jamaica. *New Scientist* 3 (1502): 33–37.

Bellwood, D.R., T.P. Hughes, C. Folke and M. Nystrom. 2004. Confronting the coral reef crisis. *Nature* 429: 827–33.

Bisson, R.A., and U. Maharaj. 2003. *Megawatersheds of the Caribbean: New water from old rocks*. Available at: http://www.earthwaterglobal.com/pubs/Megawatersheds%20of%20the%20Caribbean%20New%20Water%20from%20Old%20Rocks.pdf.

Blouet, B.W. 2006. Population: Growth, distribution and migration. In *Latin America and the Caribbean: A systematic and regional survey*, ed. B.W. Blouet and O.M. Blouet, 124–56. New York: Wiley.

Challenger, B. 2002. *Linking adaptation to climate change and disaster mitigation in the eastern Caribbean: Experiences and opportunities.* Available at http://www.onu.org.cu/havanarisk/papers_cchange3/Challenger.pdf.

Church, J.A., N.J. White, R. Coleman, K. Lambeck and J.X. Mitrovica. 2004. Estimates of the regional distribution of sea level rise over the 1950–2000 period. *Journal of Climate* 17 (13): 2609–25.

Claude Davis and Associates. 2002. *Technical support document for the regulatory impact analysis for air quality regulations developed by the national environment and planning agency.* Prepared for the National Environment and Planning Agency (NEPA) by Claude Davis and Associates, Mississauga, Ontario.

Cooper, B., and V. Bowen. 2001. *Integrating management of watersheds and coastal areas in small island developing states of the Caribbean.* Environment Division, Ministry of Tourism and Environment, Government of Antigua and Barbuda.

daCunha, L.V. 1989. Climate change and water resources development. In *Climate and geosciences*, ed. A. Berger, S. Schneider and J.C. Duplessy, 639–60. NATO ASI Series C: Mathematical and Physical Sciences, vol. 285, Dordecht: Kluwer.

Dale, V.H. 1997. The relationship between land-use change and climate change. *Ecological Applications* 7 (3): 753–69.

Day, M.J. 1978. Engineering hazards in tropical karst terrain. *Applied Geography Conferences* 1: 288–98.

———. 1979. The hydrology of polygonal karst depressions in northern Jamaica. *Zeitschrift fur Geomorphologie N.F.*, Supplement-Band 32: 25–34.

———. 1984. Predicting the location of surface collapse within karst depressions: A Jamaican example. In *Sinkholes: Their geology, engineering and environmental impact*, ed. B.F. Beck, 147–51. Rotterdam: A.A. Balkema.

———. 1993a. Human impacts on Caribbean and Central American karst. In *Karst terrains: Environmental changes and human impact*, ed. P.W. Williams, 109–25. Catena Supplement 25, Cremlingen-Destedt, Germany.

———. 1993b. Resource use in the tropical karstlands of central Belize. *Environmental Geology* 21 (3): 122–28.

———. 1996. Conservation of karst in Belize. *Journal of Cave and Karst Studies* 58 (2): 139–44.

———. 1999. Short-term decline in soil carbon dioxide concentrations upon burning of secondary vegetation in the karst of Belize. In *Essays in the ecology and conservation of karst*, ed. I. Barany-Kevei and J. Gunn, 63–69. IGU Commission on Sustainable Development and Management of Karst Terrains. *Acta Geographica* 36.

———. 2003a. An assessment of karstic collapse hazards at Mount Rosser, Ewarton, Jamaica. In *Sinkholes and the engineering and environmental impacts of karst*, ed. B.F. Beck, 40–49. American Society of Civil Engineers, Geotechnical Special Publication 122.

———. 2003b. Some observations on karst landscape ecology and conservation in Belize. *Journal of Belizean Affairs* 5 (1): 31–45.

———. 2006. Stakeholder reaction to the proposed establishment of the National Park, Cockpit Country, Jamaica. In *Ethnography of protected areas: Endangered habitats – endangered cultures*, ed. P. Simonic, 121–31. Ljubljana, Slovenia: Filozofska Fakulteta.

———. 2007a. Natural and anthropogenic hazards in the karst of Jamaica. In *Natural and anthropogenic hazards in karst areas: Recognition, analaysis and mitigation*, ed. M. Parise and J. Gunn, 173–84. Special Publication 279. London: The Geological Society.

———. 2007b. The karstlands of Antigua, their landuse and conservation. *Geographical Journal* 173 (2): 170–85.

Day, M.J., and M.S. Chenoweth. 2004. The karstlands of Trinidad and Tobago, their land use and conservation. *Geographical Journal* 170 (3): 256–66.

Day, M.J., and S. Koenig. 2002. Cave monitoring priorities in Central America and the Caribbean. *Acta Carsologica* 30 (1): 123–34.

DeFries, R., A. Hansen, A.C. Newton and M.C. Hansen. 2005. Increasing isolation of protected areas in tropical forests over the past twenty years. *Ecological Applications* 15 (1): 19–26.

Fischer, G., M. Shah and H. van Velthuizen. 2002. *Climate change and agricultural vulnerability*. Laxenburg, Austria: International Institute for Applied Systems Analysis. http://iiasa.ac.at/Research/LUC/JB-Report.pdf

Gardner, T.A., I.M. Côté, J.A. Gill, A. Grant and A.R. Watkinson. 2003. Long-term region-wide declines in Caribbean corals. *Science* 301 (5635): 958–60.

Gillieson, D. 1996. *Caves: Processes, development, management*. Cambridge, MA: Blackwell.

Heath, J., E. Ayres, M. Possell, R.D. Bardgett, H.I.J. Black, H. Grant, P. Ineson and G. Kerstiens. 2005. Rising atmospheric CO_2 reduces sequestration of root-derived soil carbon. *Science* 309 (5741): 1711–13.

Hulme, M., and D. Viner. 1998. A climate change scenario for the tropics. *Climatic Change* 39 (2–3): 145–76.

Intergovernmental Panel on Climate Change (IPCC). 2001. *Climate change 2001*. 4 vols., Geneva: UNEP.

Kleypas, J.A., R.W. Buddemeier, D. Archer, J-P. Gattuso, C. Langdon and B.N. Opdyke. 1999. Geochemical consequences of increased atmospheric carbon dioxide on coral reefs. *Science* 284 (5411): 118–20.

Kueny, J.A., and M.J. Day. 1998. An assessment of protected karst landscapes in the Caribbean. *Caribbean Geography* 9 (2): 87–100.

———. 2002. Designation of protected karstlands in Central America: A regional assessment. *Journal of Cave and Karst Studies* 64 (3): 165–74.

Lashley, B., and S.B. Bandara. 2001. *A bibliography of natural hazards in the Caribbean.* Kingston: Caribbean Disaster Information Network.

Lugo, A.E., L.M. Castro, A. Vale, T. Lopez, E.H. Prieto, A.G. Martino, A.R.P. Rolon, A.G. Tossas, D.A. McFarlane, T. Miller, A. Rodriguez, J. Lundberg, J. Thomlinson, J. Colon, J.H. Schellekens, O. Ramos and E. Helmer. 2001. *Puerto Rican karst: A vital resource.* US Department of Agriculture, Forest Service, General Technical Report WO-65.

Ma, T., Y. Wang and Q. Guo. 2004. Response of carbonate aquifer to climate change in northern China: A case study at the Shentou karst springs. *Journal of Hydrology* 297 (1–4): 274–84.

Miles, L., A.C. Newton, R.S. DeFries, C. Ravilious, I. May, S. Blyth, V. Kapos and J.E. Gordon. 2006. A global overview of the conservation status of tropical dry forests. *Journal of Biogeography* 33 (3): 491–505.

Mujica-Ortiz, B., and M.J. Day. 2001. Karst conservation and protected areas in northern Puerto Rico. *Caribbean Geography* 12 (1): 11–23.

Nurse, L.A., and G. Sem. 2001. Small island states. In *Climate change 2001: Impacts, adaptation, and vulnerability*, ed. J.J. McCarthy, O.F. Canziani, N.A. Leary, D.J. Dokken and K.S. White, 843–75. Cambridge: Cambridge University Press.

Organization of American States (OAS). 2001. *Drought risk Antigua.* Washington, DC: OAS, http://www.oas.org/pgdm/hazmap/drought/ant_map.pdf.

O'Hara, M. 1990. Flood hydrology of western Jamaica: A study in a karstic limestone environment. *Singapore Journal of Tropical Geography* 11 (2): 100–116.

Parry, M.L., C. Rosenzweig, A. Iglesias, M. Livermore and G. Fischer. 2004. Effects of climate change on global food production under SRES emissions and socio-economic scenarios. *Global Environmental Change* 14: 53–67.

Pelling, M., and J.I. Uitto. 2001. Small island developing states: Natural disaster vulnerability and global change. *Environmental Hazards* 3: 49–62.

Perry, C.T., and K.G. Taylor. 2004. Impacts of bauxite sediment inputs on a carbonate-dominated embayment, Discovery Bay, Jamaica. *Journal of Coastal Research* 20 (4): 1070–79.

Peterson, T.C., M.A. Taylor, R. Demeritte, D.L. Duncombe, S. Burton, F. Thompson, A. Porter, M. Mercedes, E. Villegas, R.S. Fils, A.Klein Tank, A. Martis, R. Warner, A. Joyette, W. Milk, L. Alexander and B. Gleason. 2002. Recent changes in climate extremes in the Caribbean region. *Journal of Geophysical Research* 107 (D21): 4601.

Proctor, G. 1986. Cockpit Country and its vegetation. In *The forests of Jamaica*, ed. D.A. Thompson, P.K. Bretting and M. Humphreys, 43–48. Kingston: Jamaican Society of Scientists and Technologists.

Sandberg, S.K., and H. Barnes. 2004. Geophysical methods applied to saltwater intrusion in Antigua. In *Symposium on the application of geophysics to engineering and environmental problems*, ed. B. Allred, 936–43. Wheat Ridge, CO: Environmental and Engineering Geophysical Society.

Secretariat of the Convention on Biological Diversity (CBD). 2006. *Global biodiversity outlook 2*. Montreal: CBD. 81pp, http://www.biodiv.org/doc/gbo2/cbd-gbo2.pdf.

Simms, A., and H. Reid. 2006. *Up in smoke? Latin America and the Caribbean: The threat from climate change to the environment and human development*. New York: New Economic Foundation.

United Nations (UN). 2002. Antigua and Barbuda: Country profile. http://www.un.org/esa/agenda21/natlinfo/wssd/antigua.pdf.

United Nations Conference to Combat Desertification (UNCCD). 2005. Draft national action plan for Antigua and Barbuda. http://www.unccd.int/actionprogrammes/lac/national/2005/antigua_and_barbuda-eng.pdf.

United Nations Environment Programme (UNEP). 2005. *Caribbean environment outlook*. Nairobi: UNEP.

United States Geological Survey (USGS). 2006. Minerals yearbook. http://minerals.usgs.gov/minerals/pubs/myb.html.

Urich, P.B., and M.J. Day. 2005. Potential impacts of sea-level rise on Belize. *Caribbean Geography* 14 (2): 73–90.

Vogel, P. 1998. The Cockpit Country workshop. *Gosse Bird Club Broadsheet* 70: 1–2.

Watson, J., E. Hamilton-Smith, D. Gillieson and K. Kiernan. 1997. *Guidelines for cave and Karst Protection*. Gland, Switzerland: International Union for the Conservation of Nature (IUCN).

Waltham, T., Bell, F. and M. Culshaw. 2005. *Sinkholes and subsidence: Karst and cavernous rocks in engineering and construction*. Berlin: Springer-Verlag.

Watts, D. 1987. *The West Indies: Patterns of development, culture and environmental change since 1492*. Cambridge: Cambridge University Press.

Webster, P.J., G.J. Hollans, J.A. Curry and H.R. Chang. 2005. Changes in tropical cyclone number, duration and intensity in a warming environment. *Science* 309 (5742): 1844–46.

Williams, C. 1994. Towards the sustainable management of freshwater resources with special reference to Antigua-Barbuda. *Water Management* 1: 9–47.

Yuan, D. 1997. Rock desertification in the subtropical karst of South China. *Zeitschrift fur Geomorphologie*, Supplement-Band 108: 81–90.

CHAPTER 6

Environmental Vulnerability and Ecosystem Services in the Jamaican Tourist Industry

ELIZABETH THOMAS-HOPE AND ADONNA JARDINE-COMRIE

Introduction

The Caribbean tourist industry is heavily reliant on the physical and social environment for survival within the existing globally competitive market. Despite this, environmental destruction brought about by the direct and indirect impacts of tourism on coastal and marine ecosystems is widespread throughout the Caribbean (de Albuquerque 1991; Bacon 1995; Pattullo 1996; Goodbody and Smith 2002). Much of the destruction has been caused by the removal of vegetation, including mangroves, for the construction of hotels; the use of fresh water in excess of supply; and the contamination of land and water by solid waste and sewage effluent. Less obvious, but also critical, are increased levels of carbon dioxide emissions that occur at tourism resort locations due to greater electricity usage, along with the reduction in the compensatory vegetation cover. The "carrying capacity threshold" of a tourist industry is the point where the services provided by the ecosystem to restore the quality of the required resources cannot meet the demand. The cost of ignoring the significance of this threshold is the decline or even collapse of the industry itself, for, as Pattullo puts it, this threshold is the point at which "tourists vote with their feet and start to go elsewhere" (Pattullo 1996, 106).

This chapter addresses issues surrounding the vulnerability of the physical environment not only in terms of its direct role as a provider of resources for tourism, but also its indirect role in the provision of ecosystem services. Ecosystem services include restoration of the chemical balance of air, water and land through the treatment and removal of waste matter. Where the ecosystem's capacity to absorb waste becomes seriously exceeded to the point that the natural balance can no longer be maintained, environmental degradation – whether of air, water or land – will result. If the process continues unchecked, it will eventually lead to the collapse of the industry, since tourism is so fundamentally dependent upon the quality of the environment. Thus, the extent to which an industry or other activity depends upon ecosystem services for its survival is an indication of its environmental vulnerability.

The purpose of this chapter is not only to raise this important issue but also to advance a methodological approach for factoring in ecosystem services as a cost in the tourist industry, measurable in economic terms. As an example, ecosystem services in the Jamaican tourist industry are examined with respect to the absorption of liquid waste (sewage and waste-water) and sequestration of CO_2. This chapter will provide estimates of the environmental and economic cost of waste-water treatment and carbon dioxide sequestration for the hotel sector of the Jamaican tourist industry. Data on the volume and type of solid waste generated by the hotel sector were not available, so that the calculation of environmental services required for this aspect of waste absorption was omitted from the study. Although the focus here is on tourism, the implications are far broader, and could apply to other industries and activities that depend on the quality of the environment.

The continued viability of the tourism industry depends upon the sustained high quality of all forms of capital: natural, human and human-made. Thus there needs to be an integrated approach to the planning and implementation of tourism expansion in Jamaica and other small island developing states (SIDS) that takes into account both the direct and indirect economic, social and ecological impacts of the industry, many of which have been extensively discussed elsewhere (Holder 1991; Momsen 1998; Romm and Browning 1998; Kempadoo 1999; World Bank 2001; Clayton 2002; Dunn and Dunn 2002; Yaw 2004; Henry 2003; Rhiney 2005; Thomas-Hope 2005; Thomas-Hope and Jardine-Comrie 2007a). Although human and human-made forms of capital are also critical factors in the sustainability of the tourism industry, they are outside of the scope of this present study.

Tourism in Jamaica

In the early decades of the development of Jamaica's tourist industry, from the 1940s to the 1970s, a small, elite overseas market was targeted. Partly as a measure to include Jamaican clients, the trend towards lower-cost, mass tourism was actively encouraged from the 1980s onwards, with the establishment of "all-inclusive" hotels. As this shift to mass-tourism proceeded, so increasing emphasis was placed upon the larger hotels rather than villas and guest houses, and accessibility to beaches became all-important.

Recent Tourism Developments

This new thrust in the tourism product has had the effect of reinforcing the spatial concentration of the industry in specific coastal locations. The environmental impact of this concentration in a few locations has been partly the result of the density of the tourist facilities – hotels, shops, entertainment venues – but partly the concentration of residences for those servicing the industry through various formal and informal activities. Thus there has been a major increase in environmental impacts, including those relating to squatting, as a result of the demand for low-cost housing solutions for workers and informal operators in the industry. Additionally, a number of unemployed persons gravitating to the tourism centres, in the hope of obtaining work, have exacerbated the pressures on surrounding land areas and water courses.

A master plan for sustainable tourism development was prepared by the Government of Jamaica to guide the development of the industry

Table 6.1 Master Plan Targets 2000–2010

Indicator	Baseline 2000	Growth Rate	Target 2010
Stop-over Arrivals (000)	1,320	5.5	2,200
Cruise Passengers (000)	907	10.0	2,200
Visitor Expenditure (US$ Million)	1,362	8.4	2,935
Room Stock	24,000	4.0	35,000
GDP Contribution	8%	8.50%	15%

Source: Government of Jamaica 2002.

and it included plans to aggressively accelerate expansion over the period 2000 to 2010 (see table 6.1) (Government of Jamaica 2002).

The urgency for tourism to be based on the principles of sustainable development was articulated in the master plan, emphasizing the need for the protection of both natural and cultural resources at the national and regional levels. The report states: "Tourism must take a lead role in achieving environmental sustainability, ensuring its own negative impact on the environment is minimized and where necessary, mitigated" (Government of Jamaica 2002, 6). In the report, concern focused upon resource degradation, in particular the quality of coral reefs and water resources, as well as the impact of recreational usage on beach erosion and the depletion of fish stocks through over-fishing.

Thus, while the master plan reflected awareness of environmental issues, and pointed out that there were serious and growing concerns over the decline in the quality of the environment taking place in resort centres, it supported plans that could undermine such stated objectives. In particular, the master plan recommended the expansion of the all-inclusive segment of the hotel industry rather than diversifying the industry as a whole. This has meant that larger resorts have been supported and built in fragile coastal locations than those that existed before.

Guided by the master plan, though contrary to both the views indicated by the Government of Jamaica's review team and issues raised by local environmental lobby groups, investments in seven major hotel developments on the north coast of the island by international (mostly Spanish) chains were agreed and the recent wave of hotel construction commenced in 2005. The large scale of the facilities would approximately double the current number of hotel rooms in Jamaica, thereby adding very significantly to demands for water, sewage treatment, solid waste absorption and energy.

The cruise ship industry is also a major and growing component of Jamaican mass tourism. The total number of cruise ships arriving in Jamaica in 1999 was 269 (Jamaica Tourist Board 2003). The wider Caribbean area is the largest cruise destination in the world and Bryan, citing figures from the Cruise Lines International Association (2005), stated that the number of cruise passengers increased by 13.9 per cent during 2004 (Bryan 2007). He also reported that, despite the hotel developments that are currently taking place in Jamaica, the figure for projected new cruise ship berths is higher than the number of hotel rooms in Jamaica, the Bahamas, Barbados, Trinidad and Tobago, St Lucia, and Antigua and Barbuda combined.

Environmental Implications of the Tourism Developments

Despite the apparent economic successes of the tourist industry, it is important to note that promoting tourism as the way forward for Jamaica, as for other SIDS, brings to the fore the challenge of balancing the increased use of natural resources and the resulting environmental degradation with the goals of sustainable development (Thomas-Hope 2002a, 2002b). Tourism expansion has major implications for marine and fresh water supply as well as quality. There is also a concern for air quality, and both air and water are dependent upon such ecological services as solid waste and sewage assimilation and neutralization.

What many of these impacts of recreational use of natural resources reflect is the excessive stress placed upon ecosystem services that ultimately lead to contamination and pollution of terrestrial and marine ecosystems. The physical impact on terrestrial ecosystems as a result of the construction of large hotels, together with access roads, other facilities, and the new formal and informal housing developments that will be concentrated in the vicinity of the resorts, can hardly be fully assessed.

The environmental impact of tourism development in Jamaica is also evident from the comparative usage of essential resources such as fresh water, as well as the greater propensity for water and air pollution by the tourism sector as compared with the domestic averages. It is known that the tourism industry makes heavy demands on water resources through its use of water for domestic purposes, swimming pools, golf courses and landscaping. For example, it was estimated that local residents use 50 gallons (227.5 litres) of water per day while tourists use 80 to 150 gallons (365 to 680 litres) of water per day (World Bank 2001). In a study conducted in the Montego Bay area of Jamaica, it was calculated that while tourists accounted for 4 per cent of Montego Bay's population, the hotels consumed nearly 15 per cent of the water supplied through the National Water Commission (NWC). The all-inclusive hotels consumed disproportionately more water per person than did the small hotels and villas (Sam 2001).

Calculations by the Jamaica Promotions Corporation (JAMPRO) estimated that the projected hotel developments would require an additional 7.5 million gallons (34 million litres) of fresh water per day (mgd), of which the NWC would be required to supply 3.5 mgd (16 million litres per day) (Government of Jamaica 2005). This occurs against a background of water supply shortages in two of the parishes (St Ann and St Mary) where the new developments are sited. Further supplies will be needed for the additional housing for local residents that will accompany the resort developments. Although the hotels will pay a utility cost for the water consumed, there is evidence that this is not commensurate with the replacement cost of water should the supplies be depleted in the long run. For if that were the case, alternative fresh water supplies would be necessitated through desalinization of sea water or other technological solutions. A study has demonstrated that the replacement cost of fresh water currently used by hotels is approximately eighty times more than the amount paid as a utility to the NWC (Thomas-Hope and Jardine-Comrie 2007b).

Additionally, there are strong positive correlations between volume of tourist arrivals and CO_2 emissions and water pollution (as measured by biological oxygen demand [BOD]) (Thomas-Hope 2001). Calculations show that 1 kilogram of solid waste per person per day is produced on average by local residents as compared with 3.79 kilograms per person per day by tourists (based on 2001 data from the Statistical Institute of Jamaica [STATIN] and the National Environment and Planning Agency [NEPA]). The figures for waste produced per tourist are calculated per bed night of hotel stay and thus exclude the additional amount produced by cruise ship day visitors. Furthermore, those hotels that have sewage treatment plants treat the effluent only up to the secondary level, and even then, some of these plants have been deemed unsatisfactory with respect to the NEPA guidelines for secondary treated sewage effluent.

There will undoubtedly be a substantial increase in the volume of sewage produced as well as waste-water generated by laundries of the new hotels. Inadequate sewage treatment and disposal facilities are primary causes of contamination of the rivers, gullies and coastal waters. Large, new sewage treatment plants are planned for the Falmouth and St Ann's Bay areas and also at the sites of a number of hotels. However, even despite the secondary treatment plants, the additional sewage and waste-water that will be generated will bring with it the necessity for a vast increase in the amount of sewage that will have to be treated in the environment. This sewage will be from the secondary treated level as well as much that will not have been treated at all prior to deposition in the surrounding terrestrial and marine environments. Already there have been sewage problems at one of the new hotel properties, where it was reported that waste-water was disposed of in deep wells intended only for relatively small quantities in excess of that used for irrigation of the hotel property (*Sunday Herald*, 22–28 April 2007).

If the required level of treatment of effluent is not met, the probable outcome will be increased nutrient loading leading to eutrophication. This, along with over-fishing and *diadema antillarum* (black long-spined sea urchin) die-offs, would result in coral reef degradation. As a consequence, there would be reduced habitat for fishes, reduction in the areas for fish nurseries, erosion of the coastline, and, in addition, limited attractions for tourist viewing as well as sports such as diving and snorkeling. Discharge of the effluent from any treatment facility is a source of potential danger that needs very serious attention. Even where sewage is partially treated before discharge into the environment, its further treatment requires ecosystem services which, once exceeded, would have to be provided through additional technological input with its commensurate costs.

Not visible, but also of major significance, is the increased production of CO_2, and at the same time, the reduced capacity of the environment to absorb it, as trees are removed for resort construction. The implications of this are far reaching because of the significant effect of carbon emissions on atmospheric warming both of the immediate locality and, cumulatively, at the global level. CO_2 emissions into the atmosphere will increase commensurate with the extent of energy consumed from fossil-fuel-derived electricity. There are no mandatory stipulations that the hotels should establish installations for alternative sources of energy. Nor are the large, all-inclusive hotels being built according to specifications designed to save energy either through the enhancement of natural lighting or cooling.

There needs to be an integrated approach to any expansion of the tourist industry taking the direct and indirect economic, social and ecological implications into account. This approach should involve continuous assessment and monitoring of the sector to ensure that it operates within the carrying capacity of the island's environment. The establishment of appropriate sustainability indicators then becomes important. These indicators would need to be of both a qualitative and quantitative nature, and would require the monetary valuation of environmental resources and services (even unpriced ones) used in the tourism industry. This would allow for comparisons to be made between the economic benefits of the sector and its natural resource use, while providing the information that would allow environmental use to be measured in terms of monetary cost, since it is on the basis of such criteria that most policy decisions are made.

Ecosystem Services in the Tourism Industry

Costing Ecosystem Services

For the purpose of this chapter, the economic contribution of the tourism sector is represented by gross domestic product (GDP). In addition, as indicated above, the ecosystem services selected are sewage treatment and the sequestration of CO_2. It will be, of necessity, a partial model of ecosystem services demand. Nevertheless, it is a model that includes two critical elements with respect to the industry. A comprehensive model of environmental vulnerability would include all indicators of tourism ecosystem services as well as resource use. The development of such a model is prevented by the lack of comprehensive data.

Limitations

In this study, the attempt to value ecosystem services encountered many conceptual and empirical challenges. In particular, the limitations of data necessitated the use of the hotel sector as a proxy for the tourism industry. As a consequence, the very significant cruise ship and transportation aspects of tourism, among others, were not included. This meant that, for example, CO_2 produced during air and ground transportation of tourists was not part of the calculation. This was the case even though they are two of the most significant sources of this greenhouse gas associated with tourism and, therefore, of far greater environmental burden than that associated with the stay in the hotel. Furthermore, sewage treatment and CO_2 sequestration were the only elements for which time series data existed.

The data limitations meant that calculations of environmental uses provided a very conservative view of the true level of environmental vulnerability of the industry. This serves to highlight the fact that the situation can only be more severe than the calculations made in this study. It therefore underscores the need for improved monitoring of the environmental implications of tourism (as well as other economic sectors) so that they can be included in national accounts and thus provide the basis on which assessments of contributions to sustainable development can be made.

The absence of data over an extended time period also meant that only limited numbers of years could be used to calculate water use and electricity use in the tourism sector. As a consequence, extrapolations had to be made from point estimates in some instances; many categories of ecosystem services used in the sector were omitted, as were the value of tourism implications for biodiversity loss and landscape change or ecosystem destruction. In spite of these limitations, valuation estimates were calculated if only to determine a crude initial measure of the magnitude of the cost of two critical ecosystem services with respect to some of the demands of the industry – namely the liquid and gaseous waste assimilation functions provided by the environment.

Valuation Techniques Used to Assess the Cost of Ecosystem Services

Waste-water absorption and carbon dioxide emissions are two of the ecosystem services which, along with solid waste absorption, are required by Jamaica's hotel sector to function efficiently. Sewage absorption and carbon dioxide sequestration are selected for attention in this chapter, based on the authors' research first published in *International Development Planning Review* (see Thomas-Hope and Jardine-Comrie 2007b).

The methodology for the valuation of ecosystem services involved the collection of estimates of the production of sewage and other waste-water effluent and carbon dioxide (CO_2) emissions. For the calculation of sewage and waste-water effluent, the assumption was made that the volume of water used would be equivalent to the volume of effluent produced. CO_2 emissions, both from direct electricity use and from the production of water for the hotel sector, were calculated from data obtained from the Jamaica Public Service Company (JPSCO) and from the NWC.

The replacement cost valuation technique was used for the valuation of the two ecosystem services selected. This technique has been identified as a useful form of valuation where limitations would prevent the application of contingent valuation methods. It involves the estimation of the value of the resource or damage to the resource, based on the cost to restore, rehabilitate or replace the relevant resource or ecosystem service (Ulibarri and Wellman 1997). The result is a surrogate measure of value that is as consistent as possible with the economic concept of "use value".

Sewage Treatment

Secondary treatment has been found to remove only about 10 to 30 per cent of the nitrogen and phosphorus compounds present in sewage. It is estimated that excessive growth of algae occurs when phosphorous levels exceed 0.1 parts per million (ppm), if phosphorous is limiting, or when nitrogen levels exceed 0.3 ppm, if nitrogen is limiting (Mele n.d.). Yet, secondary treated effluent typically contains 20 ppm nitrogen and 9 ppm phosphorous. The environment, therefore, must assimilate and neutralize the excess nitrogen and phosphorous in this treated effluent for the integrity of the ecosystem to be maintained. Should the status of the environment deteriorate to the point that it is unable to function in this way, the equivalent service would have to be provided by further technical intervention along with relevant costs.

Various criteria have been used for estimating these costs (Gordon 2001). The cost associated with the conversion of secondary treated sewage effluent to tertiary level, that is the stage of being reusable irrigation water for agriculture, was used here as a crude estimate of this service provided by the environment. The volume of sewage effluent released by hotels in Jamaica was obtained from the NWC. The approximate value of the waste assimilation service was calculated as the cost of increasing the level of sewage effluent treatment from the secondary to tertiary level based on a rate of US$0.15 per cubic metre. (calculations by Haruvy, Yaron and Shalhevet 2001; cited in Gordon 2001). The cost of tertiary treatment or value of environmental waste assimilation services was calculated as US$2,570,000, in 2004, based on an estimated 17.13 million cubic metres of sewage effluent. The amount of effluent increased in volume from 13.67 million cubic metres in 1996, with an estimated US$ 2,050,000 cost of assimilation. These costs represent conservative estimates, as other calculations have ranged from US$1.45 per cubic metre to US$4.00 per cubic metre (Gutteridge, Hoskins and Davey 2002; de Rooy and Engelbrecht 2003).

Sequestration of Carbon Dioxide

CO_2 is a greenhouse gas produced in the generation and use of energy derived from fossil fuels. Since trees and other plants provide a critical service in the sequestration of CO_2, the extent to which vegetation cover is depleted in combination with electricity produced and consumed are major factors causing the excessive accumulation of CO_2 in the atmosphere. The occurrence of such atmospheric pollution reflects the extent to which the capacity of the environment to provide the necessary ecosystem services has been exceeded.

The international estimate for the sequestration of CO_2 by vegetation cover is 1 hectare of forest needed to sequester every 5.2 tonnes of CO_2 (Barrett et al. 2002). On this basis, this chapter used levels of CO_2 emission from hotel operations to calculate the biological productive area (forested land) that would be needed to absorb this greenhouse gas so that there was no increase in atmospheric CO_2. The replacement cost of forest was therefore used to estimate the value of the hotels' use of the atmosphere to provide the ecosystem service necessary to clear excessive CO_2 (Moulton and Richards 1990; Parks and Hardie 1995; Stavins 1995; Sedjo et al. 1995).

The (Jamaica) Forestry department data were used for estimates of the cost of forest establishment and maintenance, and data from the Jamaica Public Service Company (JPSCO) provided measures of electricity used by hotels (JPSCO 2004). January to December 2003 was taken as the period for which the calculations were made. The total number of tourist bed nights for the same period was then used to calculate the electricity usage per tourist per night (Jamaica Tourist Board 2003). This information was used to calculate the KWh usage of electricity by hotels per annum from 1996 to 2003, both in direct electricity usage and also the consumption of electricity used in the production of water used by the hotels.

Data on the CO_2 emissions per KWh of electricity generated from all power plants were obtained from the JPSCO. The information indicated that, on average, 1.11 kilograms of CO_2 is released per KWh of electricity produced in Jamaica. It should be pointed out that this is not simply a reflection of the consumption of electricity, but it is compounded by the high level of CO_2 produced in the conversion of oil to energy used for the production of electricity in Jamaica. This contributes significantly to the environmental costs of the hotel industry as also to all other users of electricity. Data from the NWC indicated that, in 2002, 9034.24 KWh of electricity were needed to supply one megalitre of water, which translated into 9,034,240 KWh per million cubic metres. From this, the total electricity used in the hotel industry, both directly and in the production of water for the sector, was calculated. It should be noted here that water supplies are chiefly obtained by pumping from aquifers, and that this requires large amounts of electricity, greatly exceeding the amount of electricity directly used by the hotels.

The total CO_2 emissions of the hotel sector for one year, for example, from January to December 2003, would have amounted to 152,077.74 tonnes. This figure would have increased since that time, commensurate with the increase in the number of hotel bed nights sold. Although data for electricity use are not yet available for the years since 2003, the upward trend is evident in the previous years, when in 1996, with

5,012,398 hotel bed nights sold, the total CO_2 emissions were calculated to have been 125,631 metric tones; in 1999, with 5,687,446 hotel bed nights sold, the total CO_2 emissions had risen to 135,717.30 metric tonnes. This annual rate of increase in numbers of bed nights sold and in carbon dioxide emissions has continued (see figure 6.1).

The economic cost of compensating for these levels of CO_2 emission was calculated on the basis of the cost of planting and managing forests on the scale that would be required for the sequestration of the volume of CO_2 emitted. This cost would be very high in the first three years of

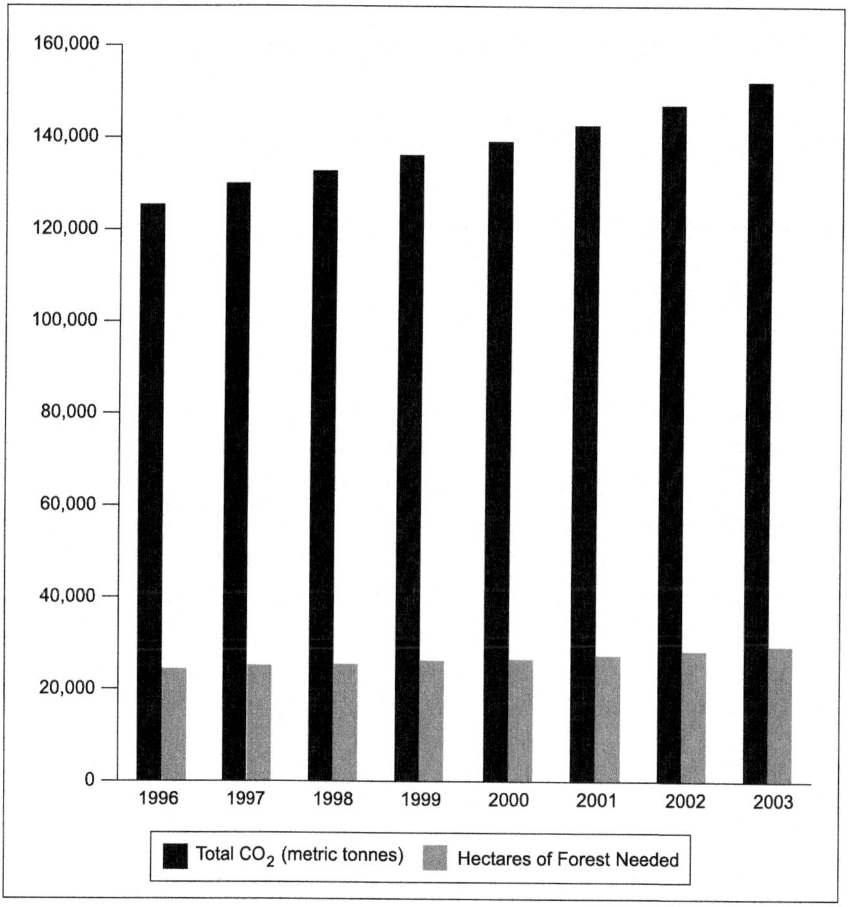

Figure 6.1 Carbon dioxide produced by hotels and potential for forest sequestration.
Source: Based on data from the Jamaica Public Service Company and the National Water Commission, Jamaica.

any such project and then would decrease while only management, as opposed to planting and management, was required. Should the year 2003 be taken, for example, as the first year of forest regeneration, the more than 29,000 hectares of forest calculated as required to compensate for the CO_2 emissions in that year would be at a cost in the order of US$ 400 million per year for about three years (Thomas-Hope and Jardine-Comrie 2007b).

Value of Tourism Environmental Services Relative to Contribution to GDP

As a further indication of the cost of ecosystem services provided in the treatment of sewage and sequestration of CO_2, the estimated economic costs of each of these were compared with the economic contribution of the hotels to gross national production. Taking the year 2003 as providing the baseline for calculations, the combined estimated economic cost of sewage treatment and CO_2 sequestration amounted to approximately US$ 20,000 million, whereas the total contribution of the tourist industry was approximately US$ 380 million (Thomas-Hope and Jardine-Comrie 2007b).

First, the direct payments made for utilities were subtracted and adjustments made for annual inflation, but even then, it appeared that the contribution of the tourism industry to GDP fell short of the economic value of the environmental services that the tourism industry received. This was particularly the case with respect to sequestration of CO_2. This reflects the fact that while environmental resources were paid for as utilities, the additional functions of the environment that allowed sewage treatment and CO_2 sequestration to take place were effectively treated as being “free”. It must be noted that this is a common situation and by no means confined to the tourism industry.

Conclusion

In the case of the present study, it is important to stress that only two of the environmental goods and services used by hotels are included here and that, even then, they exceed the cost of the total GDP contribution of the overall industry. If the cost of high quality fresh water and air as well as the cost of solid waste absorption had also been included, the discrepancy between GDP contribution and cost would have been the ratio of cost to contribution, and would undoubtedly be very high indeed. This means therefore that the actual demand for ecosystem services by the tourist industry is much greater than these measurements show. The economic benefit of tourism is reflected in the contribution to GDP of the industry as a whole, but the costs are rarely quantified.

The results of the study indicated that CO_2 sequestration is an ecosystem service of major significance that is provided free to the hotel industry. Furthermore, unlike the other services for which hotels currently bear some cost (for example, the treatment of sewage up to the secondary level), there is no charge for the emission of CO_2 into the atmosphere. The method used to calculate the value of CO_2 sequestration services could also have contributed to the high value of this service, but cheaper methods of achieving the same goal are not yet available, as already indicated by the environmental cost of production of electricity in Jamaica. The only way to establish the true picture would be to do a complete life-cycle analysis for the industry, which is outside the scope of this chapter.

It is important to note that the calculations made with respect to the use of environmental services in the Jamaican tourist industry are applicable to the Caribbean more generally. As far as the authors are aware, no similar studies exist for other countries in the region, and thus, a lack of accounting of environmental services for the tourist industry occurs in all Caribbean states. Therefore, the gap that was calculated to exist for Jamaica between the cost of environmental services in the form of sewage treatment and carbon dioxide sequestration, as well as the economic returns of the industry, would be expected to exist in other regional examples.

In all Caribbean countries, the full picture of the impact of tourism on the environment is a complex one and the demands that the industry makes on the environment will undoubtedly continue to grow, given current expansion plans. To address this issue, there needs to be a full accounting of all natural resources and environmental services utilized by the industry, with these incorporated into the national tourism planning strategy. Obtaining the support of the major players, in particular hoteliers, in pursuit of this process is unlikely to occur spontaneously, and a system of incentives and penalties would be required.

Where the costs and benefits of an industry are analysed on the basis of direct resource use, and exclude indirect use through services provided by the environment, then only short term and localized effects are taken into account. Where the assessment of national, sustainable development is the objective of calculating costs and benefits, both direct and indirect uses of the environment in goods and services need to be included. Only when payments are made that could potentially replace depleted resources, can they be managed so that the environment will continue to provide the required services for future generations. Since many of the resources used by the industry are not replaceable, it is vital that the monetary cost of such environmental resources and ecosystem services that maintain their good condition serve as an indication of the environmental vulnerability and thus the extent of caution that needs to be exercised in the usage of the relevant resources. This information should be used in the formulation of plans aimed at reducing such usage through the development of a new orientation for the industry.

References

Bacon, P.E. 1995. Wetland resource rehabilitation for sustainable development in the eastern Caribbean. In *Environment and development in the Caribbean: Geographical perspectives*, ed. D. Barker and D.F.M. McGregor, 46–56. Kingston: The Press, University of the West Indies.

Barrett, J., H. Vallak, A. Jones and G. Haq. 2002. *A material flow analysis and ecological footprint of York: Technical report*. Stockholm: Environment Institute.

Bryan, A. 2007. Sustainable Caribbean tourism: Challenges and growth to 2020. In *No island is an island: The impact of globalization on the Commonwealth Caribbean*, ed. G. Baker, 44–72. London: Chatham House, Royal Institute of International Affairs.

Clayton, A. 2002. Strategies for sustainable tourism development: The role of the concept of carrying capacity. *Social and Economic Studies* 51 (1): 61–98.

de Albuquerque, K. 1991. Conflicting claims on the Antigua coastal resources: The case of the McKinnons and Jolly Hill Salt Ponds. In *Caribbean ecology and economics*, ed. N. Girvan and D.A. Simmons, 195–206. Kingston: Institute of Social and Economic Research.

de Rooy, E., and E. Engelbrecht. 2003. *Experience with residential water recycling at Rouse Hill.* CD-ROM, Water Recycling Australia, second national conference 1–3 September, Brisbane. Australian Water Association, Sydney.

Dunn, H. S., and L.L. Dunn. 2002. *People and tourism: Issues and attitudes in the Jamaican hospitality industry.* Kingston: Arawak.

Goodbody, I., and D. Smith. 2002. Recreational use of natural resources. In *Natural resource management for sustainable development in the Caribbean*, ed. I. Goodbody and E. Thomas-Hope, 389–425. Kingston: Canoe Press.

Gordon, D. 2001. Incorporating environmental cost into an economic analysis of water supply planning: A case study of Israel. MSc thesis, McGill University, http://www.rem.sfu.ca/pdf/gordon.pdf

Government of Jamaica. 2002. *Master plan for sustainable tourism development.* Kingston: Government Printing Office.

———. 2005. Tourism/bauxite expansion study framework progamme of action: Tourism component. Report, Planning Institute of Jamaica, Sustainable Development and Regional Planning Unit. Kingston.

Gutteridge, G., G. Haskins, and G. Davey. 2002. *Urban growth areas water recycling strategy report.* Melbourne: Australian Academy of Technological Sciences and Engineering.

Haruvy, N., D. Yaron and S. Shalhevet. 2001. *Irrigation with treated waste water: Assessment of agriculture and environmental aspects.* (English translation of title, original in Hebrew.) Jerusalem: Rehovet.

Henry, M. 2003. The prospects for using hotels as a niche market for agricultural produce the Lower Rio Grande Valley, Portland, Jamaica. MSc thesis, Integrated Urban and Rural and Environmental Management, University of the West Indies, Kingston, Jamaica.

Holder, J. S. 1991. Pattern and impact of tourism on the environment of the Caribbean. In *Managing tourism*, ed. S. Medlik, 278–85. Oxford: Butterworth-Heinemann.

Jamaica Tourist Board. 2003. *Annual travel statistics 1996–2003.* Kingston: Ministry of Tourism, Government of Jamaica.

Jamaica Public Service Company (JPSCO). 2004. *Report of the Energy and Service Department of the Jamaica Public Service Company Limited.* Kingston: Jamaica Public Service Company.

Kempadoo, K. 1999. *Sun, sex and gold: Tourism and sex work in the Caribbean.* Lanham, MD: Rowman and Littlefield.

Mele, A. J. n.d. *Assessing impact of sewage effluent.* http://www.users.nac.net/jmele/AITSE.html.

Momsen, J. H. 1998. Caribbean tourism and agriculture: New linkages in the global era? In *Globalization and neoliberalism: The Caribbean Context*, ed. T. Klak, 115–33. Lanham, MD: Rowman and Littlefield.

Moulton, R., and K. Richards. 1990. *Costs of sequestration carbon through tree planting and forest management in the United States, general technical report* WO-58. Washington, DC: US Department of Agriculture and Forest Service.

Parks, P.J., and I.W. Hardie. 1995. Least-cost forest carbon reserves: Cost effective subsidies to convert marginal agricultural land to forest. *Land Economics* 71 (1): 122–36.

Pattullo, P. 1996. *Last resorts: The cost of tourism in the Caribbean.* Kingston: Ian Randle.

Rhiney, K. 2005. Tourism and its linkage with the local food supply network: A case study of Negril, Jamaica. Paper presented at the Annual Meeting of the South-Eastern Division of the Association of American Geographers (SEDAAG), West Palm Beach, Florida.

Romm, J.J., and W.D. Browning. 1998. Greening the building and the bottom line: Increasing productivity through energy-efficient design. Typescript. Rocky Mountain Institute, Snowmass, Colorado.

Sam, S. 2001. The level of efficiency and social equity that are associated with water consumption in the Montego Bay tourism sector. MSc thesis, Integrated Urban and Rural Environmental Management, University of the West Indies, Mona, Jamaica.

Sedjo, R.A., J. Wisniewski, V.A. Sample and J.D. Kinsman. 1995. The economics of managing carbon via forestry: An assessment of existing studies. *Environment and Resource Economics* 6 (2): 139–65.

Stavins, R. 1995. *The costs of carbon sequestration: A revealed preference approach*. Discussion Paper 95–21. Washington, DC: Resources for the Future.

Thomas-Hope, E. 2001. The role of environmental management in Caribbean economic development. *Integration and Trade* 5 (15): 83–104.

———. 2002a. Resource valuation techniques in promoting the sustainable use of coastal resources: The case of Ocho Rios, Jamaica. In *Ten years after Rio: Steps towards sustainable use and development of coastal and marine areas*, ed. M. Foth et al., 138–43. Bremen, Germany: Internationale Weiterbildung und Entwicklung.

———. 2002b. Managing nature as resource. In *Environmental management for sustainable development in the Caribbean*, ed. I. Goodbody and E. Thomas-Hope, 1–12. Kingston: Canoe Press.

———. 2005. "No nook of ground secure from rash assault?" Ecotourism, heritage tourism and the Jamaican Maroons. In *Caribbean tourism: More than sun, sand and sea*, vol. 3, ed. C. Jayawardena, 193–204. Kingston: Ian Randle.

Thomas-Hope, E., and A. Jardine-Comrie. 2007a. Caribbean agriculture in the new global environment. In *No island is an island: The impact of globalization on the Commonwealth Caribbean*, ed. G. Baker, 19–43. London: Chatham House, Royal Institute of International Affairs.

———. 2007b. Valuation of environmental resources for tourism in small island developing states: Implications for planning in Jamaica. *International Development Planning Review* 29 (1): 93–112.

Ulibarri, C.A and K.F. Wellman. 1997. *Natural resource valuation: A primer on concepts and techniques*. Washington, DC: US Department of Energy.

World Bank. 2001. *Tourism and the environment in the Caribbean: An economic framework*. Paper no. 80, Environmental Economics Series. Washington, DC: World Bank.

Yaw, F. 2004. Cleaner technology as an aspect of sustainable tourism: Caribbean case studies. PhD diss., University of Southern Mississippi.

CHAPTER 7

Framing Vulnerability in Jamaica's Cockpit Country

Economic and Political Constraints on Scientific Claims

KEMI GEORGE

Introduction

In studying the politics of environmental management, one of the central questions to arise is: How can concerned actors turn scientific knowledge into policy to reduce environmental vulnerability? The *epistemic communities* approach to the study of international environmental politics argues that networks of scientists who can gather consensual knowledge about an environmental problem can then persuade states to adopt effective environmental management. However, studies of environmental policy formulation in the Caribbean, and small island developing states (SIDS) in general, indicate that scientific knowledge is not simply presented to policy makers and turned into action. Pressures for economic development encourage policy makers to endorse short term development projects that are highly environmentally stressful; researchers then have to contend with elite economic and political interests who have competing environmental visions (for example, Goodbody and Thomas-Hope 2002; Pugh and Potter 2003, 6–7). This chapter indicates how the use of frames, or symbols and cognitive cues that order actor preferences, constrains scientific consensus and knowledge in a case study of environmental management in Cockpit Country, Jamaica.

Epistemic Communities and Framing

Epistemic communities are knowledge-based networks of individuals, perceived of as authoritative in their field, who share a consensus on causal beliefs, normative concerns, appropriate policy recommendations and scientific validity claims (see, *inter alia*, Haas 1989, 1992, 2001; Keck and Sikkink 1998; Ruggie 1998; Bernstein 2001; Dimitrov 2003). The epistemic community approach is drawn from the constructivist approach of international relations theory, which argues that ideas determine how states and other social actors in international relations behave, and attempts to identify how ideas are created and transmitted (see, *inter alia,* Waltz 1979; Wendt 1987; Sprinz and Vaahtoranta 1994; Burchill et al. 1995; Cox 1996; Finnemore 1996; Katzenstein 1996).

Epistemic communities are generally most effective in policy areas characterized by causal complexity and scientific uncertainty. As supposed truth seekers and custodians of advanced knowledge, scientists may, and arguably should, have a decisive role in influencing decision-making in these cognitively complex areas, particularly when they can generate consensus on the relevant causal arguments (Haas 1989, 1992; Andresen et al. 2000, 10; Thomas 2003; Dimitrov 2003). Scientific consensus reduces uncertainty, de-legitimates competing claims and clarifies appropriate courses of action (Haas 1992; Andresen et al. 2000; Dimitrov 2003). The greater the level of overall agreement, so to is the greater the effectiveness of arguments from epistemic communities.

However, when scholars present the relationship between knowledge and policy as a "linear sequencing of problem definition", they may ignore the role that sociopolitical context plays (Social Learning Group 2001, 349). Strategic *issue-framing*, or the set of metaphors, symbolic representations and cognitive cues employed by policy advocates, limits the policy responses presented to policy makers, and it precludes certain alternatives (Litfin 1996; McAdam, McCarthy and Mayer 1996; Keck and Sikkink 1998; Bernstein 2001; Schreurs et al. 2001).

Consequently, selecting the appropriate frame to present policy arguments will impact the ability of an actor to achieve its preferences. Strategic framing is further constrained by other actors seeking to impose their own agendas on policy discussion, as well as the "master frame": the dominant discourse informing any given issue area (Snow et al. 1986; McAdam, McCarthy and Mayer 1996; Diani 1996). In SIDS, the master frame that informs environmental management interprets environmental policy through the lens of economic development: development is the answer to problems such as resource scarcity and overconsumption, and should be promoted even where the impacts seem environmentally unfriendly (Founex 1972; WCED Report 1987, cited in Bernstein 2001, 65). Of course, SIDSs are also vulnerable to the transboundary effects of environmental degradation caused by heavily industrialized countries (HICs), including ozone depletion, air and water pollution, and global climate change, but this chapter is concerned primarily with those processes SIDS can address.

Methodology

This chapter investigates how an epistemic community attempted to influence protected area management in Cockpit Country so as to preserve biodiversity. Although local and transnational stakeholders were ultimately important in effecting policy change, this research is primarily concerned with how scientific networks use information. Over thirty, open-ended, elite interviews were conducted in this qualitative case study, and this included follow-up interviews and questionnaires of individuals in populations of policy makers and the scientific community. The interviewing took place over June–August of 2005 and 2006 in order to bound the populations, measure scientific consensus and determine how the actors were thinking about the issue of biodiversity management. Where quotes or points are attributed to specific actors, they will be identified, unless respondents requested anonymity. In the interest of space, points derived from multiple actors will be attributed to the social group of which the actors were members.

In order to triangulate the measures and fill in informational gaps left by non-respondents, policy documents, internal and external project reports, grant applications and memoranda pertaining to the issue were analysed. Findings were subsequently coded to measure consensus and framing of the information available on the issue. Consensus is considered a scalar variable, with possible scores of *high*, *low* and *no* consensus; the indicators of consensus are: *causes*, *consequences* and *extent* of a problem (Neumann 2000; Dimitrov 2003; George 2006). Issue-framing is a nominal variable, and indicators of this concept are *framed impacts* and policy options (Neumann 2000; Schreurs et al. 2001; George 2006).

Biodiversity at Risk in Cockpit Country

Cockpit Country is a 450-square-kilometre area of karst limestone, covering the parishes of Trelawny, St Elizabeth and St James (Day 2006). Cockpit Country is a site of high endemism, as many species therein are endemic not only to Jamaica, but to the area itself (ENACT 2004, 20; UNEP 2003, 14; Morgan 2005, 2; Koenig et al. 2000, 11–12; NEPA 2003, 4). The geological history and geomorphology of Cockpit Country has contributed, over time, to the biological diversity present today. The isolated conical hills and depressions characterizing the area have, in combination with poor species dispersal capability, led to the creation of numerous microhabitats and specialized evolution (Koenig et al. 2000, 30; see also NEPA 2003, 4). For example, according to National Environment and Planning Agency (NEPA) (c.2000), over 100 species of plant are endemic to Cockpit Country (many of which are threatened). One of the two surviving populations of the Homerus Swallowtail butterfly occurs in the area, twenty-seven of Jamaica's twenty-eight endemic bird species and thirty-three endemic amphibians are found in the area.

The primary anthropogenic environmental pressures referred to by all members of the epistemic community are bauxite mining activities and small-scale agriculture. Other threats that have been addressed by the epistemic community include hunting and amateur collection, recreational tourism, charcoal production, and the introduction of alien species (Nature Conservancy 2005, appendix c; Forestry Department 2000). In bauxite mining, environmental degradation is caused by the standard open-cast method, which entails removing the entire layer of topsoil and vegetation covering a deposit, then mining the material beneath, causing deforestation and forest degradation (Morrison and Mitchell 1999; Evelyn and Camirand 2003, 355; NEPA 2003; Owen Evelyn [Forestry Department], interviews). The construction of access roads for the transportation of bauxite equipment contributes indirectly to environmental degradation by allowing greater human entry to formerly inaccessible areas (Forestry Department 2001a, 66).

There are also significant secondary impacts emerging from the depletion and degradation of forest cover. The forest provides food sources and habitats to both local and migratory birds, as well as fauna in both terrestrial and aquatic ecosystems. Forest loss, particularly in regards to endemic species, could therefore negatively impact on the food supply of such highly specialized biota, again leading to diminishing populations (Koenig et al. 2000, 30–32; UNEP 2003, 15).

The above problems may also emerge from small-farm agriculture and logging caused by communities residing in the buffer zone. Crops produced in Cockpit Country include yam, marijuana, dasheen and pumpkin, for subsistence farming and sale on local markets, while coffee and pimento are grown for export (Spence 2000; ENACT 2004, 15). Some types of yam production are particularly stressful on the environment. In order to grow yams, farmers require "yam-sticks". These are poles of 3 to 4 metres in length used to support plant biomass and aid in photosynthesis (Forestry Department 2001, 100; Barker and Beckford 2003).

One common means of harvesting yam-sticks is by cutting and stripping saplings of hardwood trees, such as Logwood and Burneye. This prevents regeneration and exacerbates degradation caused by forest conversion for monoculture crop production (Spence 2000; ENACT 2004, 19; NRCA 1999, 5–6; Thomas-Hope and Spence 2006). Cockpit Country is especially vulnerable to yam-stick production, as the parish of Trelawny is the primary yam producing area in the country, with the eastern and southeastern zones providing approximately 42 per cent of Jamaica's national output (Spence 2000, 15; Barker and Beckford 2003; NEPA 2003, 30).

Alien fauna, such as the Shiny Cowbird, and flora, such as the Asian fern, gain access to vulnerable areas when the forest is disturbed (STEA 2003, 15; Forestry Department 2001b). These invasives initially may have no natural predators and can dramatically disrupt the ecological cycles of areas to which they gain access. In addition, the use of chemical pesticides and fertilizer in agriculture can lead to the run-off of toxic chemicals that have leached into soils. The downstream effects of chemical run-off vary, but include soil and water pollution, which could have deleterious effects on human health as well as biodiversity (Bass and Geoghegan 2002, 7).

Finally, general deforestation from these combined activities can cause soil erosion and downstream sedimentation. Combined with increased light in cleared areas, these processes could cause eutrophication in aquatic ecosystems (Kimberly Johns, Nature Conservancy, interview). The relationship between human activity and environmental degradation in Cockpit Country is thus quite complex, moreover since the environmental resources of this area are highly interdependent.

Economic Development and Environmental Vulnerability

The economic rationale behind mining and agricultural activities is fairly clear. Jamaica's bauxite production, in 1999, stood at 13 million tones per annum, or 7 per cent of the total world supply. Bauxite mining in total contributes between 8.5 per cent and 10 per cent of the GDP (Morrison and Mitchell 1999, 2–3; STATIN 2007).

Agriculture contributes less to GDP, 7 per cent in 2006 (CARICOM 2007; FAO 2007). However, agriculture is far more important in utilizing the massive surplus labour of the country, contributing approximately 21 per cent of total employment (CARICOM 2007; Vogel 1999: 3–4). These figures incorporate a dramatically unequal land distribution regime in Jamaica. Ownership is highly concentrated, with 3 per cent of landowners controlling 62 per cent of available farmland. The majority of farmers, both nationally and locally, are therefore smallholders, and virtually all farms in Cockpit Country are less than four hectares (Spence 2000, 14–15; Weis 2000, 302; JSDN 2007).

Consequently, economic development contributes to environmental vulnerability in slightly different ways for both bauxite mining and agriculture. In mining, concentrated industrial interests benefit directly by exploiting natural resources, while in agriculture, poverty and underemployment contribute to a peasantry that increasingly extends agricultural practices into ecologically sensitive areas (Weis 2000; NEPA 2003; Thomas-Hope and Spence 2006).

Policy Makers, Advocacy Networks and Epistemic Communities

Another problem affecting environmental management in Cockpit Country became apparent after conducting research to identify the relevant policy makers responsible for environmental management. The environmental policy framework is characterized by overlapping, occasionally competing, legislation, and a multiplicity of agencies all possessing jurisdiction. The Jamaican government attempted to address these complexities in the regulatory framework through bilateral projects implemented with Canada (ENACT 2004). Despite a reduction from niety-nine articles of environmental legislation, the regulatory framework in Jamaica remains fairly incoherent, with a current count of fifty-two environmentally related articles (NEPA 2003, 13; Webster and Daye 1999, 7; NRCA 1999, 9–10; see also Forestry Department 2001a, II-9 to II-12).

These competing sources of authority, along with a lack of an overarching explicit environmental framework, make it very difficult to discern a coherent direction in environmental management. Indeed, it is not always clear which policy makers are most relevant, a situation common to post-colonial SIDS in the Caribbean (Pugh 2006, 13). Environmental management in one sector may be obviated by another: leases granted under the Mining Act can supersede protection established in Forest Reserves (interviews with Donna Blake [Ministry of Local Government and Environment], Marilyn Headley [conservator, Forestry Department], Susan Koenig [Windsor Research Centre]).

In terms of methodology, this also means that it is inadequate to analyse the state's response to environmental advocacy as if it were a monolithic actor. The identified actors indicated that the most relevant state stakeholders with the authority to implement policies affecting the holistic management of the area are the Ministry of Agriculture and its agencies in the Forestry Department, Jamaica Bauxite Institute (JBI) and the Mines and Geology Division; the Ministry of Local Government and Environment; NEPA; and the prime minister's cabinet (Torres 1998, 18.1; Forestry Department 2001a, 2001b; NRCA Act; interviews and questionnaires with all respondents).

The epistemic community consists of a collection of diverse actors drawn from civil-society groups, the scientific academy and a network of environmental non-governmental organizations (ENGOs). ENGO members are the Nature Conservancy, the Windsor Research Centre (WRC) and the Southern Trelawny Environmental Association (STEA). Researchers from University of the West Indies (UWI), the Institute of Jamaica (IOJ) and foreign universities have been identified, and also self-identify, as members of this scientific community. Although a state agency, the Forestry Department has staff members that share research efforts, findings, and methodological tools with the epistemic community, blurring boundaries between policy maker and epistemic community (Thomas 2003).

In addition to the epistemic community, a transnational advocacy network of stakeholders in Cockpit Country emerged. This network developed through a variety of horizontal linkages: local stakeholders among community based organizations of Accompong Maroons and domestic ENGOs, as well as international actors such as the UK-based Birdlife International, shared a concern about potential resource loss and degradation in the area (UNEP 2003; Spence 2000). The network became formally institutionalized in October 2006 with the creation of the Cockpit Country Stakeholder's Group (CCSG), which currently consists of over thirty organizations and activists.

For conceptual clarity, it should be noted that although the CCSG has become a key actor in environmental advocacy, the absence of a unifying scientific research programme distinguishes it from the narrower epistemic community. The Maroons, for example, justified their participation in Cockpit Country's policy advocacy on the basis of its appeal as a cultural and historic site (Spence 2000). Despite the "unscientific" base of the broader network, the coalition established between these actors and the epistemic community aided in later public mobilization in support of increased environmental protection in Cockpit Country.

Identifying and Framing Threats to Cockpit Country

Two specific threats to Cockpit Country were focused upon in the course of this research: bauxite mining and agricultural degradation. In terms of responses from the epistemic community, these threats were examined in terms of perceived causation, consequences and extent. These are regarded here as indicators of the overall threat to Cockpit Country. An assessment of the degree of consensus has been made, and this is taken forward into a discussion of "framing", both internal among the epistemic community members, and then in terms of their presentation of their arguments, "externally" to policy makers.

Bauxite Mining

Causal Consensus

The causes of the bauxite mining pressure are recognized by the epistemic community: industrial bauxite companies, specifically Alcoa and Clarendon Aluminium Production, conduct standard open-cast mining in Cockpit Country, a process which leads to environmental degradation. The actors responsible for this type of anthropogenic environmental degradation are easily identified, and the causal process is short and direct. When the causal chain in an identified problem is short and possesses identifiable catalysts, policy solutions should become easier to determine and causal arguments more persuasive (Keck and Sikkink 1998, 27–28).

Consequences

The epistemic community shares an agreement about the environmental consequences of bauxite mining. Members are familiar with the process and terminology of standard open-cast mining, and referred to it in interviews and policy documents (Morrison and Mitchell 1999;

Table 7.1 Consensus on Bauxite Mining

Indicator	Score
Causal Consensus	*High*
Consequences	*High*
Extent	N/A
Overall	**High**

NEPA 2003; interviews with epistemic community). All epistemic community members believed mining was the most significant anthropogenic environmental threat to Cockpit Country (Taylor 1998, 7; Evelyn and Camirand 2003, 360; STEA 2003, 15; Nature Conservancy 2005, appendix c; CCSG 2007; interviews and questionnaires with epistemic community).

Extent

At the time of writing, there is no mining in the core of Cockpit Country. Nevertheless, members shared maps indicating the extent of bauxite reserves in Cockpit Country, as well as a belief that mining was inevitable: "[The] whole issue that they will face in the near future, as you probably would have heard, is that [Cockpit Country is] supposed to be a huge bauxite reserve. And so if we do run out of areas of bauxite outside the Cockpit Country, they're going to actually head in there" (Kurt McLaren [UWI researcher], interviews). Overall consensus on the relationship between bauxite mining and environmental change in Cockpit Country is summarized as high (table 7.1).

Agricultural Degradation

Causal Consensus

The causes of agricultural degradation are more wide ranging, compared with the causes of bauxite mining, and result from the collective actions of a large number of frequently anonymous individuals acting clandestinely (Barker and Miller 1995; Evelyn and Camirand 2003, 355).

Consequences

There is minimal agreement on the consequences of agriculture-related environmental degradation: the terms used to classify and evaluate the different activities, particularly "agriculture", "yam-stick cultivation" and "forest conversion", are not used in a consistent manner across policy documents. This complicates attempts to assess and evaluate the environmental significance of the various agriculturally related activities. In addition, the different actors in the epistemic community do not rank the severity of agriculture-related degradation in a consistent manner (Taylor 1998, 7; Nature Conservancy 2000a, 2000b, 2005; Spence 2000, 3; NEPA 2003; interviews with epistemic community).

Extent

The extent of environmental degradation from these activities is unclear, as the minimal hard data available for illustrative purposes have been contested on scientific merit. One primary indicator of environmental degradation is the rate of deforestation, but figures for the national deforestation rate have varied dramatically, from a low estimate of 0.03 per cent to a high of 11.3 per cent of loss of forest cover per annum, in part because of differences in methodology used in forest taxonomy. This is a problem mirrored in attempts to establish the rate of loss in Cockpit Country (Koenig et al. 2000; Forestry Department 2001a, 23–24; Evelyn and Camirand 2003, 355–56; ENACT 2004, 42).

The overall consensus on the relationship between agricultural degradation and environmental change in Cockpit Country is summarized as low (table 7.2).

Table 7.2 Consensus and Agriculture

Indicator	Score
Causal Consensus	*Low*
Consequences	*Low*
Extent	*Low*
Overall	**Low**

Framing I: Internal Framing

In framing their arguments, epistemic community members adopted one type of frame for internal communication, and another for communication with the policy makers. The internal frame adopted by the epistemic community emphasized the *ecological value* of biodiversity management, while the frame used in communication with policy makers emphasized *economic value*. In *ecological value* frames, environmental management was necessary to preserve vulnerable ecosystems and species endemism for intrinsically valuable reasons.

Framed Impacts

The internal arguments used by epistemic community members portrayed biodiversity in Cockpit Country as integrated with the overall environmental health of the island. The standing limestone forest is presented as an important habitat for local and migratory fauna, which display a high degree of habitat sensitivity, endemism and hence vulnerability to human activity (Forestry Department 2001a, 2001b, II–2). Consequently, when explaining their interest in biodiversity management, epistemic community members emphasized a holistic perspective on the role of the area and expressed concern about the figuratively downstream effects of degradation on environmental health: "If you're looking at it as a scientific point of view, if you lose elements of biodiversity, a particular segment of biodiversity, it's going to impact the entire biodiversity. Whether it's trophic levels, relationships, you're going to impact the entire thing" (Dayne Buddo [Jamaica Clearing House Mechanism], interview).

Policy Options

Concomitant with their ecological perspective, epistemic community members expressed a belief that the best management approach to biodiversity in Cockpit Country is to dramatically reduce, if not eliminate, agriculture and bauxite mining in the Cockpit core. Sustainable agriculture, if managed effectively, could take place in the buffer zones.

Currently, bauxite companies are legally obligated to restore mined-out lands. However, companies do not literally have to replace the former vegetative cover of mined areas, which may leave the topography and flora of restored lands significantly altered (Barrett 2005, 5). In fact, studies of land restoration indicate that companies have tended to replace natural forest with cash crops or grassland (Koenig et al. 2000; NEPA 2003). Further, only an estimated 76 per cent of land disturbed by mining is restored annually, with a backlog of land dating to the 1970s (NEPA 2003, 26). Citing ecological concerns, the epistemic community has asserted that restoration is an inadequate response to biodiversity loss and ecological disruption, and they preferred to prohibit mining: "But you know, you can't really replace the Cockpit Country. You can put a forest somewhere else, you can pay for a forest to be planted somewhere else, but you can't replant the Cockpit Country" (Marilyn Headley, interview).

Framing II: External Arguments

However, despite this internal belief that biodiversity was important, primarily due to its ecological function, the epistemic community deliberately adopted economic language in presenting their arguments to the regulators of mining, believing that this language would be more persuasive to governmental actors: "We're trying to communicate to [policy makers] in dollars and cents, which is a language that they more understand than to say, 'This is a particular species that is only found in Jamaica' " (Dayne Buddo, interview).

Framed Impacts

One of the most apparent ways in which the epistemic community attempted to frame their arguments to policy makers strategically was their emphasis on the economic impacts of environmental mismanagement in Cockpit Country. In 2003, the epistemic community attempted to construct a water valuation study of Cockpit Country's hydrological system, which encompasses several Watershed Management Units, supplying tourism centres on the north coast (Spence 2000, 12). This analysis would have established a "compelling cost-benefit analysis when compared with high-value activities such as bauxite mining" (Nature

Conservancy 2005: 2). "What we were going to do is attach a cost, that there's a price associated with the clean water that's supplied by the Cockpit Country. . . If you compare water with bauxite mining . . . [compare] the benefits to the society and the economy of maintaining the resources" (Kimberly Johns, interview).

Policy Options

Where the epistemic community used the water valuation study to portray an economic disincentive to mining, the members also recommended certain policies to illustrate the economic benefits of biodiversity management. The epistemic community argued that the government could create a national park under IUCN classification in Cockpit Country (Spence 2000, 5; NEPA 2003, 9). Under this system, management of the area would be funded in part by the Jamaica National Park Trust Fund (JNPTF) and by international donors. Currently, it is believed that as a result of the complexity of environmental legislation, national park status under existing Jamaican environmental legislation would not amount to much *de facto* additional protection against mining. A further point is that designation of the area as a national park would inhibit the ability of many of the local communities to pursue an agricultural livelihood, reflecting the "fortress conservation" approach to protected area management in SIDS (Spence 2000, 5; Few 2003, 24).

Declaration of the area as a national park would possibly increase its marketability as a destination for ecotourism (CCSG 2007, 7). Ecotourism in sites such as the forest corridor around Burnt Hill Road would alleviate some pressure on coastal tourism and generate revenue through user access fees (Forestry Department 2001a; UNEP 2003, 54; Nature Conservancy 2005, appendix a).

Management strategies aimed at curbing decentralized agricultural activity have presented "carrot-and-stick" incentives to local communities by combining Forest Reserve enforcement with the provision of economic alternatives to environmentally damaging activities. For example, the Forestry Department is attempting to establish plantations of fast-growing tree species to be coppiced for yam-sticks in order to alleviate unmanaged sapling harvesting in Cockpit Country (Spence 2000, 26; Forestry Department 2001a, 51). STEA has also contributed to promotion of the use of the "hedgerow alley" system for growing yams in southern Trelawny, whereby carefully spaced hedgerows aid in curbing soil erosion and the inevitable decrease of agricultural productivity over time (Spence 2000, 23).

Mixed Success

If scientific consensus alone determined policy adoption, it would be expected that the epistemic community would have greater success in persuading the government to adopt an environmentally friendly approach to managing bauxite mining in Cockpit Country. Yet, while there has been some success in promoting environmental protection in Cockpit Country, it has mostly occurred in managing damage caused by agriculture rather than in preventing bauxite mining.

In late 2006, major newspapers reported allegations by residents in Cockpit Country that bauxite companies had begun prospecting and relocating residents (*Jamaica Observer*, "Cockpit Country worry", 19 November 2006; *Jamaica Gleaner*, "Cockpit Country issue may be taken to court", 15 December 2006). The epistemic community, through a letter writing campaign to the Ministry of Agriculture and the Jamaica Bauxite Institute (JBI), as well as to media outlets, sought information about the plans of bauxite companies and continued to ask Minister Clarke to rescind the issued leases (epistemic community, interviews and questionnaires). In December of 2006, the civil society actors appeared to have obtained access to the decision-making process when Schwartz of the WRC, Dixon of STEA and McCauley of JET were all invited to participate in a mining policy consultation with Minister Clarke.

At this consultation, however, the minister and the JBI revealed that the mining leases had been issued weeks before, leading to public recrimination from the epistemic community (*Jamaica Observer*, "Cockpit row boils", 15 December 2006). Minister Clarke claimed the following: "[Prospecting] will not affect the water or damage the environment. If you follow the environmentalists, you would never mine anything" (ibid.; JBI 2006). As of the time of writing, attempts to establish an IUCN national park in the area were postponed indefinitely.

In response, the CCSG mobilized. Actions included issuing media releases, petitions and threatening to conduct protests, and taking a more confrontational stance in arguing for the cessation of mining leases (Wendy Lee [Jamaica Environmental Advocacy Network], Mike Schwartz [WRC], Ann Hayes-Sutton [Nature Conservancy/JET], questionnaires). Shortly after, Minister Clarke suspended the leases (CCSG 2007). Though this achieved one of the primary goals of the epistemic community, this was brought about through the political pressure affected by the advocacy network, not through scientific persuasion by experts. The research conducted by the epistemic community nevertheless played an important role in galvanizing public opinion.

This indicates that the state agencies responsible for regulating mining did not learn from scientific expertise alone. Further analysis indicates that the inability of the community to persuade can be attributed to the fact that the frames were ineffectively constructed. Although the community was using economic language to advocate for policy, the members could not conclusively demonstrate either the economic opportunity loss caused by bauxite mining, or the projected windfall from ecotourism: "[We've] been told number and number of times that because we don't have those figures, we really don't have that sort of a balance to push into the economic model that says bauxite will earn so many billion US dollars" (Marilyn Headley, interview). Policy makers in the JBI argued that since the economic rationale for preventing mining lacked quantifiable data, it was methodologically suspect and unconvincing (Shanti Persaud [JBI], interview).

Where the Ministry of Agriculture and the JBI endorsed bauxite mining in Cockpit Country through the end of 2006, the Forestry Department demonstrated a willingness to learn from the epistemic community and adopt recommendations to manage the biodiversity impacts of agricultural activity. The Forestry Department incorporated management strategies, such as threat identification methodology, from the Nature Conservancy's Conservation Action Plan, and staffers received training from the WRC (Kevin Porter [ranger, Forestry Department], Susan Koenig, Marilyn Headley, interviews). Since 1996, the department has gradually shifted focus from the management of forest resources for commerce and fuel to a more conservationist, ecological approach, incorporating the function of the forest as a species habitat and watershed protection (Forestry Department 2001a, 52). "[Biodiversity] is a new discussion, and I suppose what has changed [are] the views of managing forest for different, different reasons . . . So everybody is talking about the ecosystem, and the habitat, and we even talk about birds and bats, which we didn't do in the Forestry Department before" (Marilyn Headley, interview). In this instance, the community had a lower level of scientific consensus, but had more apparent success in promoting learning in the state. However, in this case, the Forestry Department has had extensive and institutionalized socialization with the epistemic community network, facilitating the transference of norms and information from the civil society, where other agencies have not (ENACT 2004; Jean Jo Bellamy [ENACT project coordinator], Forestry Department, interviews).

Conclusion

This chapter has attempted to analyse how political and economic considerations influence the production and use of scientific knowledge. Although the literature on epistemic communities suggests that scientists have a privileged position in policy formation, the results of this investigation in Jamaica indicate that epistemic communities may have comparatively limited options to persuade, rather than pressure, policy makers.

Epistemic communities may use economic framing to communicate with policy makers concerned about promoting national development, but their success in effecting policy change will depend on their ability to convincingly create an economic cost-benefit analysis. From an economic standpoint, it is argued here that it is less risky to exploit a resource for guaranteed and calculable short term benefits than to prevent exploitation in the hopes of long term benefits of unknown value, especially if those benefits are not guaranteed.

A demonstration of short term economic consequences to environmental management could be greatly enabled if international and local donors commit sufficient resources for establishing protected areas to offset the considerable gains of resource extraction in developing countries. In addition, local and transnational stakeholders can visibly raise the political cost of environmental inaction, as the CCSG demonstrated at the end of 2006. Political coalitions and networking should be used by scientists to construct models clearly demonstrating an economic cost to environmentally unfriendly policy.

On the other hand, this process would leave environmental management vulnerable to changes in short term economic calculations. As with the Forestry Department, internalization of ecological norms might facilitate more effective management. However, the socialization that characterized the relationship between the Forestry Department and the civil society is not a given.

In Jamaica, as in other SIDS, sectors of economic elites have a privileged position in policy-making, limiting the access of the civil society (Edie 1991). A criticism of NEPA, levelled by civil society respondents and supported by interviews with a former NEPA official, is that appointments to the body, since 2001, have been politically based, resulting in an agency unwilling to challenge environmentally unfriendly development projects. Although international donors have attempted to make environmental participatory planning a condition for aid, governments throughout the Caribbean have varied in their commitment to co-operate with other agencies or the civil society, preferring top-down implementation of policy (Pugh and Potter 2003; Pugh 2006).

This research indicates that scientific knowledge claims are subject to political and economic considerations, limiting the space researchers have to use their expertise to inform policy creation. Scientific networks may rely more on publicizing results and mass public mobilization to gain preferred policy. Future research could indicate whether interactions with the mass public further constrain the arguments and frames used by scientists.

References

Andresen, S., T. Skodvin, A. Underrdal and J. Wettestad. 2000. *Science and politics in international environmental regimes*. Manchester, UK: Manchester University Press.

Barker, D., and C. Beckford. 2003. Yam production and the yam stick trade in Jamaica: Integrated problems for resource management. In *Resources, planning and management in a changing Caribbean*, ed. D. Barker and D.F.M. McGregor, 57–73. Kingston: University of the West Indies Press.

Barker, D., and D.J. Miller. 1995. Farming on the fringe. In *Environment and development in the Caribbean: Geographical perspectives*, ed. D. Barker and D.F.M. McGregor, 271–92. Kingston: The Press, University of the West Indies.

Barrett, B. 2005. *Cockpit Country Parks-in-Peril Project: Conservation action planning strategies workshop report*. Kingston: Nature Conservancy.

Bass, S., and T. Geoghegan. 2002. *Incentives for watershed management in Jamaica: Results of a brief diagnostic*. Technical Report no. 314. Laventille, Trinidad and Tobago: CANARI.

Bernstein, S. 2001. *The compromise of liberal environmentalism*. Ithaca: Cornell University Press.

Burchill, S., A. Linklater, R. Devetak, M. Paterson and J. True, eds. 1995. *Theories of international relations*. Cambridge: Palgrave Macmillan.

Caribbean Community (CARICOM). 2007. Country statistics: Jamaica. http://www.caricomstats.org.

Cockpit Country Stakeholders Group (CCSG). 2007. Save Cockpit Country: Fact sheet. Prepared by the Cockpit Country Stakeholders Group, with a grant from the Jamaica Environmental Advocacy Network (JEAN). http://www.jamaicancaves.org/Save_Cockpit_Country_Fact_Sheet_2007.pdf.

Cox, R. 1996. *Approaches to world order*. Cambridge: Cambridge University Press.

Day, M.J. 2006. Stakeholder reaction to the proposed establishment of the National Park, Cockpit Country, Jamaica. In *Ethnography of protected areas: Endangered habitats, endangered cultures*, ed. P. Simonic, 121–31. Ljubljana, Slovenia: Filozofska fakulteta.

Diani, M. 1996. Linking mobilization frames and political opportunities: Insights from regional populism in Italy. *American Sociological Review* 61: 1053–69.

Dimitrov, R. 2003. Knowledge, power and interests in environmental regime formation. *International Studies Quarterly* 47: 123–50.

Edie, Carlene. 1991. *Democracy in the Caribbean: Myths and realities*. Westport, CT: Praeger.

Environmental Action Plan (ENACT). 2004. The ENACT Programme, Kingston: Stories in sustainability. Prepared by C. Easton and Associates for the ENACT Programme. Kingston, Jamaica.

Evelyn O.B., and R. Camirand. 2003. Forest cover and deforestation in Jamaica: An analysis of forest cover estimates over time. *International Forestry Review* 5 (4): 354–63.

Few, R. 2003. Participation or containment? Insights from the planning of protected areas in Belize. In Pugh and Potter 2003, 23–44.

Food and Agriculture Organization (FAO). 2007. FAO statistical yearbook country profiles: Jamaica. Rome: FAO. http://www.fao.org/es/ess/yearbook/vol_1_2/pdf/Jamaica.pdf.

Finnemore, M. 1996. *National interests in international society*. Ithaca: Cornell University Press.

Forestry Department. 2000. *National implementation support partnership*. Kingston: Government of Jamaica.

———. 2001a. National forest management and conservation plan. Government of Jamaica. http://www.forestry.gov.jm/PDF_files/ForestPlan.pdf.

———. 2001b. Forest policy 2001. Government of Jamaica. http://www.forestry.gov.jm/PDF_files/ForestPolicy2001.pdf.

Founex. 1972. *Report on development and environment*. New York: Carnegie Endowment for International Peace.

George, K. 2006. Economic and political constraints on scientific claims in the developing world. Paper presented at the joint panel How the Leviathan Got Its Sight, thirty-ninth annual conference of the Northeastern Political Science Association. Boston, MA.

Goodbody, I., and E. Thomas-Hope, eds. 2002. *Natural resource management for sustainable development in the Caribbean*. Kingston: Canoe Press.

Haas, P.M. 1989. Do regimes matter? Epistemic communities and Mediterranean pollution control. *International Organization* 43: 377–403.

———. 1992. Introduction: Epistemic communities and international policy coordination. *International Organization* 46: 1–35

———. 2001. Policy knowledge: Epistemic communities. In *International encyclopedia of the social and behavioral sciences*, ed. N.J. Smelser and P.B. Baltes, 11578–86. New York: Elsevier.

Jamaica Bauxite Institute (JBI). 2006. No plans to mine in the Cockpit Country. JBI press release. Kingston, Jamaica.

Jamaica Social Development Network (JSDN). 2007. Jamaica Social Development Network project. http://www.jsdnp.org.jm/.

Katzenstein, P.J., ed. 1996. *The culture of national security: Norms and identity in world politics*. New York: Columbia University Press.

Keck, M., and K. Sikkink. 1998. *Activists beyond borders*. Ithaca: Cornell University Press.

Litfin, K. 1996. *Ozone discourses*. Ithaca: Cornell University Press.

Koenig, S., A. Haynes-Sutton, G. Proctor and P. Vogel. 2000. Cockpit Country conservation report: Biodiversity assessment. Report prepared for NRCA/NEPA. Kingston, Jamaica.

McAdam, D., J.D. McCarthy and N.Z. Mayer, eds. 1996. *Comparative perspectives on social movements*. New York: Cambridge University Press.

Morgan, P. 2005. Organising for large-scale system change: The environmental action (ENACT) programme, Jamaica. Case study prepared for the project Capacity, Change and Performance. European Centre for Development Policy Management. Maastricht, Netherlands.

Morrison, D., and M. Mitchell. 1999. *National biodiversity strategy and action plan development project: Sector assessment reports – Mining*. Kingston: National Environment and Planning Agency.

National Environment and Planning Agency (NEPA). 2003. *National strategy and action plan on biodiversity in Jamaica*. Kingston: National Environment and Planning Agency.

Nature Conservancy. 2000a. *Cockpit Country Conservation Project preparation report*. Kingston: The Nature Conservancy Cockpit Country Conservation Project.

———. 2000b. *Cockpit Country Conservation Project report: Annex – Institutional and financial aspects*. Kingston: Nature Conservancy Cockpit Country Conservation Project.

———. 2005. *Cockpit Country conservation action plan: A summary*. Kingston: Nature Conservancy Cockpit Country Conservation Project.

Natural Resources Conservation Authority (NRCA). 1999. Jamaica: Towards a watershed policy. Green paper no. 2/99. Kingston: Natural Resources Conservation Authority. http://www.nrca.org/policies/watershed/toc.htm.

Neuman, W.L. 2000. *Social research methods: Qualitative and quantitative approaches*. 4th ed. Needham Heights, MA: Allyn and Bacon.

Pugh, J. 2006. Physical development planning in the anglophone Caribbean: The re-articulation of formal state power. In Pugh and Momsen 2006, 7–20.

Pugh, J., and J.H. Momsen, eds. 2006. *Environmental planning in the Caribbean*. Aldershot, UK: Ashgate.

Pugh, J., and R.B. Potter, eds. 2003. *Participatory planning in the Caribbean: Lessons from practice*. Aldershot, UK: Ashgate.

Ruggie, J. 1998. *Constructing world polity*. New York: Routledge.

Schreurs, M.A., W.C. Clark, N.M. Dickson and J. Jaeger. 2001. Issue attention, framing, and actors: An analysis of patterns across arenas. In Social Learning Group 2001, 349–64.

Snow, D.A., K. Burke, S.K. Rochford Jr, K. Steven, R. Worden and D. Benford. 1986. Frame alignment processes, micromobilization, and movement participation. *American Sociological Review* 51: 464–81.

Social Learning Group. 2001. *Learning to manage global environmental risks*. Vol. 1. Cambridge, MA: MIT Press.

Southern Trelawny Environment Agency (STEA). 2003. *Biodiversity manual*. Albert Town, Jamaica: Southern Trelawny Environment Agency with the Canada Jamaica Green Fund/World Bank Small Grants Programme.

Spence, B. 2000. GEF Cockpit Country management report. Prepared for NRCA/NEPA, Kingston, Jamaica.

Sprinz, D., and T. Vaahtoranta. 1994. The interest-based explanation of international environmental policy. *International Organization* 48: 77–105.

Statistical Institute of Jamaica (STATIN). 2007. Jamaican statistics. http://www.statinja.com/stats.html.

Taylor, K. 1998. *Report on the legal imperatives and implications of the Cockpit Country Conservation Project*. Kingston: Nature Conservancy Cockpit Country Conservation Project.

Thomas, C. 2003. *Bureaucratic landscapes: Interagency cooperation and the preservation of biodiversity*. Cambridge, MA: MIT Press.

Thomas-Hope, E., and B. Spence. 2006. Challenges to promoting agro-biodiversity in Caribbean small farming systems: A Jamaican case study. In Pugh and Momsen 2006, 33–51.

Torres, I. 1998. *The mineral industry of Jamaica*. Washington, DC: United States Geological Survey. http://minerals.usgs.gov/minerals/pubs/country/1998/9515098.pdf.

United Nations Environment Programme (UNEP). 2003. *Sustainable conservation of globally important Caribbean bird habitats: Strengthening a regional network for a shared resource*. Global Environment Facility Project Document, UNEP. http://www.gefonline.org/projectDetails.cfm?projID=1604.

Vogel, P. 1999. *National biodiversity strategy and action plan development project: Sector assessment reports – Terrestrial fauna*. Kingston: National Environment and Planning Agency.

Waltz, K. 1979. *Theory of international politics*. New York: McGraw-Hill.

Webster, D., and M. Daye. 1999. *National biodiversity strategy and action plan development project: Sector assessment reports – Tourism*. Kingston: National Environment and Planning Agency.

Weis, A. 2000. Beyond peasant deforestation: Environment and development in rural Jamaica. *Global Environmental Change* 10 (6): 299–305.

Wendt, A. 1987. The agent-structure problem in international relations theory. *International Organization* 41: 335–70.

CHAPTER 8

Building Capacity and Resilience to Adapt to Change

The Case of the Blue and John Crow Mountains National Park

SUSAN OTUOKON AND SHAUNA-LEE CHAI

Introduction

Global environmental, economic and social changes are being reported, monitored and studied throughout the world. Although the Caribbean contributes relatively little to the causes of these changes, like other regions of small island developing states (SIDS), it is highly vulnerable to their impacts. Mitigation and adaptation strategies geared at coping with issues like climate change and market liberalization must involve improved environmental management, such as maintaining and restoring ecological buffers, building human and social capacity, and economic re-structuring (Tompkins et al. 2005). In Jamaica, climate change is recognized as a threat to the Blue and John Crow Mountains National Park, however, effective management of the park can make a significant contribution to mitigating climate change and adapting to the inevitable consequences (JCDT 2005). This chapter presents the work of the Jamaica Conservation and Development Trust (JCDT) and other organizations in managing the Blue and John Crow Mountains National Park from the perspective of coping with global changes and their impact on vulnerable natural and social environments. It documents experiences and lessons learned, as well as recommended approaches to plan, implement, monitor and evaluate adaptation strategies.

Background

The park protects about 500 square kilometres of forested land in eastern Jamaica and consists of three mountain ranges – the Port Royal Mountains to the west, the Blue Mountains in the centre and the John Crow Mountains in the east. The maximum height attained in the Port Royal Mountains is 1,539 metres at Catherine's Peak, and they link to the Blue Mountains in the Hardwar/Silver Hill Gap area, where a paved road connects Kingston to Buff Bay in Portland. The Blue Mountains comprise over 60 per cent of the park's area and rise very steeply to the Grand Ridge, a series of peaks including the highest elevation in the island, Blue Mountain Peak (2,256 metres). The John Crow Mountains rise gently to a plateau with a maximum height of 1,140 metres, are separated from the Blue Mountains by the Rio Grande Valley, and converge at Corn Puss Gap (see figure 8.1).

The northeast trade winds influence rainfall which averages 2,700 millimetres per annum. The mountains cause a marked rain shadow effect which results in greater levels of precipitation on the northern slopes and in the Rio Grande Valley. The higher elevations are often covered in a blanket of mist in daylight hours. Mean monthly temperature is between 18.5 and 20.5 degrees Celsius. The park protects the uppermost portions of ten of the twenty-six watershed management units in the island, and the Port Royal Mountains and Blue Mountains give rise to nine major rivers in the east. An estimated 40 per cent of the island's population, including most of Kingston, derives their potable water from the park (JCDT 2005).

There is wide variation in geology and physical characteristics; the Blue Mountains are dominated by Cretaceous volcanic and igneous rocks while the John Crow Mountains are mainly white limestone. These

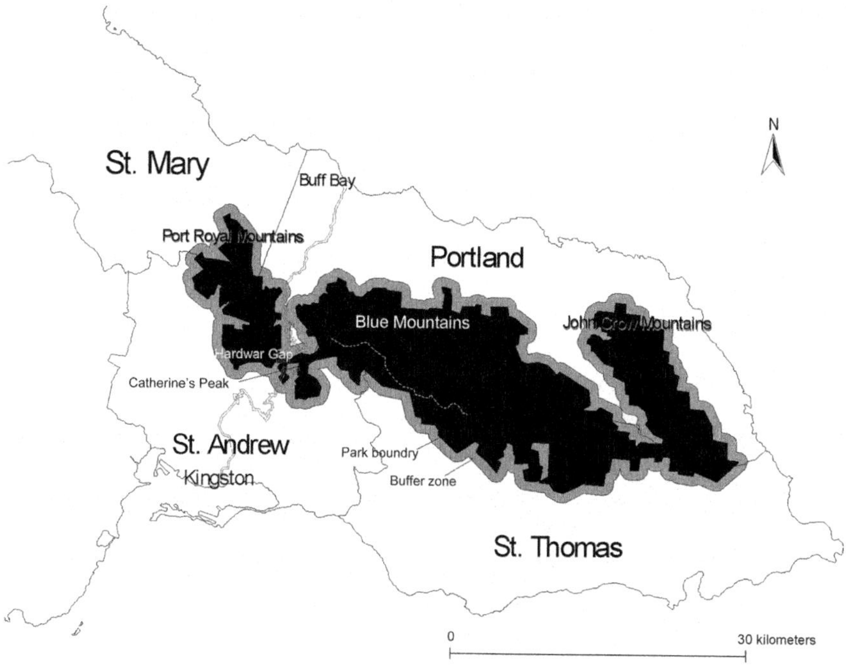

Figure 8.1 The Blue and John Crow Mountains National Park.

environmental conditions have resulted in a variety of ecosystems, with an exceptionally high level of biological diversity (JCDT 2005). The level of plant endemism is over 40 per cent (Grubb and Tanner 1976; Muchoney et al. 1994), and the site is included in the WWF-IUCN list of Globally Important Sites for Conservation of Plant Biodiversity (JCDT 2005). The park protects the largest contiguous area of closed broadleaf forest in Jamaica, covering about 30 per cent of such remaining forest, and the relative inaccessibility of this lush, natural forest creates a refuge for many of the island's endemic, threatened and endangered species.

The area was declared a national park in 1993, and the boundaries were based on those declared by the Forest Reserve in 1951. It is an IUCN Category II Protected Area, managed primarily for protection of its ecosystems, biodiversity and recreational opportunities. In 1998, the JCDT became the delegated management authority and the Natural Resources (National Park) Regulations of 1993 provide the main legislative basis for its operation. The park is managed collaboratively by JCDT, the Natural Resources Conservation Authority (through the National Environment and Planning Agency) and the Forestry Department, with the former playing the lead role in operational management. The 2005–2010 management plan has six programmes geared at addressing the root causes of the threats to the park's ecosystems:

- Conservation
- Enforcement and Compliance
- Education and Public Involvement
- Recreation and Tourism
- Monitoring and Evaluation
- Governance and Administration (JCDT 2005)

The core of natural forest, comprising more than 50 per cent of the park, is designated the preservation zone, where the management objective is strict conservation. The recovery zone lies between this core area and the boundary where the focus is mainly on forest restoration. A 1-kilometre-wide, inhabited band outside the park boundary is maintained as a buffer zone, and has about 50 communities and 26,000 residents (STATIN 2001). Much of the work of park management is focused in these buffer zone communities on defending the park boundary and restoring degraded and deforested areas within the Recovery Zone (Smith 1995; JCDT 2005).

Natural and Economic Context

The Caribbean region of SIDS is particularly vulnerable to the impacts of global changes. It relies, to a large extent, on imported goods and services, along with tourism and a very limited range of agricultural produce to generate the foreign exchange to pay for these imports. The islands are constrained by limited land and human resources, and are at risk from environmental hazards such as hurricanes and flooding (Bass and Dalal-Clayton 1995; Tompkins et al. 2005). Soil erosion and associated land degradation are Caribbean-wide phenomena, and climate change is likely to increase their negative impacts through increased seasonal rainfall following long spells of drought (McGregor 1995; McGregor, Barker and Campbell, chapter 9)

The Blue and John Crow Mountains region is particularly vulnerable because of the steep slopes, heavy rainfall and rapidly draining soils (Davis-Morrison 1998). The rural communities of eastern Jamaica, located in the buffer zone, are farming communities. Bananas were a major income-earning crop, but with the termination of preferential trade agreements, production and income are declining and, while coffee farming is increasing, the majority of households in the buffer zone are dependent on subsistence farming and farm work. In addition, bad road conditions caused by inadequate drainage and maintenance make marketing of produce very difficult (Smith 1995; Davis-Morrison 1998; Weis 2000; JCDT 2005). Many communities have declining and ageing populations because young people migrate to urban areas seeking employment. Those that remain are hampered by a lack of economic opportunities, and low levels of literacy and education affect their ability to create their own opportunities.

Physical conditions make the area vulnerable to natural hazards such as high winds, flooding and soil erosion. The heavy rainfall and wind associated with tropical storms and hurricanes often results in damage to trees and property as well as landslides, which affect homes, roads and property (McGregor 1995; Wade 1996). Deforestation, along with intensified and inappropriate land use all act as principal factors influencing the increasing rates of soil erosion. These factors result in serious land degradation, particularly on the southern slopes of the Blue Mountains, and are directly related to both small-scale farmers and large-scale coffee farmers. Many small-scale farmers occupy poor lands on steep slopes as a result of historical and socio-economic factors related to land tenure issues in the pre- and post-emancipation periods (McGregor 1995; Wade 1996; Edwards 1995; Weis 2000). Continued deforestation and land degradation will eventually lead to decreased biodiversity, decreased water supply, decline in economic production, increasing rural poverty and increasing natural disasters, among other negative impacts. The solution to the problem of deforestation and inappropriate land use needs to be multi-faceted and may include a variety of options suggested by several authors (McGregor 1995; Wade 1996; Edwards 1995; Davis-Morrison 1998; Thomas-Hope and Spence 2006):

- maintaining of natural forest cover on slopes above twenty-five degrees through management for conservation, including enforcement of relevant legislation relating to forest conservation and watershed management;
- promoting agroforestry and agro-biodiversity;
- working with farmers to develop and implement farming techniques that provide both adequate income generation and effective watershed protection; and
- developing viable, non-farming, income generating alternatives.

Park Management and Coping with Climate Change

The park's management plan (2005–2010) recognizes climate change as a factor likely to increase natural hazards, particularly through increased rainfall from an increased number of tropical storms and hurricanes (Gamble chapter 2; McGregor, Barker and Campbell, chapter 9). The likely result is increased deforestation, land degradation and concomitant loss of wildlife, seriously affecting the livelihoods of buffer zone communities, as well as income from recreation and tourism. In addition to their importance for biological diversity and water supply, the forests of the Blue and John Crow Mountains play an important role in mitigating climate change and, perhaps more importantly, buffering against the impacts of natural hazards. They help to break up storm-force winds associated with hurricanes, and also help prevent flooding by absorbing excess water produced by heavy rains. However, the forests are seriously threatened, particularly by the interlinked problems of clearance for agriculture, overgrowth of invasive species and fires (JCDT 2005).

Current national park management approaches require special attention to buffer zone communities, including tangible benefits accruing to these communities from the park. The aim is to ensure support for the protected area in terms of compliance with legislation, and protection of the park's resources. Potential management approaches, including participatory planning and management, along with integrated conservation and development projects, have met with mixed results (Brandon, Redford and Sanderson 1988; Cleaver 2001; MacKinnon 2001). Despite this, some authors suggest inadequate community participation and benefits have limited the success of conservation in protected areas (Ghimire and Pimbert 1997; Martin 2006).

An objective in preparing the park's new management plan was to ensure it was feasible. Since it is not possible to reduce or mitigate climate change *per se*, a range of interrelated strategies was developed, geared at building adaptive capacity, reducing vulnerability to hazards, and, to a certain extent, mitigating against natural disasters. Adger et al. (2003, 181) note that the vulnerability of a system is "determined by its exposure, by its physical setting and sensitivity, and by its ability to and opportunity to adapt to change", while the vulnerability of individuals and societies is "determined not only by the likely responses of the resources on which they depend but by the availability of resources and their entitlement to call on these resources".

From its inception in 1993, efforts were made at involving communities living around the boundary of the park through the creation of local advisory committees (Smith 1995). However, the 2005–2010 management plan approach is to work with existing community groups to reduce economic and social vulnerability by building capacity to plan and implement activities that will result in improved environmental management and sustainable development. All the management plan programmes are integrated and aim to involve local community members (table 8.1). However, the education and public involvement programme most directly addresses the issue of community participation and reducing human vulnerability by increasing capacity for adaptation.

Reducing the Vulnerability of Buffer Zone Communities: The Education and Public Involvement Programme

The first objective of this programme is "to facilitate capacity building (including education and training) of communities around the Park, with respect to: using natural resources with minimum impact, taking advantage of more environmentally friendly income-generating

Table 8.1 Results of Monitoring the Bowden Pen Farmers Association Blue and John Crow Mountains National Park Action Plan

Activities/Outputs	Monitoring Results: Quarter 1 & 2
Preparatory Phase	
Park Management Plan presented BPFA in agreement	Yes – Meeting held with BPFA & members indicated that their concerns were adequately addressed
BPFA agrees to collaborate in BJC-MNP management	Yes – JCDT and BPFA agree to draft and sign a co-management agreement
Training of relevant staff in participatory methodologies	Yes – staff training conducted, particularly for education & community outreach officer
Resources available for implementation of action plan	Above-mentioned officer assigned responsibility JCDT sought funding; approved at the end of 2nd qtr
Organizational Capacity	
Executive elected and working	Yes – BPFA already had these in place
Minutes kept of meetings	Yes, and training conducted
Record keeping systems	Yes, and book-keeping training conducted
Funding for projects	JCDT assisted BPFA with successful preparation of a proposal to the GEF Small Grants Programme.
Co-management Organization Formation	
Co-management agreement/MOU drafted and signed	By end of 2nd qtr, a memorandum of understanding had been drafted and agreed on. (The MOU was signed in May 2006.)
Action plan developed	Yes
Sub-committees established	Yes (and Steering Committee)

Table 8.1 continues

Table 8.1 Results of Monitoring the Bowden Pen Farmers Association Blue and John Crow Mountains National Park Action Plan *(cont'd)*

Activities/Outputs	Monitoring Results: Quarter 1 & 2
Technical and Management Indicators (Based on BJCMNP Management Programmes)	
1. Enforcement & Compliance Programme – Community Members	
Share information with rangers trained in environmental law, use of cameras and GPS units	BPFA call JCDT to advise of illegal activities or environmental concerns Training conducted, but limited by resources
Assist rangers on patrol (monthly)	BPFA members involved in joint patrols with rangers and assist police in catching a river poisoning perpetrator
2. Education & Public Involvement Programme – Community Members	
Involved in educating other community members and visitors	BPFA makes presentation to Comfort Castle A/A School and hosts students on Cunha Cunha Pass Trail
3. Conservation, Monitoring & Research Programme – Community Members	
Trained in identifying invasive species	The focus, instead, was on bird identification and members assisted park rangers in monitoring activities
Trained in plant propagation techniques and reforestation	Park rangers assisted BPFA in establishing tree nursery
4. Recreation & Tourism Programme – Community Members	
Begin to share economic benefits from business opportunities	Cunha Cunha Trail re-opened after hurricane repairs BPFA ran a successful food stall at Misty Bliss, Holywell, March 2006 Increase in visitors to Ambassabeth (lodge run by BPFA) provides income from accommodation, meals and tours

Source: Adapted from Bedasse 2006.

activities, and being involved in Park/natural resources management". This objective guides the community conservation and sustainable livelihoods sub-programme using "action learning" to plan, implement, monitor and evaluate the strategies (Bedasse 2004, 2006; JCDT 2005; Otuokon 2006a, 2007). The new approach was brought about by decreases in park funding which resulted in loss of the staff needed to maintain functioning local advisory committees and their subsequent demise, prompting an evaluation (Bedasse 2004).

In 2004, work began on preparation of the 2005–2010 management plan and the opportunity was used to develop a mechanism for involving community members in planning and implementing park management. It was felt that there was a need to interact more closely with smaller groups to have greater impact, and the local advisory committees had been somewhat contrived. Liaison with extant groups formed around particular interests, such as farming or community development, was felt to be more sustainable. As Cleaver (2001) suggests, there is a tendency for participatory projects to seek to establish "formal" organizations within communities, whereas institutions within communities often already exist, though perhaps are less bureaucratic.

Preparing the management plan included mobilizing people from specific geographical areas to attend meetings, and implementing participatory approaches to obtain feedback from attendees with respect to management strategies and activities. The process resulted in the integration of community recommendations into the management plan and a three-month action planning, implementation and monitoring cycle for involving community groups in park management. The Bowden Pen Farmers Association in Millbank, Portland, in the Upper Rio Grande Valley, was selected as the pilot group because of their enthusiasm, willingness to participate and interest shown (Bedasse 2005a, 2005b; Otuokon 2006b). Interestingly, this group grew out of the remnants of the area's local advisory committee, with the assistance of one of the park's former community outreach officers (Bedasse 2006). Under the pilot project, a one-year action plan, intended to benefit their community socio-economically, was developed to start in October 2005. The first two quarters of the year were monitored and evaluated according to a plan developed for the purpose (table 8.1). The park's education and community outreach officer (ECOO) provided assistance to the farmers group to implement their action plan, liaising where necessary with other programme officers. This action learning approach was judged successful, as the group planned and implemented several activities that contributed to management of the park's forest resources. The overall approach was adopted and the ECOO sought to revitalize relationships with three other communities: around Holywell at Hardwar Gap (formerly involved in another Local Advisory Committee); Woodford, Irish Town/Redlight, and Cascade.

The Blue and John Crow Mountains were badly impacted by hazards resulting from Hurricane Ivan in September 2004 through to Hurricane Katrina in September 2005. Many roads were blocked and damaged; for example, Cascade was completely isolated from Kingston, as the Buff Bay Road totally broke away at more than one point. Fourteen homes were destroyed by mudslides in the communities of Millbank and Moore Town in Rio Grande Valley, Portland. These events heightened people's awareness of forest conservation, and buffer zone communities became interested in implementing activities that could reduce their physical vulnerability. The ECOO was able to use this increased interest to catalyse environmental management planning and action among local, community-based organizations. For example, the 2005 National Labour Day theme "Prepare for Disaster – Recover Faster" provided an excellent opportunity to begin watershed management activities and, in the four selected communities, over 350 trees were planted in 2005 between Earth Day and Labour Day on sloping lands or home/farm plots. This enhanced awareness was used to develop another theme: "Watershed Management for Disaster Mitigation". Also, it led to the identification of appropriate watershed management activities, such as reforestation and soil conservation, as well as appropriate farming practices, including agroforestry. In addition, alternative income strategies related to recreation and tourism were identified.

Outreach to schools started in September 2005 and funds were obtained from two sources for two-year community outreach projects which started in March 2006. The plan was to use the experience gained during the pilot project with the Bowden Pen Farmers Association to guide the action planning and learning mechanism with the three additional communities. A fifth community at Hayfield, in St Thomas, was to be used as a "control", with no management interventions, to further aid in assessing the success of the mechanism for community involvement. By the end of 2006, the action planning mechanism had suffered several setbacks, and while the focus was still on the four targeted communities, efforts became more ad hoc, and less centred around a detailed action plan. In Millbank, the Bowden Pen Farmers Association became more focused on implementing a reforestation project, for which JCDT helped obtain funds. The ECOO found it difficult to organize meetings to address action planning and to work with four communities, one of which had almost non-existent organizational capacity. Activities eventually targeted individual members from all four communities through an existing youth training programme funded partially by a regional UNESCO project. A further setback involved a change in outreach officer and early closure of a project, highlighting the problem of insufficient resources.

The other three objectives of the education programme address environmental education: implementation of elements targeting school children in the buffer zone communities; interpretation for visitors to the park; and dissemination of information through publications and the media. The park's assistant education officer visits all the schools in the buffer zone annually so as to raise environmental awareness and knowledge on specific issues (Gentles 2005), and organize a poster competition. Currently, interpretation is offered mainly at Holywell, through educational tours for groups from primary, secondary and tertiary level institutions. An interpretive signage plan has been designed and funds are being sought to construct a series of signs at Holywell. Other information can be obtained from the JCDT's website (www.greenjamaica.org.jm), an in-house library, and a range of brochures and media releases.

Reducing Vulnerability Through Other Programmes

Conservation Programme

This programme was developed using the Nature Conservancy's 5-S Site Conservation Planning Process (2005) methodology. Eight conservation targets were identified: (1) montane forest on shale, (2) montane forest on limestone, (3) epiphytic communities, (4) headwater ecosystems, (5) montane forest birds, (6) Jamaican coney, (7) Jamaican boa and (8) the giant swallowtail butterfly. Montane forest on shale and montane forest on limestone were the most important of these targets because their conservation offered a holistic approach to conserving the park's resources. They are ecosystem-level targets, rather than community or singlespecies targets, and, due to their large size, form the habitat for all the other conservation targets. So, to an extent, once these forests are protected and healthily maintained, the flora and fauna of the park ought to be conserved. This important assumption was verified for montane forest birds in an analysis of a seven-year bird monitoring data project (Vogel 2004). Thus, in its first years, the conservation programme is intended to focus on protecting forests rather than single-species conservation. This latter would require in-depth knowledge of the range of species present, currently not fully researched. A focus on protecting forests also addresses the main threat to the park, namely conversion of forest to agriculture, and gives park management the most cost and time effective "value for money".

Activities to address forest conservation have focused on rehabilitating degraded areas via control of invasive plant species and replanting with native forest tree species, activities new within Jamaica's conservation arena. The following outlines two pilot projects conducted between 2005 and 2007.

1. *Invasive control along the Blue Mountain Peak Trail (August 2005 to December 2007)*

Invasive species, regarded by an experts workshop (1999) as the second most critical threat to the park's ecosystem, include the following: *Pittosporum undulatum*, *P. viridiflorum*, Molasses or Wynne grass (*Melinus minutiflora*), *Gleichenia* sp (fern), *Hedychium gardnerianum* (ginger lily), and *Polygonum chinnense* (red bush). The first invasive species control project focused on controlling *P. undulatum* (wild coffee or mock orange), regarded as the most threatening invasive plant species (Tanner *pers.comm.* 2004). Originally from Australia, and introduced into the forests from Cinchona Botanical Gardens about thirty years ago, this plant grows as a shrub or tall tree up to 23 metres in the Blue Mountains (Goodland and Healey 1996). It may lead to the competitive exclusion of many native species, and a strong negative correlation has been found between the dominance of *Pittosporum* and both the density and diversity of native tree seedlings (Healey and Hall 1994), strongly suggesting that *Pittosporum* is causing or contributing to a major decline in the seedling layer native flora.

Further, there is evidence that hurricane disturbance accelerates the invasion of Blue Mountains montane rain forests by *Pittosporum* (Bellingham, Tanner and Healey 2005). Controlling *Pittosporum* is highly recommended, but control programmes involving forest disturbance must proceed with caution because this invasive regenerates readily and competitively in tree gaps. The disturbance derived from cutting *P. undulatum* could trigger increased recruitment of the invasive (Healey and Hall 1994; Goodland and Healey 1996). To contain the spread of *P. undulatum*, all three life stages were given control treatments over an area of 4.8 hectares along the Blue Mountain Peak Trail. Subsequently, the focus was on the adult life-stage, producing thousands of seeds for dispersal. Control treatments involved uprooting *P. undulatum* seedlings, cutting back saplings and painting the cut stump with herbicide, and injecting adult trees with a herbicide that slowly kills the plant. Native species have been planted near the root of dead trees to replace those removed. In total, 237 adult *P. undulatum* were treated along the Trail. The mortality was 80 per cent (sample plot results), so 190 trees were actually eradicated (Chai 2007). If treated trees showed signs of springing back, a booster treatment was administered. Injected trees died standing upright, avoiding the sudden creation of forest gaps. Based on the success of this pilot project, the control method was replicated in Holywell, where eighty adult trees were eradicated. A total of 250 tree seedlings of seven indigenous species have been planted with a 90 per cent survival rate, replacing *Pittosporum* in rehabilitated areas (Chai 2007).

2. Forest rehabilitation at Mount Horeb and Catherine's Peak (March 2006 to May 2007)

Numerous studies have highlighted the role of plantation forests (especially indigenous tree species) in accelerating the rate of regeneration of species-rich natural forest stands in degraded sites. The following indigenous trees, based on recommendations by McDonald et al. (2003), were used in replanting projects: *Alchornea latifolia* (dovewood), *Clethra occidentalis* (soapwood) and *Podocarpus urbanii* (mountain yacca). Also used were *Hibiscus elatus* (blue mahoe), *Schefflera sciadophyllum* (umbrella tree), and *Sapium harrisii* (milkwood) due to their ease of propagation and availability. These trees were propagated from wildings in a nursery located at Holywell, except for *S. sciadophyllum* which was propagated from cuttings. Areas were planted out using a line contour system developed by the Forestry Department. Lines were cut 10 to 12 feet (3 to 3.5 metres) apart, and cleared of weeds. Seedlings were pegged and planted 10 to 12 feet (3 to 3.5 metres) apart along cleared lines, with 250 seedlings per acre of land (approximately 100 per hectare). In total, four hectares were planted out on Catherine's Peak, after illegal forest destruction by a coffee farmer. Lines were re-cleared approximately four months later, seedlings were circle weeded and dead seedlings replanted. This is scheduled twice per year for at least two years.

The second most threatening invasive plant, *Hedychium* sp. (wild ginger), grows heavily along the Fairy Glade Trail on Mount Horeb in the Port Royal Mountains, and was controlled over a 1-hectare area, divided into 10-by-10-metre plots. Manual labour was used to dig up and remove the ginger. No replanting is required once the ginger is removed, as existing native seeds in the soil seed bank grow quickly and out-compete the ginger. The plots need to be revisited as Wild Ginger has a rhizomatous root system and will regenerate from pieces left behind. However, since March 2006, areas cleared of ginger appear to be regenerating naturally without re-growth (Chai 2006a).

Enforcement and Compliance Programme

This programme focuses on reducing encroachment across the park boundary and concomitant destruction and degradation of the forest. The main approach, particularly since implementation of the new management plan, was to increase and maintain a higher level of presence of park rangers and other enforcement officers in and around the park. In 2004, overseas migration of two rangers early in the year left

Table 8.2 Offences Recorded for 2006 in the Blue and John Crow Mountains National Park

Offence	Number of Incidents	Places Observed
Dumping coffee pulp in river	2	Wallenford Coffee Plant, Mavis Bank Coffee Plant
Farming	13	Bangor Ridge, Buff Bay Valley, Catherine's Peak, Cinchona, Clifton Hill, Farm Hill, Fruitful Vale, Green Hill/Middleton, Greenwich Trail, Johnson Mountain Leith Hall, Mount Telegraph, Long Road, Hardwar Gap, Shirley Castle
Logging	5	Durham, Mount Telegraph/Long Road, Penlyne, Shirley Castle, Cinchona
Hunting/shooting/capture	4	Hayfield, Packi River, Roaring River/Mount Felix, Section
Washing in river	2	Buff Bay Valley, Packi River
Clearing forested land	2	Bangor Ridge, Hardwar Gap
River poisoning	3	Buff Bay Pencar River, Morant River, Bucket River
Fire	3	Cave Bottom, Cinchona, Garbrand Hall/Mullet Hall
Dumping garbage	4	Bellevue, Buff Bay Valley, Mount Charles, Rio Grande River
Charcoal burning	1	Bangor Ridge
Sand mining	1	Buff Bay Valley

only three officers, who averaged 1.25 recorded patrols per month. In 2005, regular patrols moved from an average of 1.7 per month between January to July 2005, to an average of 6.4 per month between August and December 2005, following the employment of a chief of corps (Otuokon 2006a). In 2006, 105 patrols were carried out (an average of 8.75 per month) with 40 offences recorded (JCDT 2006) (table 8.2).

If an illegal activity is observed and the offender is present, they are given a verbal warning, and for repeat offences, a written warning is served. In the case of repeated offences, park rangers liaise with the Island Special Constabulary Force who accompany them to the site for warning and, if necessary, arrest. If offenders cannot be identified, rangers seek information from persons in the area. Detailed monthly reports are sent to the Forestry Department and the National Environment and Planning Agency, and follow-up conducted with these agencies if further action is necessary.

Park rangers are often in the field conducting activities other than patrols, such as bird monitoring, seedling collection and visitor management. Their approach is one of interpretive enforcement, which aims to ensure that people understand the reasons for the park's rules and regulations, and what they are expected to do in order to conserve the park's forest resources and live in a more sustainable way. For example, rangers talk to community members, provide brochures and ask for information on breaches of environmental legislation. In turn, community members have expressed concern about several environmental issues, including poisoning of rivers with pesticides to catch crayfish and the existence of yellow snakes for which most Jamaicans hold a dreadful, yet unfounded, fear. In one case, discussions between the JCDT, the Bowden Pen Farmers Association and the Nature Conservancy resulted in funding for a project aimed at decreasing the practice of river poisoning through public education, community action and investigation into the feasibility of crayfish rearing. In a second case, reports from community members led to site visits by the park's conservation science officer, and to community meetings regarding the non-poisonous nature and conservation value of the yellow snake (*Epicrates subflavus*), along with other biodiversity conservation issues, all organized by the park's education officers.

Monitoring and Evaluation Programme

This programme is a combination of target-based and threat-based monitoring, which allows an assessment of both the threat status and the ecological integrity of the park's natural resources.

The numerous threats and illegal activities include the conversion of forest to agriculture, growth of invasive alien species and harvest of non-timber products, as well as logging, fires and hunting. The most damaging, however are the conversion of forest and the growth of invasive plants (JCDT 2000). Monitoring threats can reveal whether they are changing in severity or geographic range as a result of conservation strategies. This type of monitoring is referred to as threat-based or ranger-based because rangers are the eyes and ears of protected area management and are closest to situations on the ground. Rangers have the power to effect immediate action against unacceptable changes and are the first line of defence against offenders of protected area legislation (JCDT 2005). Surveillance, stopping illegal activities and monitoring the progression of mitigation are important activities, and immediate action is essential to stop new offences (such as deforestation). Threats-based monitoring involves rangers taking photographs and GPS readings to geo-reference any threats they encounter. This information is entered into an ArcView GIS database. This effort started in September 2005 and, in the absence of previous records, the first visit to a particular area is used as the baseline against which further visits are compared.

Target-based monitoring assesses the impacts of management activities on the protection of biodiversity, and provides the opportunity to manage adaptively and to demonstrate effectiveness to donors and other stakeholders, thereby providing the accountability needed to sustain conservation investments. The park implemented systems for monitoring birds and stream water quality, between 1996 and 1999. Results show that, for the most part, freshwater ecosystems are being maintained in a stable condition (Chai 2006b) while bird monitoring has shown an important dependency connection between forests and endemic birds (Vogel 2004).

Annual photographic records of certain forest areas are taken for comparison and, in one such case, showed an increase in vegetation cover and a decrease in farming, associated with ranger patrols and warnings to farmers who had encroached into the park.

Recreation and Tourism Programme

Tourism is a major income generator for most Caribbean islands but development is mainly coastal and thus vulnerable to sea-level rise.

Increasing the sustainability of tourism is an important adaptation to climate and other global changes. To do so, several approaches should be used, including diversification of tourist locations away from coastal areas, encouraging visitors to stay longer and spend more particularly in rural communities and promoting ecotourism that invests in the preservation of buffering ecosystems rather than exploiting natural resources (Tompkins et al. 2005). The park's recreation and tourism programme uses ecotourism principles in its provision of recreational opportunities for local and international visitors, with the goal of generating income and support for conservation of the park's resources (JCDT 2005). Plans for a Blue Mountains sustainable tourism programme include infrastructural development, training, packaging tours and marketing in the buffer zone communities surrounding selected park recreation areas, such Holywell and Cunha Cunha Pass Trail.

Working with local communities to develop opportunities and implement successful projects in sustainable heritage-based community tourism can strengthen community support for conservation (MacKinnon 2001; Martin 2006). The collaboration with the Bowden Pen Farmers Association in the Upper Rio Grande Valley includes working with park management to enhance opportunities at the community level for income generation through recreation and tourism. These activities are an important component of the Bowden Pen Farmers Association's action plan for co-management. In collaborative work between 2003 and 2005, park rangers provided technical assistance to the farmers group in re-building the Cunha Cunha Pass Trail, a heritage trail once used by Maroons, and later used as a market route between St Thomas and Port Antonio, before the coast road was constructed. The Bowden Pen Farmers Association felt the project would benefit their visitor accommodation at Ambassabeth in Bowden Pen, and provide an alternate access route via Bath and Hayfield in St Thomas. Since 2005, marketing and capacity building assistance provided by park management has enhanced the farmers group's ability to liaise directly with relevant government agencies, donors and the private sector. Further, the quality community-based tourism product offered at Ambassabeth has resulted in an increasing number of visitors, referred mainly by satisfied clients.

Lessons Learned

Anticipating negative impacts from climate and other global changes, the approach to managing the Blue and John Crow Mountains National Park is to decrease the vulnerability of natural systems by increasing their resilience, through conservation, enforcement, education, public involvement, recreation and tourism. Since 2005, conservation and enforcement have increased and become more effective at maintaining and expanding natural forest. Community involvement has changed from the approach described by Smith (1995) to one which seeks to liaise more directly with targeted communities through participatory approaches, including action learning, and capacity building for project planning and implementation by working with community groups and individuals (JCDT 2005).

As discussed earlier, participatory approaches to conservation and sustainable development are not a simple, "cure-all" solution, as they are sometimes portrayed. Experience with a successful community-based organization (Bowden Pen Farmers Association) encouraged a similar approach with several other community organizations, though with limited success. The Bowden Pen Farmers Association continues to do well and liaises with the JCDT in natural resources management and sustainable development activities, though not strictly according to an action plan. In Woodford, the Anglican Church and All Age School has been assisted with an agroforestry project on its property, establishing a variety of fruit trees on a hectare of degraded land. Activity in the Irish Town area has been limited, but involved each of the three main community development groups there. In the Cascade/Section community, despite several meetings, there has been limited action, and there is no community-based organization despite efforts of young people to form one. JCDT has continued to work with Woodford and Cascade farmers in particular, and a sustainable agriculture project based on the lessons learned over the past three years and aimed at reducing vulnerability to climate change was initiated in November 2008. Several factors were identified that reduce the likelihood of success of participatory approaches, including some that are external to the facilitating organization and more difficult to address. These are as follows:

1. *Limited capacity of community-based organizations:* Thomas-Hope and Spence (2006) report "inter-community suspicion, lack of

leadership" and a "lack of confidence and organizational framework" among communities in the Rio Grande Valley, and our experience is that this can be extended to include most of the park's buffer zone communities. In communities with active organizations, there are individuals with leadership qualities that form the core of the group and help to maintain continuity. However, organizations often form and dissolve abruptly over an internal disagreement, suspicion of an individual benefiting more than others, or an externally induced challenge, and they may or may not re-form around issues. The communities of Cascade and Section have had difficulty forming a long-standing community group, apparently because of suspicions regarding the activities of various individuals and organizations. Further, the landslides occurring along the Section to Buff Bay Road and the very slow responses from politicians and relevant agencies have resulted in a sense of hopelessness. Edwards (1995) refers to the paternalistic system of patronage, in Jamaica and Barbados respectively, that tends to dominate the relationship between politicians, civil servants and project managers, and small farmers and citizens. Pelling (2003) notes the tendency for reliance on state actors that "potentially undermines self-reliance and community-level social capital". In the national park's buffer zone, community members have been waiting for government or others to solve not only infrastructural challenges such as poor roads, but also economic hardships. While there are certainly things that the state should provide, community and individual self-help is important. JCDT's approach to community involvement in conservation and sustainable development projects is ultimately aimed at strengthening the capacity of community-based organizations and their members to implement and evaluate projects, directed towards even small, short term achievements that can be built upon in order to develop a sense of accomplishment.

2. *Land tenure and livelihood issues:* Edwards (1995), Wade (1996), Weis (2000) and Mills (2006) describe the historical complexity of the relationship between small farmers and land in the Caribbean. In the Blue Mountains, this land is of the lowest agricultural quality – steep slopes and poor soils. Frequently, land is not owned by the farmer, though many claim family ownership but have no

title. Davis-Morrison (1998) did not find an intrinsic link between private land ownership and sustainable resource use in her study in the Rio Grande Valley, but noted that this may be related to Maroon values regarding land use. It may also be due to differences between soils and slopes in the valley compared to the southern slopes of the Blue Mountains, where farmers practice a more shifting form of agriculture. This latter practice, along with low levels of employment in many rural communities, is resulting in the clearing of land on ever-steeper slopes sometimes within the national park/forest reserve boundary, threatening to change the forests of the Grand Ridge irrevocably. Despite the contested relationship between land ownership and good land management practices (Davis-Morrison 1998; Mills 2006), lack of land tenure is an issue for small farmers in the buffer zone communities and needs to be addressed for social, economic and environmental reasons.

Edwards's (1995) study on small farmers and the protection of watersheds indicates that, while farmers are aware of the negative consequences of their farming practices, soil conservation structures and techniques are too costly to implement and maintain, both in terms of financial and human resources. The farmers have to use their limited resources cautiously and therefore continue to use tried and practiced methods rather than adopt new practices which would be too risky for them, even when weighed against the vagaries of reaping crops which can be lost through disease, bad weather and praedial larceny. Thus, in a dilemma between short term economic needs and protecting long term requirements through environmental management, farmers opt for the former.

3. *Stakeholder benefits and incentives:* Krishnarayan, Geoghegan and Renard (2002) point out the challenges in committing resource users to conservation practices where a threat to livelihoods is perceived and suggest that stakeholders are more likely to adopt practices where benefits have been demonstrated. In synthesizing lessons learned from our pilot project in Bowden Pen, Bedasse (2006, 40) noted that "the opportunities to earn during the process of implementation serves as a catalyst for people involvement"

and further that, "public recognition and incentives to community groups (for example, donor funding for programmes) assuming the role of environmental stewards is an important tool to motivate change in behaviour and attitudes about the environment"

Cleaver (2001, 48) recommends care in the assumptions regarding the types of incentives and the targeted beneficiaries, because motivating factors may not be easy to identify. Further, she highlights the tendency to focus on economic benefits and not "personal psychological motivations, for the needs of individuals for recognition, respect or purpose, which may be independent of other material benefits". Bedasse (2006, 39) noted that the opportunities provided for the community members in Bowden Pen to educate visitors about their cultural and natural heritage have "aided in the development of self worth." In their account of lessons learned which was presented at a workshop, the Bowden Pen Farmers Association (2006, 2) remarked that "The JCDT respected our views on what actions should be taken to improve natural resources management in the area and acted on recommendations that were made by members of the [Bowden Pen Farmers Association]", "it made us feel good about ourselves" and "if you take yourself seriously, others will take you serious". This response supports the importance of the principle of respect, in participatory research and planning, as described by Dodman and Dodman (2006).

4. *Limited capacity of facilitating NGOs and importance of timescale:* Bedasse (2004) indicated a number of factors which had decreased the impact of the park's educational programmes with respect to conservation, including insufficient resources and the need for staff trained in participatory approaches and capacity building. More generally, Krishnarayan, Geoghegan and Renard (2002) point out the need for capacity building for all natural resource management partners, with respect to their world view, culture, structure, adaptive strategies, material resources, skills and linkages. There were many challenges working with four rural communities (some consisting of several districts and more than one formal organization which tend to meet on weekend evenings) situated along roads

with poor conditions and located in high rainfall regions (Gentles 2005). Funding from donors is usually for a maximum of two years, often with limited operational funding for items such as salaries and vehicles/transportation, and focuses on tangible deliverables which often prove more difficult to produce in practice than during design. This results in a management dilemma, and more time tends to be invested on the *products*, to ensure the likelihood of further funding, than on the *processes*, which would have longer-term benefits for community partners.

Krishnaryan et al. (2002, 17) conclude that "the more durable participatory resource management arrangements have been developed over a period of years, allowing stakeholders to grow and make transitions". JCDT's experience, working in partnership with the Bowden Pen group, whose growth floundered when funding for the position of community outreach officer ceased, but is now able to stand on its own, supports this conclusion. Bedasse's (2006) evaluation also supports this long term approach. Unfortunately, funding for this type of method is difficult to access and maintain, and, as Nettles (2003, 198) states in her case study of Red Thread in Guyana, reflects a "continued reliance on NGOs . . . in the 'Third World' to pick-up the slack on an eroded civil society" and, in the case of the JCDT and Jamaica, an eroded environmental management system.

Conclusion

The factors described are much the same as those that increase the economic and social vulnerability of individuals and communities, putting them at greater risk of disaster in the face of natural and other hazards that are likely to increase with climate and other global changes. JCDT's work, reported above, in building capacity of the Blue and John Crow Mountains National Park buffer zone communities to plan, implement, monitor and evaluate activities (that result in improved environmental management, increased economic benefits and reduced vulnerability to natural hazards) suggests an approach which can have positive results. Limited institutional, social and human capacity was however found to be a major hindrance in efforts to involve local stakeholders. Research conducted by the Caribbean Natural Resources Institute over the last twenty years indicates that limited capacity at all levels and the need for greater political and policy level support have constrained the effectiveness of participatory approaches to natural resources management in the Caribbean (Geoghegan et al. 1999; Krishnarayan, Geoghegan and Renard 2002; CANARI 2003).

Building capacity for improved environmental management is a necessary component of any change adaptation strategy, and will require more than skills training and the provision of material resources. The capacities of government, non-government, community-based and private sector organizations will need to be strengthened and even adjusted, in terms of their world-view, culture and structure with respect to participatory approaches to planning and management of natural resources (Krishnarayan, Geoghegan and Renard 2002). While there are principles that can guide this process, there is no "one-size-fits-all" plan and therefore each area and group of stakeholders will have to learn from and build on their own experiences (and those of others) in a continuous process. Such an action-learning approach cannot be implemented successfully as short term projects, but rather as ongoing programmes that address both internal and external factors. This *project* versus *programme* approach remains one of the difficulties of natural resources management in the Caribbean, whether NGO or government led, because effective environmental management requires a long term programmatic approach, whereas funding is available mainly for short term (two- to five-year) projects.

JCDT has had success in strengthening the resilience of the natural systems within the Blue and John Crow Mountains National Park, particularly through forest rehabilitation and increased activity of enforcement officers. Increasing the resilience of human communities, however, is more difficult and requires a practical, empowering, participatory and long term approach. Conservation of forest ecosystems will therefore require a two-pronged effort: strengthening key institutions in the short term for management activities within protected area boundaries, and, over the longer term, strengthening local communities to improve environmental management beyond these boundaries.

References

Adger, W.N., S. Huq, K. Brown, D. Conway and M. Hulme. 2003. Adaptation to climate change in the developing world. *Progress in Development Studies* 3 (3): 179–95.

Bass, S., and B. Dalal-Clayton. 1995. *Small island states and sustainable development: Strategic issues and experience.* Environmental Planning Issues, no. 8. London: International Institute for Environment and Development.

Bedasse, J. 2004. *An evaluation of the contributions of educational programmes to conservation within the Blue and John Crow Mountains National Park.* Kingston: Jamaica Conservation and Development Trust (JCDT).

———. 2005a. *A plan for on-going stakeholder involvement in the management of the Blue and John Crow Mountains National Park.* Kingston: JCDT.

———. 2005b. *Monitoring and evaluation of on-going involvement of community stakeholders in the Blue and John Crow Mountains National Park.* Kingston: JCDT.

———. 2006. *Report on the involvement of the Bowden Pen Farmers Association in the management of the Blue and John Crow Mountains National Park.* Kingston: JCDT.

Bellingham, P.J., E.V.J. Tanner and J.R. Healey. 2005. Hurricane disturbance accelerates invasion by the alien tree *Pittosporum undulatum* in Jamaican montane rain forests. *Journal of Vegetation Science* 16: 675–84.

Bowden Pen Farmers Association. 2006. *Lessons learned.* Kingston: JCDT.

Brandon, K., K. Redford and S. Sanderson, eds. 1998. *Parks in peril: People, politics, and protected areas.* Washington, DC: Island Press.

CANARI. 2003. *Building institutions for sustainable development: The role of participatory processes.* CANARI Policy Brief no. 3, UNDP/DFID. Laventille, Trinidad and Tobago: CANARI.

Chai, S-L. 2006a. *Reports 1–3, Luis Kennedy forest conservation project.* Kingston: JCDT.

———. 2006b. *Report on freshwater monitoring in the Blue and John Crow Mountains National Park for 2005/6.* Kingston: JCDT.

———. 2007. *Report on Rufford Small Grants Project: Rehabilitation of the tropical montane rainforests along the Whitfield Hall/Blue Mountain Peak Area.* Kingston: JCDT.

Cleaver, F. 2001. Institutions, agency and the limitations of participatory approaches to development. In *Participation: The new tyranny?* ed. B. Cooke and U. Kothari, 36–55. London: Zed Books.

Davis-Morrison, V. 1998. The sustainability of small-scale agricultural systems in the Millbank area of the Rio Grande Valley, Portland, Jamaica. In *Resource sustainability and Caribbean development*. eds. D. McGregor, D. Barker and S. Lloyd-Evans, 296–316. Kingston: The Press, University of the West Indies.

Dodman, D., and J. Dodman. 2006. Nuff Respec'? Widening and deepening participation in academic and policy research in Jamaica. In Pugh and Momsen 2006, 93–107.

Edwards, D. 1995. *Small farmers and the protection of the watersheds: The experience of Jamaica since the 1950s*. UWI Centre for Environment and Development, Occasional Paper no. 1. Kingston: Canoe Press.

Gentles, M. 2005. *BJCMNP Environmental Education Programme report: September, 2004–September, 2005*. Kingston: JCDT.

Geoghegan, T., Y. Renard, N.Brown and V. Krishnaryan. 1999. *Evaluation of Caribbean experiences in participatory planning and management of marine and coastal resources*. CANARI Technical Report no. 259. Laventille, Trinidad and Tobago: CANARI/UNEP/DFID.

Ghimire, K.B., and M.P. Pimbert. 1997. Social change and conservation: An overview of issues and concepts. In *Social change and conservation: Environmental politics and impacts of national parks and protected areas*, ed. K.B. Ghimire and M.P. Pimbert, 1–45. London: Earthscan and United Nations Research Institute for Social Development.

Goodland, T., and J.R. Healey. 1996. *The invasion of Jamaican montane rainforests by the Australian tree Pittosporum undulatum*. Bangor, UK: School of Agricultural and Forest Sciences, University of Wales.

Grubb, P.J., and E.V.J. Tanner. 1976. *The montane forests and soils of Jamaica: A reassessment*. Bangor, UK: School of Agricultural and Forest Sciences, University of Wales.

Healey, J.R., and J.B. Hall. 1994. *The impact on forest biodiversity of an invasive tree species and the development of methods for its control. Annual Report 1993–4 Overseas Development Agency Project R4742*. Bangor, UK: School of Agricultural and Forest Sciences, University of Wales.

Jamaica Conservation and Development Trust (JCDT). 1999. *Report on the experts' workshop for the Blue and John Crow Mountains National Park*. Kingston: JCDT.

———. 2000. *Site Conservation Plan for the Blue and John Crow Mountains National Park*. Kingston: JCDT.

———. 2005. *Blue and John Crow Mountains National Park Management Plan 2005–2010*. Kingston: JCDT.

———. 2006. *GIS analysis of enforcement patrols*. Kingston: JCDT.

Krishnarayan, V., T. Geoghegan and Y. Renard. 2002. *Assessing capacity for participatory natural resource management*. Caribbean Natural Resources Institute Guidelines Series, no. 3. Laventille, Trinidad and Tobago: CANARI.

McDonald, M.A., A. Hofny-Collins, J.R. Healey and T.C.R. Goodland. 2003. *Evaluation of trees indigenous to the montane forest of the Blue Mountains, Jamaica for reforestation and agroforestry*. Bangor, UK: School of Agricultural and Forest Sciences, University of Wales.

McGregor, D. 1995. Soil erosion, environmental change and development in the Caribbean: A Deepening Crisis? In Pugh and Momsen 2006, 189–208.

MacKinnon, K. 2001. Editorial: ICDPs working with parks and people. *Parks* 11 (2): 1–5.

Martin, G. 2006. Conservation and recreation planning on the Caribbean coast: Cahuita, Costa Rica. In Pugh and Momsen 2006, 129–54.

Mills, B. 2006. "The bad old days look better": Enlightened colonial land management practices and land reform in the British Windward Islands. In Pugh and Momsen 2006, 21–31.

Muchoney, D.M., S. Iremonger and R. Wright. 1994. *A rapid ecological assessment of the Blue and John Crow Mountains National Park, Jamaica*. Kingston: Nature Conservancy/JCDT.

Nature Conservancy. 2005. *5-S Site Conservation Planning Process*. Kingston: Nature Conservancy.

Nettles, K. 2003. "Learning but not always earning": The promise and problematics of women's grassroots development in Guyana. In *Particpatory planning in the Caribbean*, ed. J. Pugh and R.B. Potter, 173–202. Aldershot, UK: Ashgate.

Otuokon, S. 2006a. *Report on 2005 workplan*. Kingston: JCDT.

———. 2006b. *Lessons learned: Community stakeholder involvement in protected area management*. Kingston: JCDT.

———. 2007. *Report on 2006 workplan*. Kingston, JCDT.

Pelling, M. 2003. Capacity building for urban poverty reduction: The pinelands creative workshop, Barbados. In *Participatory planning in the Caribbean*, ed. J. Pugh and R.B. Potter, 98–117. Aldershot, UK: Ashgate.

Pugh, J., and J.H. Momsen, eds. 2006. *Environmental planning in the Caribbean*. Aldershot, UK: Ashgate.

Smith, D. 1995. Implementing a national park system for Jamaica: The PARC Project. In *Environment and development in the Caribbean: Geographical perspectives*, ed. D. Barker and D. McGregor, 249–58. Kingston: The Press, University of the West Indies.

Statistical Institute of Jamaica (STATIN). 2001. *The population census 2001: Country report*. Kingston: Statistical Institute of Jamaica.

Thomas-Hope, E., and B. Spence. 2006. Challenges to promoting agro-biodiversity in Caribbean small farming systems: A Jamaican case study. In Pugh and Momsen 2006, 33–51.

Tompkins, E.L., S.A. Nicholson-Cole, L-A. Hurlston, E. Boyd, G.B. Hodge, J. Clarke, G. Gray, N. Trotz and L. Varlack. 2005. *Surviving climate change in small islands: A guidebook*. Norwich, UK: Tyndall Centre for Climate Change Research.

Wade, B. 1996. The environmental imperative. *Jamaica Journal* 26 (1): 7–11.

Weis, T. 2000. Beyond peasant deforestation: Environment and development in rural Jamaica. *Global Environment Change* 10: 299–305.

Vogel, P. 2004. *Report on bird monitoring in the Blue and John Crow Mountains National Park*. Kingston: JCDT.

PART 3

Vulnerability and Domestic Food Supply

CHAPTER 9

Environmental Change and Caribbean Food Security

Recent Hazard Impacts and Domestic Food Production in Jamaica

DUNCAN McGREGOR, DAVID BARKER AND DONOVAN CAMPBELL

Introduction

It is now almost universally accepted that global warming is taking place (IPCC 2007; Gamble, chapter 2). It is also generally accepted that the net effects of global warming on agriculture, through a combination of global temperature increases, regional variations in rainfall and global increases in CO_2 levels, will be negative on developing areas as a whole (for example, see Reilly 1996; Reilly and Graham 2000; Adger et al. 2003; Parry et al. 2004).

The effects of global warming on Caribbean agriculture were postulated (for example, McGregor 1995; Watts 1995; McGregor and Potter 1997) against a background of structural weaknesses in the profile of Caribbean agriculture (Barker 1993). There are two effects to be considered; first, the effects across the Caribbean Basin of more gradual changes in environmental parameters such as temperature and rainfall.

Second, there is the more obvious effect of the damage caused in specific locations by single high-magnitude events such as hurricanes. This chapter examines the implications for food security, defined here specifically as the viability of agricultural production systems. A case study of a distinctive marginal farming system in rural Jamaica (southern St Elizabeth) is considered in terms of the effects of recent, but cumulative, environmental changes upon it. The focus is on agricultural marginalization within a complex of regional physical and local societal forces.

Food Security and the Vulnerability of the Small-Farmer Sector

Food security in developing countries is intimately connected to problems associated with globalization and the diversification of a country's economy: disasters such as famines, hurricanes, floods, droughts and the general state of agriculture, rural livelihoods, and household incomes. Further, food security policy envisages that "all people, at all times, have physical and economic access to sufficient, safe and nutritious food to meet their dietary needs and food preferences for an active and healthy life" (FAO 1999). It follows that to improve food security by increasing food availability at the national level, countries have two options: accelerate domestic agricultural production or increase imports. Domestic production remains an extremely important source of food for developing countries, since an increasing dependence on food imports through trade liberalization has contributed to increased vulnerability to external shocks in the global economy.

Kasperson et al. (2005, 249) describe vulnerability as "the degree to which a system or unit (such as a human group or a place) is likely to experience harm due to exposure to perturbations or stress, and the ability (or lack thereof) of the exposure unit to cope, recover, or fundamentally adapt (become a new system or become extinct)". This definition highlights the generally accepted idea that vulnerability has an *external component* of stresses, shocks and perturbations to which a system is susceptible; and an *internal component* which represents the resilience and capacity to cope with the external stressor (Chambers 1989; Brooks 2003). Adger (2006) argues that vulnerability has become a powerful analytical tool for describing susceptibility to harm, powerlessness and marginality of both physical and social systems. In this chapter, we conceptualize vulnerability as comprising both internal and external components. There are external risks, shocks and stresses to which an individual or a community is exposed, and an internal component of being relatively unable to adjust and cope without incurring damage and losses.

The "internal" components of vulnerability are significantly influenced by access to resources by farmers and by the ability of society, community or region to adapt to changing circumstances. A fundamental problem for Caribbean small farmers is poor access to appropriate amounts of good agricultural land. This is a significant factor in rural poverty, and the short term farm management planning horizons that rural poverty generates. Small farmers must plan for the next harvest, and do not have the financial means to plan for the longer term on their own, thwarting efforts to ensure agricultural sustainability. Government intervention can be crucial in short term disaster relief and in longer term adjustments, but the region's history of intervention is not encouraging (for example, see Edwards 1995).

Current Climate Change Projections and Caribbean Agriculture

Gamble (chapter 2) examines in detail climate change projections for the Caribbean region, and we summarize here the specific implications of these projected changes for agriculture. The longer period of data collection and collation since the early IPCC estimates of climate change (IPCC 2001a; IPCC 2001b) has indicated some clearer trends in environmental variables over the past decade or so. The Fourth IPCC Assessment Report (IPCC 2007) comes down more firmly than hitherto on the reality of progressive global warming and of the role of humans in climate change. The consequences for small islands, including the Caribbean, are detailed in Mimura et al. (2007). It should be borne in mind, however, that observed recent trends in these variables are not necessarily evidence of a progressive unidirectional change, but may, to some extent, reflect longer-term fluctuations.

The effects on agriculture of the Basin-wide temperature increases noted by Gamble (chapter 2) will be to reduce plant-available moisture, through increased evapotranspiration, unless accompanied by a balancing increase in precipitation. In semi-arid areas, with environmental conditions similar to those of the drier parts of the Caribbean, it has been estimated that for each one degree Celcius increase in temperature, potential evapotranspiration will rise by about 5 per cent (Le Houerou 1996). Other, less obvious effects may also become more problematic. The spread of some pests and diseases may hinder agriculture, while some crops may become progressively more sensitive to temperature increases. For example, preliminary research in Trinidad suggests that increasing night time temperatures may be associated with a decline in sugar cane yields (EMA 2000).

Taking forward Gamble's (chapter 2) analysis of rainfall patterns, trends can be distinguished in Jamaica's rainfall since the 1920s. The Initial National Communications of Jamaica (2000) showed that although Jamaica's annual rainfall rose slightly between 1920 and 1959, the trend from 1960 to 1998 (figure 9.1) was downward, from figures in the early 1960s of at least 2000 millimetres, to figures grouped around 1,500 to 1,700 millimetres from 1994 to 1998. Of course, these average figures for Jamaica smooth the significant rain shadow effect of the Blue Mountains, but nevertheless indicate a similar trend to that of the eastern Caribbean. The picture since about 2002 has, though, been less clear-cut (figure 9.1 – data from Jamaican Meteorological Service), with more variable annual totals.

Allied to this, the recent perceived increase in tropical cyclone activity detailed herein by Gamble may or may not, as he notes, be evidence of a longer-term trend. Early season droughts have plagued Jamaica in recent years (2004, 2005 and 2008), yet in each of these years the heavy

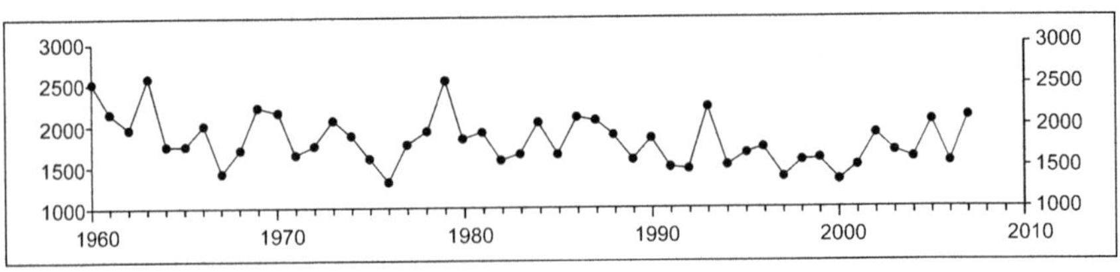

Figure 9.1 Average rainfall for Jamaica (1960–2007).
Source: Jamaican Meteorological Service.

Table 9.1 Normal Rainfall Requirement for Some Jamaican Crops

Crop	Recommended Rainfall Range (Millimetres per Year)
Sugar	1,500 of well distributed water
Citrus	1,500–1,800
Banana, coffee, yellow yam, ginger	1,300–1,800
Cassava	750–1,000
Sweet potato	1,000–1,300
Dasheen	1,500–2,000

Source: Initial National Communications 2000.

rains associated with hurricanes and tropical storm activity impacted severely on the main summer/autumn growing season in Jamaica.

With declining rainfall, or a shift towards a larger proportion of rainfall occurring in the summer (when higher temperatures would induce more evapotranspiration), already-marginal crops may become unviable locally. Losses of production are inevitable where rainfall falls below the normal requirement for typical Caribbean crops (see table 9.1 for Jamaican examples).

The influence of increasing CO_2 is variable. C_4 plants such as sugar cane, maize, millet and sorghum may be negatively affected, but there is still a need for specific research. For example, Reilly's (1996) review has wide error bands on most tropical C_4 crops, reflecting uncertainty. Singh and Maayar (1998) have attempted to model sugar cane productivity in a range of climatic change circumstances. Five models tested indicate a range of losses of between 40 and 17 per cent. However, they note that compensatory factors may operate, such as the increasing water efficiency of C_4 plants with temperature increase (due to increased stomatal resistance to transpiration), and a probable increase in yield with increasing concentrations of atmospheric CO_2.

Soil erosion rates are high on Caribbean hillsides, where the bulk of domestic food crops are grown. Visual and anecdotal evidence indicates that hillsides continue to degrade, and the effects of agriculture on erosion rates have been demonstrated experimentally for Jamaica (McGregor 1988) and St Lucia (Cox, Sarangi and Madramootoo 2006). For Jamaica, a scenario of decreasing rainfall might at first sight indicate less erosion. But a pattern of early season drought followed by a stormier wet season will ensure that soil erosion rates remain high overall. Progressive land degradation follows, thus further marginalizing already marginal agricultural systems.

With regard to "external" stresses and shocks, Caribbean small farmers are, and will be, increasingly vulnerable to the changing Caribbean climate. For much of the Basin, it appears that increasing temperatures will not be matched by concomitant increases in rainfall, but rather by decreases in water availability via increasing evapotranspiration. Some crops will become marginalized, and will require increasing and careful use of irrigation systems to remain viable. The dangers of poorly designed and improperly managed irrigation schemes are well known (for examples, see Agnew and Anderson 1992; Barrow 1999). As mentioned above, soil erosion and land degradation will probably continue to be significant. The overall effects of extreme events may increase if, as the trend in the 2000s so far suggests, both the number and magnitude of tropical storms and hurricanes increases.

Farming, Hazards and Vulnerability in Southern St Elizabeth

Jamaica suffered a series of extreme climate events in quick succession between 2002 and 2005, each of which had a significant effect on the country's agricultural production, and, in particular, domestic food production (table 9.2). The parish of St Elizabeth is one of the principal areas for domestic food production, and its fresh produce is sold in all the island's markets and major tourist resorts (see Rhiney, chapter 11).

The region of southern St Elizabeth is classed by Batjes (1994) as "dry" with relatively low moisture availability. The average annual rainfall ranges from 650 to 800 millimetres and is well below that normally required for cultivating the crops usually grown in Jamaica (see table 9.2). The main dry season lasts four to six months, while the dependable growing season (defined by Batjes as the length of the 75 per cent-dependable growing period) is only two to three months. This is one of

Table 9.2 Impact of Hurricanes and Drought on Jamaica's Agriculture, 2002–2005

	Annual Change in Agricultural GDP (%)	Annual Change Domestic Food Production (%)	Event	Damage to Agricultural Sector
2002	–8.3	–11.3	Heavy rainfall May/June & Sept	Flood damage J$1,100 million
2003	+5.7	+17.3		
2004	–10.4	–15.6	Drought first half year. hurricanes Charley, Ivan	Ivan–J$8,500 million Charley–J$90 million
2005	–7.3	–3.4	Drought first 4 months. hurricanes Dennis, Emily and tropical storm Wilma	Combined effects of storms–J$993 million

Source: PIOJ, various years.

the hottest areas of Jamaica, thus potential evapotranspiration (PET) will be relatively high.

Figure 9.2 shows the monthly rainfall totals from 2001 to 2005 for Mountainside, the nearest meteorological station to the study area (figure 9.3). The monthly totals for February and March range from highs of 155 millimetres and 83 millimetres in 2001 to nothing in 2005. Similarly, June rainfall in 2005 was 161 millimetres and nothing the previous year. November and December totals also fluctuate significantly from year to year. Rainfall is not only temporally variable in southern St Elizabeth but also spatially variable in ways that standard data collection techniques can never reflect nor verify. The uncertain volume and timing of rainfall presents a classic game theory scenario for farmers. They must consider their decision-making with respect to coping with

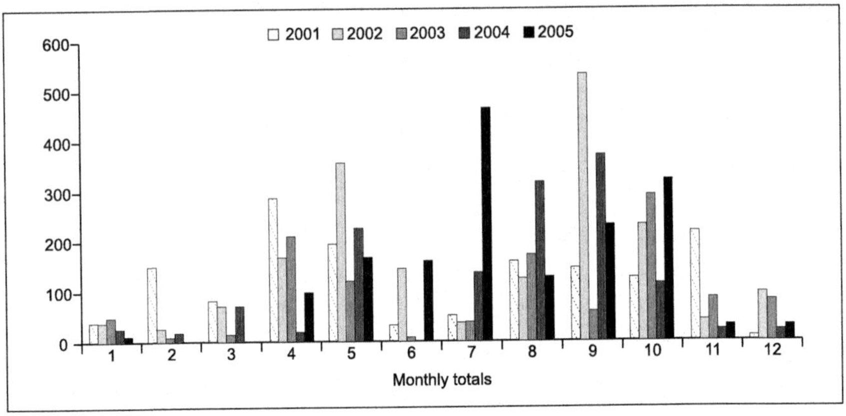

Figure 9.2 Monthly rainfall totals for Mountainside, 2001–2005.
Source: Jamaican Meteorological Service.

risk and uncertainty: should they play safe or gamble? The problem is particularly acute in deciding when to plant and when to reap. May and June is a historically short rainy period, and the sowing season in southern St Elizabeth. It is also a period adversely affected in the Caribbean Basin by the El Niño/La Niña cycle, which may produce heavier rains during the period following an El Niño event, and reduced rainfall following La Niña (Taylor, Enfield and Chen 2002).

Southern St Elizabeth experienced multiple hazards in rapid sequence between 2002 and 2005. There was a drought in the early months of 2004, followed in the same year by hurricanes Charley and Ivan, both of which tracked just south of the island. Another extended, seven-month drought ensued in late 2004 and into 2005, resulting in serious bush fires and attendant crop damage. Later in 2005, the area suffered the effects of hurricanes Dennis and Emily, as well as tropical storm Wilma. The estimated damage to St Elizabeth's agriculture by Hurricane Ivan alone has been estimated as $105 million (table 9.2).

Southern St Elizabeth, despite its low rainfall, is a major source area for Jamaica's domestic food because its farming systems are finely tuned to the environment, and rely on the ubiquitous use of grass mulch and improvised irrigation systems (figures 9.4 and 9.5). In Jamaican terms, these are prosperous, small-scale farming communities which, in the post-independence period, have offered sustainable livelihoods to

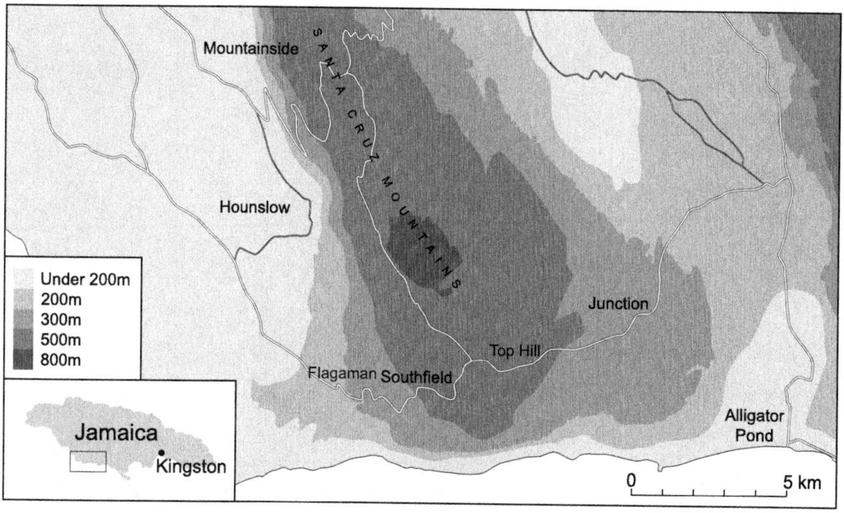

Figure 9.3 Southern St Elizabeth: survey communities.

farmers. Yet, trade liberalization and the flood of cheap food imports such as onions, carrots and red peas have compromised the area's competitive position for some domestic food crops. Onions, for example, are no longer grown in the area. A key question is: How resilient are these farming systems in the face of environmental change? In particular, to what extent will the likely increasing frequency and severity of climatic hazards further dislocate agricultural sustainability and disrupt agricultural production?

To explore some of these questions, we undertook a pilot project that focused on recent climatc hazard impacts and the response of farmers in Southern St Elizabeth. A survey of sixty small farmers investigated cropping systems, farm management practices and the perception, on behalf of the farmers, of hazard and hazard impacts in three small farming communities. The communities form an altitudinal transect from the foothills of the Santa Cruz Mountains towards the relatively narrow coastal lowlands, ranging from Top Hill (1,800 feet/550 metres), Southfield (1,600 feet/490 metres) to Flagaman (750 feet/230 metres) (figure 9.3). Twenty farmers in each community were interviewed using a detailed questionnaire survey and extensive informal discussions. The empirical data collected focused on farm size and crops grown, the use

Figure 9.4 Guinea grass being cut for mulching.
Photograph by Kevon Rhiney.

of grass mulches, crop losses due to recent extreme climatic events, the recovery process, and farmer views on the effectiveness of disaster relief, as well as their perceptions of changing weather patterns. The non-probability sample was designed for rapid data reconnaissance, though we note that a high proportion were full-time farmers (80 per cent) and 23 per cent owned a vehicle, confirming the area's status as a region of commercial farming.

Farm Management

Southern St Elizabeth is a commercial farming region dominated by small and medium-sized farms. Average farm size in the sample was about 3 acres (approximately 1.2 hectares) and, as elsewhere in Jamaica, farming systems are fragmented, with an average number of plots per farm of 2.4. The principal cash crops were carrots, scallion, watermelon,

Figure 9.5 Surface irrigation piping.
Photograph by Kevon Rhiney.

honeydew melon, cantaloupe, tomatoes, sweet peppers and cassava. There are minor variations in cropping systems by elevation; for example, carrots are grown in the higher elevations like Top Hill, while scallion and melon are grown at lower elevations such as Flagaman and Southfield.

Guinea grass was used as mulch in all 144 plots recorded (figure 9.4). The techniques of mulching are similar, and they reflect widespread and shared traditional knowledge among farmers, with only minor variations in the timing of fertilizer applications. The widespread use of a guinea grass mulch in this low rainfall area, coupled with relatively favourable soil conditions (pockets of bauxite soils, for example) are principal reasons for the area's success in the commercial production of vegetables and fruit. Farmers explain that "above ground" crops (such as melons, tomatoes and cucumber) rather than "below ground" crops (sweet potatoes and cassava) benefit more from mulching. The principal advantages of mulching listed by farmers are the retention of soil moisture, suppressing weeds, keeping plants cooler, helping germination and plant residues adding fertility to the soil. Also, they report that grass residues afford protection from rain splash (erosion) and act as a ground cushion for melon and cucumber. Farmers frequently stated that without mulching, farming would not be sustainable.

In general, farmers calculate that the ground area of live guinea grass required for mulching is approximately equal to the cropping area to be mulched. While many farmers set aside land each year for guinea grass (in effect, a planted fallow), there is a significant regional demand for sourcing additional guinea grass off-farm. Over half the farmers surveyed (57 per cent) said they needed to purchase guinea grass supplies, while only 23 per cent claimed to grow all they needed. The remaining 20 per cent grew some guinea grass and purchased the rest of their supplies. We estimate that, for a 3-acre (1.2-hectare) farm, if the farmer needs to purchase all his guinea grass, his production costs will increase by as much as US$1,000 (approximately J$70,000 at time of survey).

The study area has no organized public irrigation system, so farmers have had to adapt and respond to low rainfall conditions not only by mulching, but also by improvising irrigation systems to sustain their crops (figure 9.5). Irrigation systems appear idiosyncratic, ranging from labour-intensive and painstaking hand-watering of individual plants in fields, to field drip systems with pipes snaking out from large plastic water tanks. Farmers have hitherto dealt with their water supply problems on an individualistic basis rather than co-operating among themselves. While individual solutions often reflect personal ingenuity, such an ad hoc system is unsustainable in the face of a prolonged drought, and it increases farmers' vulnerability to high cost and unpredictable supplies from water trucks. It should be noted that households in the study area normally have two water tanks, one for domestic uses (connected to the public water supply) and the other for agricultural use (supplied by water trucks and/or rainfall).

The Effects of Recent Climatic Events

Farmers reported that all crops were affected by the storms of 2004 and 2005, but that tomatoes, scallion, melons and carrots were most affected. The list of crops most affected by drought was similar, though carrots were reportedly less affected and melons more affected. Fully 80 per cent of farmers claimed they lost all their crops as a result of the biggest single event, Hurricane Ivan, while 10 per cent said that they had lost over half their crops. More than 50 per cent also reported they lost all their crops again in the drought that occurred after Ivan, and a third of farmers (35 per cent) said 2005 was worse than 2004, combining a lengthy drought followed by several storm events (though it should be noted that 50 per cent felt Ivan was worse – these to some extent contradictory figures being perhaps an artefact of the relatively small sample size). Just over half (53 per cent) of farmers said it had taken them up to 6 months to recover from these events, but 32 per cent said it had taken more than a year to recover. It was frequently stated, however, that impacts are highly variable and localized.

It is acknowledged that more detailed empirical field research is needed to disentangle these damage reports from farmers, but what these preliminary data confirm is a complex scenario of successive and severe hazard impacts. A hypothesis worthy of further investigation is whether multiple "low"-impacts hazards may have a cumulative effect on farming that is greater than a single isolated hazard of greater magnitude (such as Hurricane Gilbert in 1988).

Response of Farmers to Hazards

Farm prices increase dramatically after storm events and during prolonged drought events. Farmers' responses to hazard impacts, whether storm or drought, are necessarily short term, precipitated by the urgent need for cash. A common strategy reported here was to plant a rapidly maturing crop like scallion to generate quick cash flow three months hence. The downside, acknowledged by many farmers, is that when a large number of farmers choose the same strategy, a market glut results, which in turn depresses prices. Sensitivity to market prices is a significant feature of the decision-making process for small farmers in this area, and it partly explains why the area no longer produces onions.

Another adjustment to Ivan and the subsequent drought, reported by 40 per cent of the sample, was to alter farm planning and management practices. The two most common responses were to delay the timing of planting (and thus reaping) and to scale down operations by taking land out of production. Both responses were influenced (in southern St Elizabeth at least) by the high cost of inputs – water for irrigation and guinea grass for mulching. An additional factor is precipitated by their own innovative, adaptive response to high input costs – planting guinea grass for their own use in order to reduce production costs. Since the guinea grass crop is also adversely affected by drought conditions, the problem of obtaining supplies of their most critical farm input is compounded, and their resilience to adversity is further weakened. Whether scaling back of production is a short term or permanent response will need longer-term monitoring.

Farmers' Perception of Relief Efforts

Most farmers were negative and highly critical of the government's relief efforts after Hurricane Ivan and during the succeeding drought. Sixty-two per cent said they had never been visited by RADA (Rural Agricultural Development Authority), the government agency charged with organizing relief efforts among farmers. Moreover, 68 per cent stated emphatically that they had received no government assistance. Financial assistance from government was perceived to be a token gesture, reportedly a J$5,000 coupon (US$71), redeemable only against carrot seeds supplied by a single large Jamaican agro-processing company and regardless of a farmer's particular cropping speciality. Farmers reported that the registration arrangements for assistance were chaotic, requiring them to travel out of the area to nearby towns, and the registration system was so loosely managed it allowed persons who were not farmers to claim assistance.

During the drought, relief assistance took the form of government-organized water trucks but, according to the farmers, this too was poorly orchestrated and expensive. Farmers were charged J$4,000 (US$57) for a small truck of 4,000 gallons and double that for a large 8,000 gallon truck. Bureaucratic procedures required farmers to visit the local National Irrigation System (NIS) office in Hounslow (see figure 9.3) to purchase the water in advance, only to wait sometimes up to three weeks for the truck to deliver, by which time many farmers said their crops had already dried up. An informal sector water supply operation readily filled these formal sector deficiencies, but farmers reported being charged an additional premium of J$2,000 (US$29) per truckload.

Conclusion

The southern St Elizabeth case study shows that the degree of small farmer vulnerability is partly dependent on the type of hazard. Though losses may be significant in a hurricane, drought is a more insidious hazard, and farmers are probably less resilient because a key aspect of vulnerability is access to critical farm inputs. In the case study, these were found to be guinea grass and water, both of which are more affected by drought than hurricanes. Taken together, hurricanes and droughts compound vulnerability, though in the St Elizabeth case we are unable to quantify these effects at the present time. Poorly organized and inadequately financed relief efforts also impinge on farmers' ability to recover from such shocks and stresses, and this was clearly evident from the data collected here. It also has a polarizing effect on communities, creating deep-seated antagonisms between local communities and government planning bureaucracies. Loss of markets is another potential vulnerability of which farmers are aware, but these preliminary data are inconclusive in this respect.

It is clear that Jamaica was improperly prepared for the impact of these extreme climatic events on the agricultural sector, and that the ability of farmers to cope not significantly improved by government interventions. As a result, farmers' perceptions of the official response were hostile and bitter. There is an urgent need for farm modernization and improved irrigation systems to enable farmers in drought-prone areas of Jamaica to deal with future events. It is clear from the southern St Elizabeth case that relief and response systems need to be radically improved if droughts and hurricanes become more frequent and/or more severe. It is also clear that southern St Elizabeth will not be the only area of Jamaica where farming systems will become marginalized. There is an urgent research need for such potentially marginal areas to be identified. This can be most easily undertaken through a thorough analysis of recent climate trends and the application of methodologies such as that of the FAO Agro-Ecological Zones approach, already outlined by Batjes (1994).

The probabilities are that many of Jamaica's existing agriculture systems will become more marginalized by environmental change, but there are also opportunities that can be explored. There are many adaptation strategies which could be employed to cope with external stresses and shocks. The feasibility needs to be explored of adaptation options such as changing management practices (forms of soil and watershed management such as irrigation, mulching, zero tillage), investigating the possibilities for new crops (though only if current crops are identified as becoming marginal), changing the annual pattern of farm activities (planting and harvesting patterns) and introducing adapted crop varieties. There is now significant case history of such adaptations elsewhere (for examples, see Adger et al. 2003; Jones 2002; Luseno et al. 2003; Meadows and Hoffman 2003; Paavola 2004), and adaptations to the Jamaican situation can be relatively rapidly simulated. But adaptations have to be appropriate and affordable in order to ensure that marginal farmers have access to them. The difficulties in ensuring that equity and justice are appropriately applied to natural resource-dependent societies are rehearsed by Thomas and Twyman (2005). In the absence of equity and justice for the small farmer, increasing physical and economic vulnerability seems inevitable. Economic environments change with climate change, and though there will be "gainers", who have the resources to seize opportunities, the livelihoods of those with less room to manoeuvre will inevitably be prejudiced.

The major pressure at present is on the small farmer, growing food for the domestic market. The small farmer is less able economically to adapt and does not have easy access to more or better land. The southern St Elizabeth experience suggests an urgent need for capacity-building, funded by government or with external support, aimed at the rural community and individual farmer level.

Research on agricultural change in the wider Caribbean focuses on the impacts of trade liberalization rather than environmental disasters, and on export crops rather than domestic food crops. In the Dominican Republic, the transformation of the rural livelihoods is linked to economic globalization and the role of multinationals in the food trade (Raynolds 1997). In the anglophone Caribbean, the fate of traditional export crops has merited considerable attention as a result of the dismantling of preferential access to the EU. The discussion on Windward

Islands bananas tends to focus on policy issues (Ahmed 2001) and there is a dearth of empirical research regarding the impacts on rural livelihoods. In Barbados, Drummond and Marsden (1995) analyse factors contributing to the decline of sugar production, while Momsen (2005) documents the transformation of peasant agriculture on the island in a unique analysis of repeat small farmer surveys from the 1960s. She concludes that, despite phasing out of sugar subsidies and a surge of food imports, small farmers in Barbados continue to adapt to changing economic conditions and respond to demands for fresh locally grown produce, reinforcing ideas on emerging linkages between tourism and agriculture (Momsen 1998).

We conclude that, in the wider Caribbean, the conceptual framework of trade liberalization and economic globalization underpins much of the research on agricultural change, identifying both negative and positive impacts. We feel a more holistic approach is urgently needed because local food production systems are not only vulnerable to the external shocks of trade liberalization, but also to the disastrous impact of agricultural hazards, what O'Brien and Leichenko (2000) refer to as the double stressors of climate change and economic globalization. Data on agricultural disasters is highly aggregated at best, and often only available through official government or international agency reports which tabulate the dollar impacts of crop losses and infrastructure damage. Detailed data on the local impacts of hazards, as reported in this chapter, are critical to understanding the internal component of vulnerability, that is, the ability to cope and adapt, because environmental hazards disrupt and dislocate farming systems, and produce unforeseen adjustments and unintended adaptations within rural communities.

The work reported here is part of a wider study of the impact of a suite of agricultural hazards in southern St Elizabeth, using detailed behavioural and ethnographic studies of farmers' local knowledge, adaptations and coping mechanisms. Our broad aim is to link environmental change and trade liberalization at the macro level with the vulnerabilities and coping strategies of farmers at the micro level. We feel this can be best achieved by focusing on a hitherto neglected aspect of the food security debate in the Caribbean region, namely the viability and vulnerability of local domestic food production systems in the face of environmental change.

References

Adger, W. N. 2006. Vulnerability. *Global Environmental Change* 16: 268–81.

Adger, W.N., S. Huq, K. Brown, D. Conway and M. Hulme. 2003. Adaptation to climate change in the developing world. *Progress in Development Studies* 3: 179–95. http://www.tyndall.ac.uk/publications/working_papers/wp16.pdf.

Agnew, C., and E. Anderson. 1992. *Water resources in the arid realm*. London: Routledge.

Ahmed, B. 2001. The impact of globalization on the Caribbean sugar and banana industries. *Society for Caribbean Studies Annual Conference Papers*, vol. 2. http://www.scsonline.freeserve.co.uk/olvol2.html.

Barker, D. 1993. Dualism and disaster on a tropical island: Constraints on agricultural development in Jamaica. *Tidjschrift voor Economische en Sociale Geographie* 84: 332–40.

Barrow, C.J. 1999. *Alternative irrigation: The promise of runoff agriculture*. London: Earthscan.

Batjes, N.H. 1994. Agro-climatic zoning and physical land evaluation in Jamaica. *Soil Use and Management* 10: 9–14.

Brooks, N. 2003. Vulnerability, risk and adaptation: A conceptual framework. Working paper no. 38. Norwich, UK: Tyndall Centre for Climate Change Research.

Chambers, R. 1989. Vulnerability, coping and policy. *IDS Bulletin* 20 (2): 1–7.

Cox, C., A.A. Sarangi and C.A. Madramootoo. 2006. Effect of land management on runoff and soil losses from two small watersheds in St Lucia. *Land Degradation and Development* 17: 55–72.

Drummond, I., and T. Marsden. 1995. A case study of unsustainability: The Barbados sugar industry. *Geography* 80 (4): 342–54.

Edwards, D.T. 1995. *Small farmers and the protection of the watersheds: The experience of Jamaica since the 1950s*. UWI Centre for Environment and Development, Occasional Paper no. 1. Kingston: Canoe Press.

Environmental Management Authority (EMA). 2000. *Republic of Trinidad and Tobago's first national communication to the United Nations Framework Convention on Climatic Change*. Port of Spain: Environmental Management Authority.

Food and Agriculture Organisation (FAO). 1999. *The state of food insecurity in the world*. Rome: FAO.

Initial National Communications. 2000. Jamaica's initial national communications. Report to the United Nations Framework Convention on

Climate Change, Kingston, Jamaica. unfccc.int/national_reports/non-annex_i_natcom/items/2979.php.

Intergovernmental Panel on Climate Change (IPCC). 2001a. *Climate change 2001: The scientific basis. Contribution of Working Group 1 to the Third Assessment Report of the IPCC.* Cambridge: Cambridge University Press.

———. 2001b. *Climate change 2001: Impacts, adaptation and vulnerability. Contribution of Working Group II to the Third Assessment Report of the IPCC.* Cambridge: Cambridge University Press.

———. 2007. *Climate change 2007: Impacts, adaptation and vulnerability. Contribution of Working Group II to the Fourth Assessment Report of the IPCC.* Cambridge: Cambridge University Press. Available at http://www.ipcc-wg2.org/

Jones, S. 2002. A framework for understanding on-farm environmental degradation and constraints to the adoption of soil conservation measures: Case studies from Highland Tanzania and Thailand. *World Development* 30: 1607–20.

Kasperson, J.X, R.E. Kasperson, B.L. Turner II, A. Schiller and W-H. Hsieh. 2005. Vulnerability to global environmental change. In *Social contours of risk*. Vol. 2: *Risk analysis corporations and the globalization of risk*, ed. J.X. Kasperson and R.E. Kasperson, 245–85. London: Earthscan.

Le Houerou, H.N. 1996. Climate change, drought and desertification. *Journal of Arid Environments* 34: 133–85.

Luseno, W.K., J.G. McPeak, C.B. Barrett, P.D. Little and G. Gebru. 2003. Assessing the value of climate forecast information for pastoralists: Evidence from Southern Ethiopia and Northern Kenya. *World Development* 31: 1477–94.

McGregor, D.F.M. 1988. An investigation of soil status and land use on a steeply sloping hillside, Blue Mountains, Jamaica. *Singapore Journal of Tropical Geography* 9: 60–71.

———. 1995. Soil erosion, environmental change and development in the Caribbean: A deepening crisis? In *Environment and development in the Caribbean: Geographical perspectives*, ed. D. Barker and D.F.M. McGregor, 189–208. Kingston: The Press, University of the West Indies.

McGregor, D.F.M., and R.B. Potter. 1997. Environmental change and sustainability in the Caribbean: Terrestrial perspectives. In *Land, sea and human effort in the Caribbean Basin*, special vol. of *Beitrage zur geographischen Regionalforschung in Latinamerika*, ed. B. Ratter and W-D. Sahr, 12–15. Hamburg: University of Hamburg.

Meadows, M.E., and T.M. Hoffman. 2003. Land degradation and climate change in South Africa. *Geographical Journal* 169: 168–77.

Mimura, N., L. Nurse, R.F. McLean, J. Agard, L. Briguglio, P. Lefale, R. Payet and G. Sem. 2007. Small islands. In *Climate change 2007: Impacts, adaptation and vulnerability. Contribution of Working Group II to the Fourth Assessment Report of the Intergovernmental Panel on Climate Change*, ed. M.L. Parry, O.F. Canziani, J.P. Palutikof, P.J. van der Linden and C.E. Hanson, 687–716. Cambridge: Cambridge University Press. Available at http://www.ipcc-wg2.org/

Momsen, J.H. 1998. Caribbean tourism and agriculture: New linkages in the global era? In *Globalization and neoliberalism: The Caribbean context*, ed. T. Klak, 115–33. Lanham, MD: Rowman and Littlefield.

———. 2005. Caribbean peasantry revisited: Barbadian farmers over four decades. *Southeastern Geographer* 45 (2): 206–21.

O'Brien, K.L., and R.L. Leichenko. 2000. Double exposure: Assessing the impacts of climate change within the context of economic globalization. *Global Environmental Change* 10: 221–32.

Paavola, J. 2004. Livelihoods, vulnerability and adaptation to climate change in the Morogoro Region, Tanzania. CSERGE Working Paper, EDM 04–12. Norwich: Centre for Social and Economic Research on the Global Environment.

Parry, M.L., C. Rosenjzweig, A. Iglesias, M. Livermore and G. Fischer. 2004. Effects of climate change on global food production under SRES emissions and socio-economic scenarios. *Global Environmental Change* 14: 53–67.

Reilly, J. 1996. Climate change, global agriculture and regional vulnerability. In *Global climate change and agricultural production*, ed. F. Bazzaz and W. Sombroek, 237–65. Chichester, UK: John Wiley. http://www.fao.org/docrep/W5183E/W5183E00.htm.

Reilly, J., and J. Graham (co-chairs). 2000. Changing climate and changing agriculture, report of the Agricultural Sector Assessment Team, US National Assessment for the US Global Change Research Programme. Washington, DC: US Global Change Research Programme. http://www.usgcrp.gov/usgcrp/nacc/agriculture/draft-report/default.htm

Raynolds. L. 1997. Restructuring national agriculture, agro-food trade, and agrarian livelihoods in the Caribbean. In *Globalizing food: Agrarian questions and global restructuring*, ed. D. Goodman and M.J. Watts, 86–96. London: Routledge.

Singh, B., and M. El Maayar. 1998. Potential impacts of greenhouse gas climate change scenarios on sugar cane yields in Trinidad. *Tropical Agriculture* (Trinidad) 75: 348–54.

Taylor, M.A., D.B. Enfield and A.A. Chen. 2002. Influence of the tropical Atlantic versus the tropical Pacific on Caribbean rainfall. *Journal of Geophysical Research* 107 (C9), 3127: 10–1 to 10–14.

Thomas, D.S.G., and C. Twyman. 2005. Equity and justice in climate change adaptation amongst natural-resource-dependent societies. *Global Environmental Change* 15: 115–24.

Watts, D. 1995. Environmental degradation, the water resource and sustainable development in the eastern Caribbean. *Caribbean Geography* 6 (1): 2–15.

CHAPTER 10

Vulnerability, Constraints and Survival on Small-Scale Food Farms in Southern St Elizabeth, Jamaica

Strengthening Local Food Production Systems

CLINTON BECKFORD AND STEVE BAILEY

Introduction

Domestic food production in Jamaica is the purview of thousands of small-scale farmers operating small holdings, on mostly marginal lands, in the hilly interior and dry southern coast of the island. Most of this food is sold, processed, resold and consumed locally, thereby providing the foundation of people's nutrition, incomes and livelihoods, and contributing to rural and national development. This is achieved despite enormous challenges faced by small-scale farmers. In this chapter we examine the challenges faced by fruit and vegetable growers in southern St Elizabeth as well as the wider implications for food availability. We discuss some of the responses of farmers to these challenges and advance an approach which might be used to strengthen local food production systems.

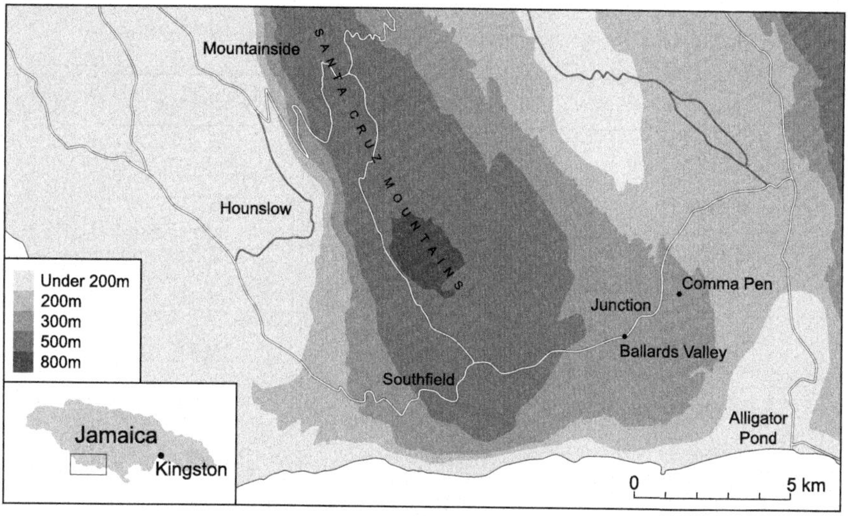

Figure 10.1 Research locations, southern St Elizabeth.

The paper is based on convenience and snowball sampling of forty-seven farmers in four communities in southern St Elizabeth in 2005: Southfield, Comma Pen, Hounslow and Ballard's Valley (see figure 10.1). The study employed a mixed methods approach in which both quantitative and qualitative techniques were systematically used. The specific mixed-methods design, where qualitative and quantitative data are collected simultaneously and both having equal status (Creswell 2003; Teddlie and Tashakkori 2003), allowed us to merge the two dominant research traditions to obtain rich data and explanations which would not have been possible using only one method. Questionnaire surveys were used to collect quantitative data while interviews and field observation were used to collect qualitative data.

Domestic Food Production in Jamaica

The significance of agriculture among rural families goes beyond satisfying household needs. Agriculture makes an indispensable contribution to national food availability and food security. In the 1970s, there was a shift in focus in agricultural policy and planning, with an increased

Table 10.1 Domestic Food Production in Jamaica, 1995–2006

Year	Production (tonnes)	% Change
1996	662,952	
1997	550,928	–16.9
1998	538,585	–2.3
1999	553,157	2.7
2000	450,530	–18.5
2001	490,296	8.8
2002	431,579	–11.9
2003	491,473	13.8
2004	414,790	–15.6
2005	391,707	–5.5
2006	467,802	19.4

Source: Compiled from Ministry of Agriculture raw data.

emphasis on domestic food production as part of a drive towards national food self-sufficiency. Even before then, the small-scale, domestic, food-producing sector made a significant contribution to both national food security and, in rural areas, household food security. It has played a key role in the economy through foreign exchange earnings, foreign exchange savings, absorption of surplus labour, contribution to tourism and contribution to agro-processing industries (Beckford 2000). Domestic food production, up to 1996, had been the foundation for growth in the agricultural sector (Ministry of Agriculture 2001). However, from 1997 to 2000, the country experienced several periods of severe drought conditions, leading to significant decreases in local food production. Domestic food production peaked in 1996 at 695,050 tonnes, but the next year there was a 25.7 per cent decrease in food production, and a 2.2 per cent decrease in 1998, followed by 2.6 per cent rebound in 1999 (PIOJ 2000). Since the turn of the century, domestic food production has fluctuated. The year 2000 saw an 18.5 per cent decrease and, despite good years in 2001 and 2003, the period from 2000 to 2005 has seen a net decrease of 13 per cent. Furthermore, production in 2006 was 29 per cent below the peak production year of 1996 (Ministry of Agriculture 2006) (table 10.1).

The decreases in food production after 1996 can be partially attributed to the aforementioned series of debilitating droughts between 1997 and 2000 and several devastating hurricanes that battered the agricultural sector between 2003 and 2005. These include hurricanes Ivan, Dennis and, most recently, Emily (see also McGregor, Barker and Campbell, chapter 9). Floods and landslides have been frequent also, with debilitating impacts on food production. The impact of extreme natural events underscores the extreme and ubiquitous vulnerability in Caribbean agriculture. In the southern St Elizabeth study area, these vulnerabilities were compounded by competition from imported food. Trade liberalization in the 1980s has led to an increase in the importation of many food items for which south St Elizabeth has traditionally been a major producer. These include foods like carrots, tomatoes, red kidney beans, onions, sweet peppers, cabbage and Irish potatoes, which are imported mainly from the United States. We will revisit this issue later in this chapter

Resource Constraints

Strengthening the capacity of local farmers to compete requires that certain problems be addressed, the first of which refers to resource constraints. Farmers in our survey identified two main issues regarding resources: access to land, which was identified by 59 per cent of the farmers, and access to financial support, identified by 83 per cent of the farmers. In our survey of farmland, 56 per cent was owned, 11 per cent leased, 6 per cent rented and just over 5 per cent squatted. Thirteen per cent of the land was operated under variety of temporary and tenuous "free" arrangements. There was a direct relationship between percentage of land owned and farm size, with percentage of land owned decreasing as farm size decreased. Insecure tenure was identified by 21 per cent of the farmers as a constraint.

Agriculture in Jamaica is characterized by deep-rooted structural dualism (Barker 1993). This is characterized by a small-scale food producing sector that primarily produces food for the domestic market, and a large-scale sector that primarily produces for export. The small-scale sector grew out of provision grounds provided for slaves during the slavery period, and the agricultural exploits of those freed who made up the peasantry in the aftermath of emancipation. Farmers in this sector grow crops which are food staples for the local population including a wide range of roots and tubers (yams, sweet potatoes, cassava, dasheen, coco, Irish potatoes), vegetables and fruits (such as carrots, cabbage, tomatoes, pumpkin, cucumber, bananas, plantains; and condiments and herbs like onions, escallion, thyme. These are usually grown for local marketing and consumption, however, small-scale farmers also grow traditional export crops like bananas, citrus and coffee, as well as non-traditional exports like yams. The large-scale sector focuses on production of the traditional export crops, mainly sugar cane, coffee, bananas, and citrus. Such dualism has historical roots and has influenced policy, creating asymmetrical relationships between small-scale farmers and centres of economic and political power (Beckford, Barker and Bailey 2007). An example is the critical issue of financial support. The degree of access to farm credit is directly related to farm size; farmers of traditional export crops have greater access to credit than small-scale domestic food farmers. In the 1990s, for example, the amount of funds allocated to the export crop sector was approximately seven times greater than to the domestic sector on average (PIOJ 2000). Farmers in the present study rarely utilized banks and other formal financial institutions. Over 27 per cent of the farmers suggested that because they did not own land they would not be able to offer collateral. Others who owned land did not have legal proof of ownership and would therefore not be able to use land as collateral. Furthermore, they were generally opposed to borrowing large sums of money from official institutions because of the risks involved to their families by mortgaging their land and houses. Forty-three per cent of the farmers did not think they would be able to afford the interest on bank loans. The point should be made, however, that only 9 per cent of the farmers had ever applied for agricultural loans, and only two farmers reported that they had been successful in securing loans. Studies of small-scale food farmers in Jamaica have consistently identified lack of access to funds as a major obstacle to production (Beckford 2000; Davis-Morrison 1998; Meikle 1998). Over 87 per cent of our farmers identified lack of finances as a major constraint. The inadequate compensation for losses from environmental hazards was a major source of angst among farmers, who reported that compensation for hurricane damage was slow and insignificant, while typically, losses from drought, fire and floods were not even addressed.

Hazards

Agriculture in Jamaica is vulnerable to a plethora of natural disasters including hurricanes, floods, landslides and droughts. The impact of these events is well known and well documented and need not be detailed here. The issue here is the extent to which farmers are assisted in their efforts to rebound from losses caused by hurricanes and other disasters. The farmers in our study were unanimous in their opinions which identified that support from government was woefully inadequate. They bemoaned the corruption, disorganization and bureaucracy which characterized the process of assessing losses and described the assistance they received in terms of tokenism (see also McGregor, Barker and Campbell, chapter 9). Next to dualism, disaster is the most ubiquitous source of vulnerability to local agriculture and is a major threat to food security through food self-sufficiency. An effective strategy for providing meaningful support to farmers after major extreme events is critical and overdue.

Farmers themselves have devised creative ways of coping with some hazards. Their response to droughts and general moisture deficiency, for example, is a study in innovation and experimentation. The communities are part of the rain-shadowed, dry, southern coastal plains, and they suffer frequent debilitating droughts and lack of rainfall. Farming systems in the area are characterized by the use of grass mulch to retain moisture, keep the ground cool and keep down weeds (see also McGregor, Barker and Campbell, chapter 9). Mulching is so widespread that as many as 69 per cent of farmers set aside some portion of their land to grow guinea grass in order to supply some of their own needs, usually on land left in fallow. This is quite ingenious as it can be used also to generate a cash income – 9 per cent reported that they sometimes sold guinea grass to neighbouring farmers.

In addition, farmers use a number of inventive homemade schemes to provide irrigation. Some have put in rudimentary sprinkler systems while others have put in quite elaborate drip irrigation systems. On most farms you will see a number of water barrels strategically placed around a farm plot from which farmers will use a can to water each plant individually. Farmers say that this is tedious and backbreaking work but they recognize that this is a very efficient way of using a scarce and critical input. Drip irrigation and the manual watering of individual plants are good examples of the ability of Jamaican small-scale farmers to efficiently manage scarce resources, a reality that has become more evident in the last twenty-five years and has been well-documented (Davis-Morrison 1998; Barker and Beckford 2003; Beckford and Barker 2007; Beckford, Barker and Bailey 2007).

Marketing and Distribution of Domestic Food Crops

Another theme emerging from the data was the lack of institutional support in the marketing and distribution of food. There is a dualism in the marketing and distribution of agricultural produce that parallels the dualistic mode of agricultural production in Jamaica. On one hand, there is an informal and inadequate system for the distribution of domestic crops, characterized by several inefficiencies, including lack of marketing information, an absence of storage facilities, unreliable transportation and lack of quality control. On the other hand, there is the relatively well-established and highly organized system for the export crop sector.

Traditional export crops benefit from structured marketing and distribution arrangements typified by commodity associations that organize production, transportation and marketing of crops like sugar cane, coffee, cocoa, citrus and banana. However, marketing of domestic crops locally is achieved through a number of unstructured arrangements. The two main ones in this study were: *farmer* and *higgler* marketing, where a farmer or his spouse takes their produce to local produce markets themselves; and higgler marketing, characterized by small-scale informal traders (mainly women) who purchase food from farmers for sale in local produce markets. In addition, sidewalk and roadside vendors often buy produce directly from farms and sell them in rudimentary stalls usually set up informally in public spaces.

Traditionally, higglers bought food directly from farms, but a recent and growing trend, which started in the late 1990s, is that farmers take their produce in bulk to local produce markets and sell directly to higglers who operate in these markets. Seven farmers were engaged in this marketing strategy. Farmers explained that, in traditional dealings with higglers, their produce is credited and payment is made to them after the higgler returns from market. Often higglers claim that they are unable to pay the farmer because they experienced a "bad market". This new arrangement typically involves up-front cash payments by the higgler to the farmer when the sale is made.

The distribution and marketing of domestic food crops was cited as a major constraint to production by 47 per cent of the farmers. Efficient marketing arrangements would encourage many farmers to produce more, despite other problems. Specific problems identified included poor roads and expensive transportation, unfair practices by export agents, the poor physical conditions of most local produce markets, inability to collect money from higglers, irregular and unpredictable orders from export agents, and the lack of control in setting prices and grading. Many farmers are still moving produce from farms on their heads, and trade with exporters is not governed by contractual arrangements. Older farmers suggested that government should play a bigger role in marketing of food, urging a return to arrangements similar to those in the 1970s with the Agricultural Marketing Corporation (AMC), on which they could depend to buy most of their produce. While farmers were concerned about receiving a fair price for their produce, 83 per cent felt guaranteed markets were more important.

Several farmers observed that local higglers, their traditional safety net, had become involved in the international trade of farm produce in Jamaica. Higglers used to buy all their food from local farmers (or market food from their own farms), but recently they have begun to buy and sell more imported food.

Competition from Foreign Food Markets

The increase in the importation of certain food items has had a marked and adverse impact on domestic food production in some areas. Jamaica's main food imports are cereals – mainly wheat, rice, flour and corn – and have accounted for an average of 80 per cent of food imports in the last decade. Rice is grown locally but only on a very small scale in the swampy areas of the western parishes of St Elizabeth and Westmoreland. The island is nowhere near self-sufficient in the production of corn, which is grown mainly as a rotational crop on very small acreages (Ministry of Agriculture 2001). Wheat is not grown in the country at all. All these cereals are critical components of local cuisine and diets. Importation of these crops has very little adverse impact on local agriculture but the same is not true for a growing list of crops which traditionally have been important cash crops for local, small-scale, domestic food farmers. These crops include onions, red peas (kidney beans), carrots, Irish potatoes, sweet pepper and cabbage – all of which, with the exception of Irish potatoes, are major cash crops in the study area.

In this study, the unregulated importation of agricultural produce was identified by 83 per cent of the farmers as the second most debilitating constraint (after shortage of water for irrigation) to sustainable and profitable farming. In Hounslow, formerly a major producer of onions, production has become almost non-existent. Of thirteen former onion growers identified in our survey, only two were growing onions regularly and they had reduced their production by over 70 per cent in the last ten years. These two farmers were able to stay in production because they had managed to secure market contracts with supermarkets, as well as small hotels and guest houses involved in south coast tourism. In Southfield, Junction and Ballard's Valley, farmers made similar complaints. They had drastically reduced their production of cabbage, sweet pepper, red peas, melons and carrots because they could not compete with the imported varieties. Seven farmers had scaled down their operations to become primary food distributors rather than farmers. They bought imported food in bulk from distributors in Kingston and resold them to supermarkets, hotels and local produce markets in Kingston, Mandeville, Falmouth and Montego Bay. The farmers identified several other local farmers who had also become mainly food distributors.

It was argued that the imported foods were so cheap that local producers were unable to compete. Two farmers, who also cultivated land located in Coleyville, Manchester, discussed the impact of imported kidney beans (red peas) on the local farmers. Red peas were a major cash crop in this area up to the 1990s, but farmers had been forced to reduce production drastically in face of competition from imported produce. Two farmers recounted how they were forced to dump barrels of peas in 2000 because of unfavourable market conditions. Planning Institute of Jamaica data indicate that there was an 80 per cent decline in the local production of red peas between 1996 and 2004. Throughout the communities studied, farmers reported their inability to sell produce and the resultant spoilage. Further, farmers identified over a dozen farms which they classified as either abandoned or which had been lying idle or in fallow for extended and indefinite periods of time.

Dependence on Food Imports: An Issue of Vulnerability

International trade has become a critical component in food security in Jamaica in the last decade. According to the Ministry of Agriculture it contributes by making up the differences between domestic production and consumption needs, reducing supply variability, and fostering economic growth (Ministry of Agriculture 2001). This interpretation is quite problematic. It suggests that the increase in food imports is driven by shortfalls in domestic supply. There is, however, no evidence of this. Food imports are becoming a progressively larger part of Jamaica's food profile, but this occurs to the detriment of local producers and puts the survival and economic fortune of many rural farm families at risk. Many of the foods being imported have been successfully grown by successive generations of local farmers, many of whom have been driven out of production by cheap imports. In 2006, Jamaica imported US$1.64 billion worth of food, which was half the country's import bill. The implications of Jamaica's over-reliance on imported food were identified by the Ministry of Agriculture (2001) and were discussed by Spence (1996).

The Jamaican situation is reflective of a general trend in developing countries that have opened up their markets in the last two decades

(Short 2000; Spitz 2002; Walelign 2002). The structural adjustment process Jamaica embarked on in the 1980s forced smallholder farmers to compete with subsidized exports, mainly from the United States. Like most poor countries, subsidies in Jamaica are insignificant, and trade barriers have been removed. At the same time, farmers in America are paid subsidies to produce food which is then "dumped" in Jamaica and other developing countries, putting local farmers out of business and making it difficult for rural farm families to provide a satisfactory livelihood for themselves (CIOEC 2003; Via Campesina 1996, 2001, 2003; UNDP 2005). In addition, export subsidies allow US farmers to sell their products at lower prices than the cost of production (Windfuhr 2002, 2003; Windfuhr and Jonsen 2005). This forces local farmers into unfavourable, often insurmountable, competitive situations.

On a global basis this problem is becoming increasingly more serious. In India, for example, imports of EU subsidized dairy surpluses have had disastrous impact on local, family operated dairy farms. Similarly, the export of pork from the United States to the Caribbean has destroyed the local Caribbean market (Windfuhr and Jonsen 2005). All forty-seven farmers in this study said that they were earning a less than satisfactory livelihood from farming.

Developments in the international global political economy have serious implications for national food security. The flood of imported food is the result of trade liberalization which drives the process of globalization. Structural adjustment from the 1980s has led to deregulation of local markets, removing protective barriers and subjecting local producers to unfettered competition from mainly subsidized food imports. The removal or reduction of barriers between the domestic and international markets made it imperative for local producers to become more efficient in order to compete. The removal of subsidies and opening up the market to international competition is perhaps beneficial to consumers through lower prices caused by more efficient production and greater access to affordable food. However, it is not economically feasible for Jamaica to grow all the food it needs, so food security realistically has to be secured through a balanced approach to international trade combined with a healthy domestic producing sector. Presently, there is an over-dependence on food imports.

There are obvious dangers associated with too great a reliance on foreign food imports, but the interdependent nature of the world economy means that any external shocks will have widespread impacts in many countries. In the context of the global political economy, a major concern is the possibility of market instability. This could cause increased prices and uncertainty of supplies – both of which leading to food shortages locally. A second, scarcely mentioned concern is the vulnerability of foreign-produced food to terrorist attack. The food system in America may become a terrorist target with implications for countries like Jamaica and its significant food imports from the United States. A third concern is that, with so much of the Jamaica's food imports originating in the United States and Canada, the contamination of food supplies is a possibility.

This latter point is important because food security refers not just to food availability, but to food safety as well. The highly centralized nature of American farms and food processing operations creates endless opportunities for contamination and spread of contaminated food (Halweil 2005). Also, the highly uniform and large-scale crop fields and livestock herds that dominate North American farms facilitate the rapid spread of disease agents. The sheer size of many operations means it is virtually impossible to monitor all farm animals and crops properly. Usually it is only after unsafe food enters the food chain and people are affected that the danger is recognized. Uniformity and large size, which are seen as signs of efficiency and economies of scale, contribute to the ecological fragility of farms (Halweil 2005). The specialized nature of many large farms makes them dependent on daily deliveries of massive amounts of inputs like feed and medicine, which present more avenues for the introduction and movement of diseases. It is estimated that, in the United States, the average food item travels between 1,500 and 2,500 miles (approximately 2,400 to 4,000 kilometres) from farm to plate (Halweil 2005), and the distance is further increased when exported offshore to countries like Jamaica. The problem here is that the farther food travels and the more frequently it changes hands, the more likely it is for it to encounter contamination.

The point should be made here that Jamaica is typical of the English speaking Caribbean in terms of its dependence on food importation.

With the exception of perhaps Guyana, which was classified as a net food exporter in the mid 1990s, the region is made up of countries that are net importers of food. As an example, the Caribbean Basin Trade Office (CBTO) reports that Barbados imports a whopping 70 per cent of its food (http://www.cbato.fas.usda.gov/efm_barbados.htm). The CBTO profiles for other countries in the region indicate a general reliance on imported food.

Supporting Local Farmers: A Food Sovereignity Approach

It follows logically from the preceding discussion that more emphasis should be placed on producing more food locally. One way to do this is to promote greater self-sufficiency. This can be done through implementation of policies consistent with the principle of *food sovereignty*. Food sovereignty is a term which was first used in 1996, by the members of the Rome-based organization Via Campesina, to describe a concept advocated by a number of farmers, peasants and fishermen's organizations: namely the claimed right of peoples to define their own food and agriculture in contrast to having food largely subject to international market forces. The concept includes support for smallholders and for collectively owned farms rather than industrializing these sectors in a minimally regulated global economy (World Forum on Food Sovereignty 2001). Windfuhr and Jonsen (2005) described food sovereignty as a platform for rural revitalization at a global level based on equitable distribution of resources, farmers having control over planting stocks and productive small-scale farms supplying consumers with healthy, locally grown food.

Via Campesina (1996) defined food sovereignty as the right of peoples to define their own food and agriculture; and that explicit in this were the necessities for people to be able to protect and regulate domestic agricultural production and trade in order to achieve sustainable development objectives; to determine the extent to which they want to be self-reliant; and to restrict the dumping of products in their markets. Via Campesina (1996) also pointed out that the concept of food sovereignty does not negate trade, but rather, it promotes the formulation of trade policies and practices that serve the rights of peoples to safe, healthy and ecologically sustainable production.

This conceptualization of food sovereignty contains several fundamental elements which have implications we consider significant in the Jamaican context. These include the prioritization of local agricultural production, ensuring that small farmers have access to productive resources (land, water, planting materials, livestock breeds and credit). Also important are the rights of small-scale farmers to produce food, the rights of countries to protect themselves from underpriced agricultural and food imports and the rights of consumers to decide what they consume, as well as how and by whom it is produced. Other important elements in this approach to food sovereignty are the participation of farmers and rural peoples in agricultural planning and decision-making (including the development of agricultural innovations), and the conceptualization of agroecology as a way to achieve sustainable livelihoods and environmental integrity as well as producing food.

Conclusion

The issues discussed in this chapter are generic to the Caribbean region as a whole. The work of many researchers in other parts of the region speaks to the importance of the local small-scale farming sector and the considerable challenges faced by farmers in this sector. Many of the seminal and classical works, including those by Demas (1970), Rankine (1973), Brierley (1976, 1991), Hills and Iton (1983), and Mintz (1985), confirm many of the homogenous characteristics of the realities of small-scale farming across the Caribbean. Collymore's work in St Vincent (1986) provides many similarities to the Jamaican context, as does the work of Thomasson (1994) in Montserrat and Brierley (1976) in Grenada.

Farmers in this study identified a number of general solutions to their problems. Thirty-two per cent mentioned the idea of more agro-processing by Jamaican manufacturers using local farm produce. Another popular suggestion was to encourage Jamaicans to buy what they termed "locally grown" produce, astutely pointing out that tourism interests and the fast food industry both should be given incentives to use locally grown food. In relation to water shortage, farmers suggested that wells and water tanks should be erected throughout the area. It was suggested that agricultural loans should be more accessible, and disaster compensation arrangements should be rethought. Farmers also argued that other hazards, in addition to hurricanes, should be included in compensation schemes.

The farmers in our study are typical of the vast majority of small-scale domestic food crop producers in Jamaica (see studies of small-scale farmers in Jamaica by Spence 1996; Davis-Morrison 1998; Meikle 1998; Beckford and Barker 2007; Beckford, Barker and Bailey 2007). They play a vital role in the economy, yet face formidable odds. We submit that, given the chance, these farmers can be even more instrumental in ensuring that the nation's food needs are more adequately met. Food self-sufficiency is of paramount importance, as it advocates meeting food needs as far as possible from domestic supplies and minimizing dependence on global trade (Pretty 2001; Friends of the Earth International 2003; Holt-Giménez 2006). This definition of food self-sufficiency represents a perfect fit with a food sovereignty approach focused on supporting local farmers and giving them every opportunity to make a decent livelihood and feed their families.

While recognizing the ongoing importance in Jamaica of international trade to meet domestic food needs, we make the case for a local strategy that honours the contribution of local farmers and reduces the vulnerability of local food availability to external events. Food sovereignty recognizes that rural peoples, who depend on agriculture for their livelihoods, need special support to overcome the problems they face in environmentally sustainable ways (Hines 2003; Schwind 2005). Such advocacy is not directed against international trade *per se* but is based on the rationale that the current rules and practices of international trade work against the interests of smallholder farmers (Kent 2001; Oxfam International 2002; IATP 2003). St Elizabeth's small-scale farmers and their families have a right to be protected against dumping and unfair competition. We conclude, therefore, that there is a need for a trade system which does not compromise local agriculture and the ability of local farmers to feed their families and supply affordable food to the local population. This requires the removal of barriers identified by farmers in this study and others, as well as an approach that recognizes their importance and commits to providing the infrastructural and financial supports they need.

References

Barker, D. 1993. Dualism and disaster on a small tropical island: Some constraints on agricultural development in Jamaica. *Tidjschrift voor Economische en Sociale Geographie* 84: 332–40.

Barker, D., and C.L. Beckford. 2003. Yam production and the yam stick trade in Jamaica: Integrated problems for resource management. In *Resources, planning and environmental management in a changing Caribbean*, ed. D. Barker and D. McGregor, 57–74. Kingston: University of the West Indies Press.

Beckford C.L. 2000. Yam cultivation, the yam stick trade and resource depletion in the yam growing regions of central Jamaica: Integrated problems for planning and resource management. PhD diss., University of the West Indies.

Beckford, C.L., and D. Barker. 2007. The role and value of local knowledge in Jamaican agriculture: Adaptation and change in small-scale farming. *Geographical Journal* 173 (2): 118–28.

Beckford, C.L., D. Barker and S.W. Bailey. 2007. Adaptation, innovation and domestic food production: Some examples of survival strategies of Jamaican small-scale farmers. *Singapore Journal of Tropical Geography* 28 (3): 273–86.

Brierley, J.S. 1976. Kitchen gardens in the West Indies with a contemporary study from Grenada. *Journal of Tropical Geography* 43: 30–40.

———. 1991. Kitchen gardens in the Caribbean, past and present: Their role in small farm development. *Caribbean Geography* 3 (1): 15–28.

Caribbean Basin Trade Office. 2008. Barbados: Country profile. http://www.cbato.fas.usda.gov/efm_barbados.htm.

Coordinadora de Integracion Economicas Campesinas de Bolivia (CIOEC). 2003. Towards a world convention on food sovereignty and trade. www.cioecbolivia.org/wgt/food_sovereignty.htm.

Collymore, J. 1986. Small farmers as resource managers: A note from St Vincent. *Caribbean Geography* 2: 92–99.

Creswell, J.W. 2003. *Research design: Qualitative, quantitative and mixed method research approaches*. 2nd ed. London: Sage Publications.

Davis-Morrison, V. 1998. The sustainability of small-scale agricultural systems in the Millbank area of the Rio Grande Valley, Portland, Jamaica. In *Resource sustainability and Caribbean development*, ed. D. McGregor, D. Barker and S.L. Evans, 296–316. Kingston: The Press, University of the West Indies.

Demas, W. 1970. The prospects for developing agriculture in small Commonwealth Caribbean territories: The role of small-scale farmers. In *Proceedings of the Fifth West Indian Agricultural Economics Conference*. Trinidad and Tobago: University of the West Indies, St Augustine.

Friends of the Earth International. 2003. Trade and people's food sovereignty. http://www.foei.org/publications/pdfs/newfinallowres.pdf.

Halweil, B. 2005. Farmland defense: How the food system can ward off future threats. In *New perspectives on food security*, Glynwood Center Conference Proceedings, 25–30. Cold Spring, NY: Glynwood Centre.

Hills, T.L., and S. Iton. 1983. An assessment of the "traditional" in Caribbean small-scale agriculture. *Caribbean Geography* (1): 24–35.

Hines, C. 2003. *A global look to the local: Replacing economic globalization with democratic localization*. London: International Institute for Environment and Development (IIED).

Holt-Giménez, E. 2006. Movimiento Campesino a Campesino: Linking sustainable agriculture and social change. *Food First Backgrounder* 12 (1), Winter/Spring. Institute for Food and Development Policy, Oakland, California.

Institute for Agriculture and Trade Policy (IATP). 2003. Towards food sovereignty: Constructing an alternative to the World Trade Organization's agreement on agriculture farmers, food and trade. International Workshop on the Review of the AoA. Geneva. www.tradeobservatory.org/library.cfm?RefID=25961.

Kent, G. 2001. Food and trade rights. *UN Chronicle*, Issue 3. http://www.un.org/Pubs/chronicle/2002/issue1/0102p27.html.

Meikle, P. 1998. Rural change and agricultural sustainability: The Rio Grande Valley Project, Portland, Jamaica. In *Resource sustainability and Caribbean development*, ed. D. McGregor, D. Barker and S. Lloyd Evans, 273–95. Kingston: The Press, University of the West Indies.

Ministry of Agriculture. 2001. Update on the state of food security in Jamaica. http://www.moa.gov.jm/agripp/int/food_sec/overview.php.

———. 2006. All island estimates of crop production 1997–2006. http://www.moa.gov.jm/agripp/agridata/data/qrtpord_estimate9706.htm.

Mintz, S. 1985. From plantations to peasantries in the Caribbean. In *Caribbean contours*, ed. S.W. Mintz and S. Price, 127–54. Baltimore: Johns Hopkins Press.

Oxfam International. 2002. *Rigged rules and double standards: Trade globalization and the fight against poverty*. Oxford: Oxfam International.

Planning Institute of Jamaica (PIOJ). 2000. *Economic and social survey of Jamaica: Annual report, 2000*. Kingston: Planning Institute of Jamaica.

Pretty, J. 2001. *Reducing food poverty with sustainable agriculture: A summary of new evidence*. Wivenhoe Park, UK: Centre for Environment and Society, University of Essex. http://www2.essex.ac.uk/ces/ResearchProgrammes/SAFEWexecsummfinalreport.htm.

Rankine, L. 1973. *Extra and intra Caribbean trade in root crops*. Occasional Paper Series no. 9. St Augustine, Trinidad: Department of Agricultural Economics and Farm Management, University of the West Indies.

Schwind, K. 2005. Going local on a global scale: Rethinking food trade in the era of climate change, dumping and rural poverty. *Food First Backgrounder* 11 (2): 1–4.

Short, C. 2000. Sustainable food security for all by 2020; food insecurity: A symptom of poverty. London: Department for International Development (DFID). http://www.ifpri.org/2020conference/PDF/summary_short.pdf.

Spence, B. 1996. The influence of small farmers' land use decisions on the status of domestic food security in Jamaica. *Caribbean Geography* 7 (2): 132–42.

Spitz, P. 2002. Food Security, the right to food and the FAO. *FIAN-Magazine* no. 2.

Teddlie, C., and A. Tashakkori. 2003. Major issues and controversies in the use of mixed methods in the social and behavioural sciences. In *Handbook of mixed methods in social and behavioural research*, ed. C. Teddlie and A. Tashakkori, 3–50. London: Sage.

Thomasson, D. A. 1994. Montserrat kitchen gardens: Social functions and development potential. *Caribbean Geography* 5: 20–31.

United Nations Development Programme (UNDP). 2005. *The Millennium Project 2005. Halving hunger: It can be done*. Final report of the Task Force on Hunger. New York: Earth Institute at Columbia University. Available at http://www.unmilleniumproject.org/who/tf2docs.htm.

Via Campesina. 1996. *Food sovereignty: A future without hunger*. http://www.viacampesina.org/imprimer.php3?id_article=38.pdf.

———. 2001. *Priorities of peoples' food sovereignty*. http://www.viacampesina.org/IMG/_article_PDF/article_34.pdf

———. 2003. *What is food sovereignty?* http:// www.viacampesina.org/IMG/_article_PDF/article_216.pdf

World Forum on Food Sovereignty (WFFS). 2001. Final declaration of the World Forum on Food Sovereignty. Havana, Cuba: World Forum on Food Sovereignty, 7 September 2001. http://www.ukabc.org/havanadeclaration.pdf.

Walelign, T. 2002. The fifth P7 summit: Food sovereignty and democracy – Let the world feed itself. *GREEN/EFA International Relations Newsletter* no. 6 (December).

Windfuhr, M. 2002. *Food security*, food sovereignty, right to food: Competing or complementary approaches to fight hunger and malnutrition? Hungry for what is right. *FIAN-Magazine* no. 1.

———. 2003. *Food sovereignty and the right to adequate food*. Heidelburg, Germany: FIAN-International.

Windfuhr, M., and Jonsen, J. 2005. *Food sovereignty: Towards democracy in localized food systems*. Rugby, UK: ITDG Publishing.

CHAPTER 11

(Re)defining the Link?

Globalization, Tourism and the Jamaican Food Supply Network

KEVON RHINEY

Introduction

In an effort to tackle the challenges of a new world economy, many island economies are turning to tourism as a cornerstone for economic growth and development. In the Caribbean, this steady growth in tourism over the past two decades has occurred alongside a progressive decline in the agricultural industry (Thomas-Hope and Jardine-Comrie 2007; Dodman and Rhiney 2008). This trend (an expanding tourism industry and a declining agricultural sector) runs counter to previous arguments that tourism's growth would, over time, lead to both an expansion and diversification of domestic cropping systems (Lundgren 1973).

A number of past researchers (Momsen 1972; Bryden 1973; Belisle 1983, 1984) have in fact shown that the tourism industry then provided only limited opportunities for domestic agriculture. These early studies pointed to tourism industry characteristics such as hotel size and ownership, as the major causes for this absence of linkage between the two sectors. For example, Belisle (1982) observed that large, high-end hotels in Jamaica displayed a greater tendency to consume food imports as opposed to smaller, low-end hotels. Similarly, Momsen's 1971 survey of St Lucia's tourism industry found that large, foreign-owned hotels relied extensively on food imports (over 70 per cent by value) compared with small, locally owned hotels, which imported only 33 per cent of their food by value (1972).

Since then, both tourism and agriculture have changed, and some contemporary scholars suggest the possibility of establishing deeper linkages between the two sectors, attributed mainly to changing attitudes about serving local cuisine in hotels, the globalization of food consumption habits, a maturing of regional tourism industries and increased government assistance to producers of locally grown foods (Momsen 1986, 1998; Torres 2003; Conway 2004). The problem, however, is that there are only a few empirical studies that have looked in-depth at agriculture-tourism linkages in the Caribbean, most of which were undertaken more than two decades ago. As such, a review of the literature on agriculture-tourism linkages in the Caribbean ends up with more questions than answers.

This chapter is part of a larger study that examines agriculture-tourism linkages in Jamaica. Jamaica, being a larger and more established Caribbean destination, provides a good opportunity to investigate the nature of the relationship existing between its tourism and agriculture industries. The study focuses on the opportunities and challenges faced by local small farmers in forging viable linkages with the domestic tourism industry against the backdrop of an increasingly globalized world economy that is characterized by free trade. The empirical research draws on fieldwork data collected from a variety of research methodologies, including a detailed questionnaire survey of hotel and restaurant food operators in the resort town of Negril, as well as in-depth interviews and focus group sessions with individual farmers in several traditional farming communities, local purveyors, food importers and two farmers co-operatives that supply the Negril area with various fruits and vegetables.

The chapter is divided into two broad sections. The first section provides both historical and economic accounts of the impacts of global trading policies and neo-liberal economic reforms on the domestic food market, and will discuss how these in turn affect the nature of the linkages that now exist between tourism and local agriculture. The second section focuses on the empirical research. Here, the relationships existing between the tourism and agriculture industries in Jamaica are examined beyond the linkage/leakage dichotomy that places emphasis on quantitative assessments such as the proportion of local to imported food purchased by tourist food establishments – an approach evident in previous studies (Lundgren 1973; Belisle 1983, 1984). Instead, the chapter extends the analysis by focusing on the major local food supply networks that exist between the two sectors and evaluating their respective strengths and weaknesses in a neo-liberal trading environment.

The Jamaican Tourism Industry

While Jamaica's tourism industry dates back to the late 1800s, it was not until the 1950s that the island started to witness a significant growth in visitor arrivals due largely to developments in transportation and communication technologies after the Second World War (Taylor 1993). Jamaica experienced a progressive increase in visitor arrivals up to the 1960s. By 1965, the island was hosting over 300,000 visitors and tourism had surpassed sugar as the major source of foreign revenue (Sealey 1982). However, by the 1970s, Jamaica saw a drastic decline in its tourism industry. This slow down came as a result of a variety of external factors including rising oil prices and increasing air fares, but largely as a result of the Manley-led administration's flirtation with democratic socialism as well as its close relations with Cuba. Manley's policies attracted a chain of negative press coverage overseas, particularly in the United States – Jamaica's chief generating market (Taylor 1993). Between 1975 and 1977, visitor arrivals to Jamaica fell by an average of 15 per cent each year (Sealey 1982, 35) and in 1977 the island's average hotel room occupancy fell to a mere 29 per cent (Sealey 1982, 43). By the end of Manley's tenure in 1980, visitor arrivals and the in-country expenditures of these visitors had reached an all time low (Sealey 1982; Taylor 1993).

The industry soon regained buoyancy during the 1980s under the Seaga-led government. The pro-US policy adopted by the Seaga-led Jamaica Labour Party, as well as the massive advertisement campaigns

Table 11.1 Summary of Main Tourism Indicators, Jamaica 2000–2005

Indicators	2000	2001	2002	2003	2004	2005
Total stopovers	1,322,690	1,276,516	1,266,366	1,350,285	1,414,786	1,478,663
Cruise ship passengers	907,611	840,337	865,419	1,132,596	1,099,773	1,135,843
Total visitor arrivals	2,230,301	2,116,853	2,131,785	2,482,881	2,514,559	2,614,506
Hotel room nights sold	2,956,315	2,903,586	2,830,604	2,971,885	3,133,963	3,253,149
Average hotel room occupancy (%)	58.6	56.3	55.5	57.9	61.4	61.9
All-inclusive hotel room occupancy (%)	70.0	69.4	66.2	69.3	71.9	70.0
Non all-inclusive hotel room occupancy (%)	39.0	35.1	38.3	38.5	39.1	42.4
Total visitor expenditure (US$ million)	1,332	1,233	1,209	1,351	1,437	1,545

Source: Jamaica Tourist Board, annual travel statistics, various years.

carried out by the Jamaica Tourist Board (JTB), helped to restore both visitor and investor confidence in the island. By 1987, total visitor arrivals had exceeded the one million mark for the first time in the island's history (Taylor 1993).

Despite slight fluctuations, the overall number of visitors to Jamaica has increased over the last two decades, and has further intensified over the last few years. In 1999, Jamaica hosted over two million visitors. Since then the industry has continued to experience a steady rate of growth in visitor numbers, room stock and attractions. Jamaica is now one of the Caribbean's leading tourism destinations, with a total of 2.6 million visitor arrivals in 2005, of whom approximately 1.5 million were stopover visitors. In the same year, gross visitor expenditure was worth US$1,545 million (table 11.1) and real gross domestic product (GDP) for hotels, restaurants and clubs grew by approximately 3.4 per cent, with most major indicators reaching record levels (PIOJ 2006). In addition, Jamaica received the top award for being the Caribbean's leading destination at the prestigious thirteenth annual World Travel Awards in 2006 (*Jamaica Gleaner*, "Jamaica corners market on tourism awards", 21 September 2006). Unquestionably, the tourism industry seems to be in prime condition compared to its past performances and compared with other sectors of the Jamaican economy.

Yet despite a revived tourism industry, a number of studies have shown that the net gains from the industry are far below gross receipts (Jayawardena and Ramjee Singh 2003; Ramjee Singh 2006). Indeed, the rapid growth in Jamaica's tourism industry has been accompanied by a drastic decline in the agricultural sector. Implicit in this observation is the fact that the island's growing tourism industry does little to stimulate domestic food production. Indeed, a recent study carried out by Ramjee Singh (2006) revealed that Jamaica experiences a 50 per cent rate of foreign exchange leakage, meaning that for every dollar spent by a tourist, half exits the Jamaican economy. This can be attributed to a variety of factors, including the industry's high import content, repatriation of profits and overseas promotional expenditures (Jayawardena and Ramjee Singh 2003). A number of past and contemporary studies have singled out food imports as a major avenue of foreign exchange leakages in island states (Belisle 1983, 1984; Pattullo 1996, 2005; Momsen 1998; Jayawardena and Ramjee Singh 2003; Hayle 2005; Ramjee Singh 2006; Timms 2006; Dodman and Rhiney 2008). This is indeed a function of the lack of linkage between tourism and local agriculture in these host economies. These arguments are very much attuned with the findings of the author's research (2006) as fieldwork data are indicating that the Negril tourism industry consumes a significant amount of food imports largely due to the disconnection between the industry and the local agricultural sector.

Data drawn from a survey of forty-four hotels and forty-one restaurants in the Negril resort area indicated that a significant amount of the food consumed in the tourism industry is imported. An independent quantitative assessment carried out by the JTB in 2005 revealed that while 70 to 90 per cent of the food used in traditional, small-scale hotels is produced locally, as little as 40 per cent of the food consumed in large, all-inclusive hotels is grown locally (Pennicook 2006). This is particularly important, as all-inclusive hotels – despite their relatively small numbers – are substantial consumers of food, both by quantity and by value. These and other similar trends, such as the tendency of larger hotels to specialize in international cuisines and employ overseas chefs, have far-reaching implications for the local economy; especially as it relates to the industry's net retention capacity.

Globalization and Caribbean Agriculture

The traditional Caribbean economy was a dependent and open system that was mainly export-oriented (Beckford 1968; Witter and Beckford 1980). Farming in most of the Caribbean has always been highly dependent on the export crop sector, depriving the domestic crop sector of much needed assistance (Beckford 1968; Witter and Beckford 1980; Mintz 1985; Barker 1993; Pemberton et al. 2002). Caribbean agriculture is characterized by a pattern of one-crop dependency or monoculture. The dominant crops, sugar cane and banana, are primarily for export to Europe (Pattullo 1996, 2005; Ahmed 2004). This is strongly related to the legacies left behind by a colonial past that has resulted in Caribbean agriculture being heavily reliant on a system of preferential access to developed markets in the former colonial powers (Momsen 1998). The consequence is that the region's local markets remain greatly underdeveloped and are characterized by labour-intensive, low-technology cropping systems that are, generally, disconnected from the food marketing network, susceptible to natural hazards and suffer from a lack of capital.

The agricultural sector in many small island economies, including CARICOM member states, is characterized by predominantly small subsistence farm holdings of five acres (approximately two hectares) and less. These are essentially high production cost units, mainly due to the diseconomies of scale associated with small-size production (CARICOM 2002). This is contrary to the market-oriented, highly mechanized and commercial agriculture practised in most developed countries which largely dominate international trade in agricultural commodities. Furthermore, developing countries, due to their open agricultural trade policies, accommodate a substantial level of food imports that directly compete with their locally grown produce (FAO 1999).

The situation has intensified in an increasingly globalized world, possessing a neo-liberal global economy dominated by a few core countries (Goodman and Redclift 1990). These imported products can often be sold at a cheaper price than their locally grown counterparts, mainly because of the ability of the governments in more-developed countries to provide significant subsidies for their agricultural producers. Additionally, developing economies are placed in a disadvantageous position due to restrictive tariffs and trading blocs such as the European Union and NAFTA, as well as trade liberalization policies promoted by a few developed nations to protect their domestic markets and advance their economic interests (Clegg 2004). This (neo-liberalism) not only presents a barrier to economic growth but significantly threatens food security and rural livelihoods in many developing countries (FAO 1999; Torres and Momsen 2004; Weis 2004).

Referring specifically to Jamaica, neo-liberalism has led not only to the reduction of the nation's long held preferential access to overseas markets, but to an opening up of the local market to an increasing proportion of cheap, imported foodstuffs (Weis 2004). The mid 1990s mark the period when the liberalization process started to dominate the local economy. Most duties on food imports had declined from over 100 per cent in the 1980s to as low as 20 per cent by the mid 1990s (Weis 2004). As early as 1993, the country started witnessing significant increases in imports of fruits and vegetables, various meats, particularly poultry parts, and skimmed milk powder. The latter led to the collapse of the Jamaican dairy industry (Campbell 1997). The nation's dairy production declined progressively from 39 million litres in 1992 to 24 million litres in 1996 – a decline of approximately 39 per cent (PIOJ various years).

Not much has changed since then. In fact, so far the twenty-first century represents a period of intense liberalization, widening trade and balance of payments deficits, looming debt burden, and a deteriorating agricultural sector. Between 1998 and 2007, Jamaica's food import bill increased from US$466.6 million to US$662 million, and its food trade deficit increased by approximately 58 per cent from a deficit of US$237.1 million to US$375.2 million over the same period (PIOJ 1999, 2008).

The combination of high input costs, limited availability and access to much-needed operational capital, and the increasing competition from highly subsidized food imports typify both the gravity and the complexity of the challenges currently faced by Jamaican farmers. To a large extent, the structural adjustment and trade liberalization policies adopted by the Jamaican government over the last twenty years have made it extremely difficult for local producers to survive in their own market. It therefore follows that the ability of small, local farmers to

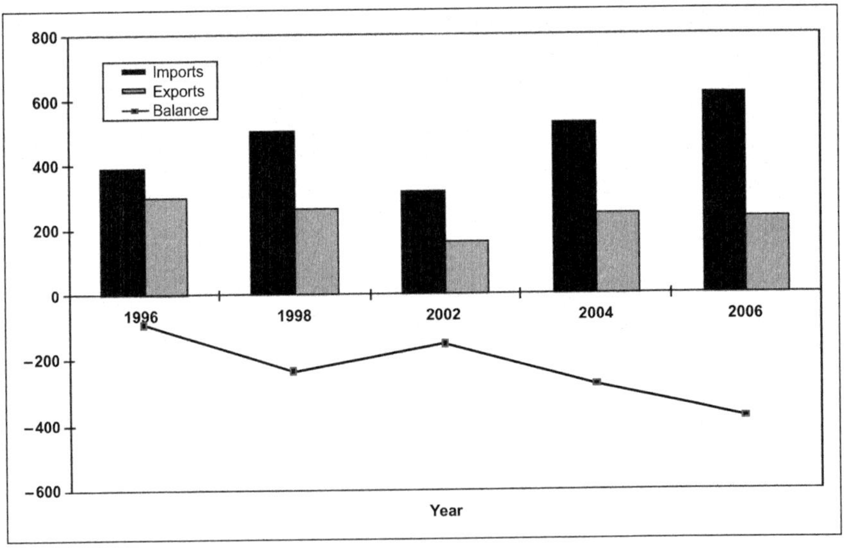

Figure 11.1 Food imports and exports, Jamaica 1996–2006.
Source: PIOJ, Economic and Social Survey of Jamaica, various years.

forge viable linkages with the Jamaican tourism industry is increasingly dependent on their ability to function in an increasingly open market economy characterized by the proliferation of cheap food imports and limited state support.

Today, tourism remains the island's leading industry, while the agricultural sector continues to experience negative trade balances (figure 11.1), high costs of production and a diminishing workforce (PIOJ various years). In 2005, real GDP for the Jamaican economy grew by approximately 1.4 per cent. This growth was reflective of all sectors of the economy with the exception of agriculture (–7.3 per cent GDP growth) and manufacturing (–1.0 per cent).

Forging the Link? The Negril Case Study

Negril is located on the western tip of the island and spans two parishes – Westmoreland and Hanover. Since the 1960s, the former fishing village has expanded to become one of Jamaica's major tourist resort towns (table 11.2). Tourism is the main contributor to the local economy in Negril today, attracting a workforce from surrounding communities.

Negril provides a distinctive opportunity to observe the interface between agriculture and tourism for several distinct reasons. First, the town is relatively small in size and is somewhat isolated from major towns such as Montego Bay and Ocho Rios (figure 11.2). Second, there is no established local produce market in the town, which allows the

Table 11.2 Selected Indicators for the Three Major Resort Towns, Jamaica, 2003

Area	Accommodation (Rooms)	Direct Employment	Stopover Arrivals by Intended Area of Stay
Montego Bay	8,019	10,117	417,274
Ocho Rios	6,777	9,288	296,344
Negril	5,851	7,773	275,124
Total for industry	24,625	30,512	1,350,285

Source: Jamaica Tourist Board, annual travel statistics, 2004.

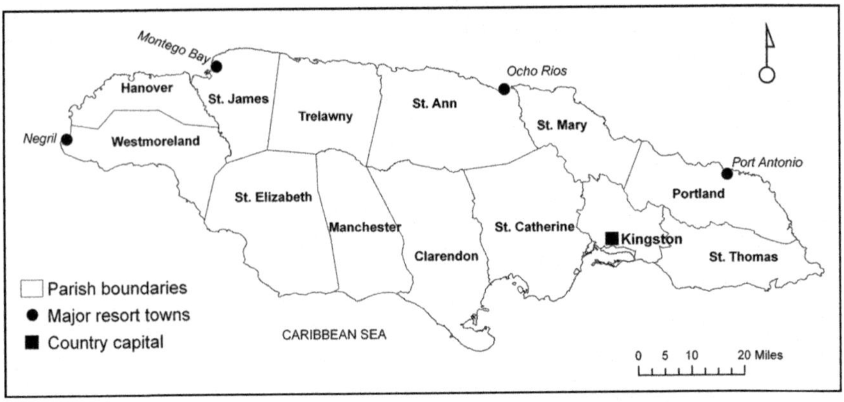

Figure 11.2 Location of major resort towns and parishes, Jamaica.

researcher to directly observe the relationship between tourism food representatives, and producers and suppliers of locally grown food items. Finally, the area is also geographically bounded by a number of farming communities spanning the parishes of Westmoreland, St Elizabeth, Manchester and Hanover.

In order to examine the factors shaping agriculture-tourism linkages in Jamaica, it is necessary to look at the constraints faced by the local food network in producing reliable and high quality food supplies on a consistent basis. The aim here is to broaden the scope of analysis by providing a detailed description of the various local food supply networks that supply the Negril tourism industry as well as to illustrate their differential abilities to forge viable and sustainable linkages within the new global political economy. The following sub-section adopts Benjamin Timms's idea of "re-conceptualising the supply chains as linkages themselves" (2006, 40). According to Timms, by "[r]edefining linkages as relations, including such characteristics as information flows between demand and supply . . . a more informed understanding of the effects that hotel demand has on agricultural production can be gained" (ibid.).

Local Food Supply Networks: An Analysis

Hotels and restaurants in Negril obtain their food supplies from several major sources. These sources can be categorized into several broad groups: local purveyors, wholesalers, retailers and supermarkets, farmers co-operatives, and the local produce market located approximately 30 kilometres to the southeast in the town of Savanna-la-Mar. Hotels and restaurants may rely on several of these sources at a time to obtain their food needs, and these may be both local and imported. The focus of the following sub-sections, however, is primarily concerned with the two supply networks that provide the bulk of local foods to the Negril tourism market. By focusing on these two local supply networks and evaluating their respective strengths and weaknesses, a greater understanding of the constraints and opportunities that exist for tourist food demand to stimulate local agricultural production in a neo-liberal trading environment can be obtained.

Local Purveyors

Local purveyors are the major suppliers of locally grown food items to the tourism sector. Each purveyor supplies several hotels and restaurants on a weekly basis. They supply mainly fruits and vegetables, seafood, and a few other products, such as eggs. Most of the establishments only have verbal agreements with their higgler (market trader) suppliers, although these relationships may have developed over a number of years. There are at least two distinct types of purveyors. First, there are purveyors who are actually farmers and supply their own produce and purchase surplus crops from other farmers to meet the required quota. The second group include those purveyors who are only involved in the marketing process.

From all indications the tourism market seemed lucrative for the majority of purveyors that were interviewed. Most of the purveyors had been supplying the Negril area for over five years. The majority of the purveyors interviewed owned their own vehicles, and a few others either rented vehicles or the vehicle belonged to a family member or the person who employed them. In fact, a few of the purveyors indicated that they had more than one vehicle and had hired other persons to assist them in the marketing process. The average purveyor works alongside someone else who assists in both the packing and unpacking of the produce – in some cases, depending on the size of the vehicle and the amount of food to be delivered, as many as three persons were observed. This is partly because the delivery process has to be a short task, yet can be

Figure 11.3 Purveyors off-loading produce at an all-inclusive hotel, Negril.

tedious and time-consuming, especially for the large all-inclusive hotels. The purveyors have to queue up outside the hotels, on specified delivery days, and wait their turn, usually between the hours of 8:30 a.m. and 12:30 p.m. and then between 1:30 p.m. and 4:30 p.m. Each purveyor is sent in one at a time and this process usually takes place on average between ten and fifteen minutes (figure 11.3). As a result, one of the purveyors takes care of the financial arrangements with the food purchasing clerk, after the produce has been inspected by the chef or someone else from the store room, and the other is mainly responsible for the unloading of the produce.

Purveyors are important in that they provide an avenue for small farmers to market their produce. Indeed, these purveyors travel across the island to source food for the tourism market. However, in the majority of cases, they usually source foodstuff from the parish in which they live. As such, the majority of the fruits and vegetables supplied to the tourism industry come from the nearby parishes of Hanover, Westmoreland, St Elizabeth and Manchester. On the other hand, this particular food-supply system is filled with a number of weaknesses in its ability to provide large quantities of high quality food on a consistent basis.

Most of the purveyors use open-back trucks and the foodstuffs are stored in a number of crates at best. Only a few vehicles have cold storage in which to transport and store the food items. Added to this factor, most of these purveyors travel significant distances to deliver these food items. This results in a substantial amount of the fruits and vegetables being delivered becoming spoiled or squeezed on delivery. The situation worsens depending on the size, shape and texture of the produce. For instance, tomatoes present a problem because their thin skins are prone to spoilage. Additionally, no formal arrangements exist between the food establishments and the bulk of purveyors supplying them with food. As such, tourism food representatives often complain about the tardiness of food suppliers and inconsistencies with the prices of products. It is also worthy to note here that purveyors are significant suppliers of both locally grown and imported foodstuffs. In fact, there are purveyors who are known to specialize in supplying only imported foods to the hotels. Aside from this, it was observed that most of the hoteliers were unaware of the exact origin of the foods purchased. This points not only to the disconnection between the local food sector and the tourism industry in terms of communication, but also to the complexities involved in analysing the material geographies of tourist food consumption practices in an increasingly globalized trading environment.

During interviews, many small farmers expressed a deep distrust of the purveyor-led marketing system. The majority interviewed indicated that they relied heavily on purveyors to market their produce in the tourism industry. This involves a complex and unplanned trading system between the farmers and purveyors where agricultural produce is usually taken on credit and then farmers are paid after the produce is sold to the different hotels and restaurants. However, there have been many cases where small farmers have been exploited and cheated through their dealings with these intermediaries. Indeed, a significant number of the farmers interviewed complained about the credit system. Their main complaint is about the lengthy period taken to get paid for their crops, and in other instances, they do not receive any payment at all. A few of the farmers indicated that they have taken purveyors to court in an attempt to secure payment that was long overdue.

Additionally, farmers also complained about the low prices purveyors offered for their crops. These purveyors, in an attempt to maximize profit, bargain for the lowest farm gate prices and then resell the produce for a much higher price in the tourism market. Due to the immense competition among small farmers in securing a market for their produce, many are forced to accept very low prices offered by purveyors, effectively lowering the farm gate prices for many local food items. This is indicative of a free market system where market forces, such as competition, dictate the prices of commodities. The end result, however, is that local farmers benefit little from the linkage formed with the tourism industry and are, in many cases, stimulated by other factors apart from tourist food demand.

Agricultural Co-operatives

Co-operatives are another form of supply network that provide a significant proportion of local food to hotels and restaurants in Jamaica. This arrangement involves a group of farmers that produce a particular crop or set of crops for a common market by pooling their productive, distributive and human resources (Timms 2006). Co-operatives operate on a formal basis, unlike that of the purveyor-led marketing system, and probably provide one of the best avenues for forging a viable link between the two sectors. The study entails detailed case studies of two farmers co-operatives – Santoy and Douglas Castle Farmers Co-operatives – the former has been able to successfully supply the tourism industry, while the latter has failed to do so.

Both co-operatives were involved in the Sandals (all-inclusive hotel chain) Farmers Programme. Douglas Castle co-operative was formed in 1996 (the year the programme started) and the Santoy group was formed two years later. Both groups were comprised of small farmers and faced similar challenges when starting out, including competition from food importers and purveyors, vulnerability to natural hazards, and a lack of funding and infrastructural support. By 1998, the Douglas Castle group had collapsed, largely due to their inability to maintain a reliable food supply. On the other hand, the Santoy group has been able to maintain its relationship with the Sandals chain and has even expanded its market to include other hotels, supermarkets, purveyors, exporters, and individual households in and around Negril. Presently, the group generates an average of J$700,000 (about US$10,000) per week from sales – some two-thirds of that sum being accounted for by the Negril all-inclusive sub-sector.

The Sandals Farmers Programme was initiated by the Rural Agricultural Development Authority (RADA) and the Sandals hotel chain, following a joint FAO and Jamaican government project where small farmers, located in the Rio Minho Valley of northern Clarendon and southern St Ann, received training in advanced post-harvest management practices. The main objectives of the Sandals programme were to improve the quality of local foods supplied to the industry, to diversify and expand the local cropping systems to suit the industry's food demand, develop a proper pricing system and improve the communication link between farmers and the hotel industry (RADA 2003).

The project began with ten farmers supplying two Sandals properties located along the island's north coast. It was later extended to five new locations in the parishes of St James, Hanover, St Mary, St Elizabeth and St Ann, involving a total of eighty farmers producing over an area of twelve hectares (RADA 2003). The arrangement was that the hotel would supply the seeds to the farmers on a crop-lien basis, and the farmers would make the necessary repayment at the end of the cropping season. The farmers supplied the hotels with various fruit and vegetable crops including watermelon, lettuce, green and red cabbage, yellow and red sweet peppers, sweet corn, zucchini, snow peas, tomatoes, yellow squash, and cantaloupe. Many of these crops were never previously planted in these communities and are thus indicative of one of the many ways the tourism sector can stimulate the domestic agricultural sector. Interviews with co-operative managers and focus groups conducted with members of both co-operatives allowed for a comparison between the two co-operative groups and an opportunity to identify the strengths and weaknesses of the co-operative system.

Agricultural co-operatives provide several distinct advantages in both developing the local agricultural sector and forging a viable link with the tourism industry. First, co-operatives allow small farmers to pool their resources together and thus tap into vital resources that may not have been accessed as individual farmers. For instance, the Santoy case study illustrates the effectiveness of the co-operative system for small farmers in terms of accessing loans and mobilizing assets in the form of access to land, training, social networks, and partnerships with large businesses and organizations such as USAID and the Land O'Lakes Foundation – one of the largest farmer-owned co-operatives in the United States.

Santoy is one of two co-operatives that have benefited from international financial aid amounting to some J$6 million (about US$90,000). This aid has been used to expand their facilities by the construction of pack houses, cold storage and nursery facilities, as well as the provision of irrigation facilities and assistance in the purchase of a tractor (RADA 2003).

Additionally, co-operatives allow for the establishment of direct and greater communication between farmers and tourism representatives. This helps to reduce the mismatch between food supply and demand, and it alleviates dependence on purveyors. Co-operatives also provide an effective means of competing with imported products because group farming allows for greater coordination and consistency in food supply (co-operative members generally stagger their production to ensure regular food supplies), as well as more competitive pricing, as the group share the costs of production and benefits from greater economies of scale.

However, co-operatives can face a number of challenges that prevent them from forging viable links with the tourism sector. Agricultural co-operatives, while they provide an effective means of transferring technologies to the local agricultural sector, are not immune to the stifling effects of cheap food imports and environmental constraints that affect domestic cropping systems in general, such as droughts and hurricanes. In the case of Douglas Castle, several factors were identified as being responsible for the group's inability to sustain the link with the tourism industry. These factors ranged from inconsistencies in food supply, particularly due to a lack of cold storage facilities and irrigation systems. Other problems cited included weak communication between hotel representatives and farmers, lengthy payment periods by the hotels, poor packaging materials for agricultural produce and weak institutional frameworks. In the latter case, poor leadership and distrust among group members were regarded as significant contributors to the group's collapse.

Apart from supply and marketing problems, the hotels were also inconsistent with their food purchases. The inconsistency of the supply orders prevented the farmers from planning production, effectively leading to many instances where production exceeded the demand, resulting in an impromptu attempt by the farmers to identify new market sources.

To make matters worse, a large amount of the food produced was exotic to the island's palate and could only be sold in the tourism market and a few grocery stores.

Conclusion

A review of the literature suggests a slight improvement in the link between tourism and agriculture in an increasing number of Caribbean islands namely, St Lucia, the Dominican Republic, the US Virgin Islands and Barbados (Momsen 1998; Conway 2004; Pattullo 2005; Timms 2006). This strengthening of the links between local agriculture and tourism is primarily attributed to changes in attitudes about serving local cuisine in hotels and restaurants, the globalization of food consumption habits, a maturing of island tourist industries and increased government assistance to producers and suppliers of local produce. In her follow up study of St Lucia, Momsen (1986) found that consumption of food imports by the largest hotels had declined to 58 per cent, down from 70 per cent in 1972. Following up on Momsen's findings, Timms (2006) discovered that tourism-agricultural linkages in St Lucia were far more variegated and dynamic than they were previously understood to be. Timms identified as many as seven major forms of linkages existing between St Lucian hotels and local food suppliers. Additionally, Timms's research found that, while small hotels and guest houses were still using considerably more local suppliers compared with larger establishments, large established hotels had started using an increasing number of local food vendors and suppliers.

Though these are some encouraging and welcome findings, they are by no means suggesting that these linkages are void of challenges; neither are they representative of the wider Caribbean region. In fact, Timms (2006) discovered in his study that there still exist many bottlenecks in supply linkages, hotels maintained flexibility in their contractual arrangements and local agricultural co-operatives were generally avoided. Additionally, a number of scholars remain sceptical about tourism's ability to forge viable linkages with local agriculture in the Caribbean (Hayle 2005; Pattullo 2005). Pattullo points out, for example, that leakage rates are generally high for Caribbean destinations – averaging between 50 and 70 per cent. Pattullo goes on to point out that the Caribbean tourism industry remains dependent on imported goods and services, which, in her estimation, is the major cause for the industry's inability to forge viable inter-sectoral linkages with other sectors of the local economy, particularly agriculture.

If Jamaica is to maximize its benefits from tourism development, ways must be found to increase backward economic linkages, including utilizing local food products in the tourism industry. The Jamaican economy strongly relies on tourism as a significant foreign exchange earner. Conversely, the agricultural sector has been plagued with poor performances over the years; this has been evident through a chronic increase in the sector's negative trade balance over the last three decades (PIOJ various years). These trends have serious implications for the hoped-for stimulation of the link between local agriculture and tourism. Indeed, if a viable and sustainable link is to be formed between tourism and local agriculture, the local food network has to develop the capacity to supply high quality food in large quantities on a consistent basis. At present, domestic agriculture is grossly underdeveloped. Inadequate help for small farmers, rising costs of living, high reliance on imported inputs and ageing farming populations are among the factors that have contributed to this continuous decline in local agricultural productivity.

If left unchecked, the collapse of the local agricultural system will lead to the demise of many rural livelihoods. While on one hand, global trading policies and neo-liberal reforms have placed the Jamaican agricultural system in a disadvantaged position, globalization has also provided tremendous opportunities for the agrarian sector to tap into (Momsen 1998). Indeed, with the increasing exposure of Jamaican cuisines in tourist generating markets and a global shift away from traditional models of mass tourism to small-scale alternative forms of tourism, new niche markets are developing rapidly (Dodman 2004). These niche markets may provide an opportunity for the increased consumption of local food in the tourism industry.

This chapter has shown the complexities involved in the production and marketing of local food crops. The study takes us into the heart of the tourist food distribution system by focusing on the two major local supply networks and evaluating their respective strengths and weaknesses in stimulating local agriculture in a neo-liberal trading environment. By highlighting the constraints and opportunities faced by local producers, the factors involved in food production and marketing processes in host economies can be better understood. The local food network is a vulnerable yet dynamic system. To improve its ability to supply high quality food on a consistent basis, the approach has to be one that is multi-faceted and incorporates every stage of the local food servicing industry, spanning the production, marketing and distribution stages. A new way of organizing and marketing local cropping systems is therefore urgently required. Indeed, there is a need to diversify the agrarian sector and tap into not only new forms of cropping systems but also new niche markets.

Though this example is from Jamaica, the nature of food networks elsewhere in the Caribbean is broadly similar (cf. Timms 2006). Therefore similar problems arise, and these problems require broadly similar solutions if the Caribbean agricultural sector is to benefit from tourism to a greater extent than it has hitherto.

References

Ahmed, B. 2004. The impact of globalization on the Caribbean sugar and banana industries. In Courtman 2004, 256–72.

Barker, D. 1993. Dualism and disasters on a typical island: Constraints on agricultural development in Jamaica. *Tijdschrift voor Economische en Sociale Geografie* 84: 332–40.

Beckford, G. 1968. Toward an appropriate theoretical framework for agricultural development, planning and policy. *Social and Economic Studies* 17: 233–42.

Belisle, F.J. 1983. Tourism and food production in the Caribbean. *Annals of Tourism Research* 10 (4): 497–513.

———. 1984. The significance and structure of hotel food supply in Jamaica. *Caribbean Geography* 1 (4): 219–33.

Bryden, J. 1973. *Tourism and development: A case study from the Commonwealth Caribbean.* Cambridge: Cambridge University Press.

Campbell, R. 1997. The Jamaican dairy and beef industries: Some recent spatial, structural and development trends. *Caribbean Geography* 8 (2): 107–19.

CARICOM. 2002. Special session of the Committees on Agriculture: WTO negotiations on agriculture – Non-trade concerns. Negotiating proposal on behalf of members of the Caribbean Community, Fourth Ministerial Conference, Doha, Qatar.

Clegg, P. 2004. The transatlantic banana war and the marginalisation of Caribbean trading interests. In Courtman 2004, 242–55.

Conway, D. 2004. Tourism, environmental conservation and management and local agriculture in the eastern Caribbean: Is there an appropriate, sustainable future for them? In *Tourism in the Caribbean: Trends, development, prospects*, ed. D.T. Duval, 187–204. London and New York: Routledge.

Courtman, S., ed. 2004. *Beyond the blood the beach and the banana*. Kingston: Ian Randle.

Dodman, D. 2004. Postmodernity or profitability? Modes of tourism in Jamaica. In Courtman 2004, 155–69.

Dodman, D., and K. Rhiney. 2008. "We nyammin": Food authenticity and the tourist experience in Negril, Jamaica. In *New perspectives in Caribbean tourism*, ed. M. Daye, D. Chambers and S. Roberts, 115–32. New York: Routledge.

Food and Agriculture Organisation (FAO). 1999. Issues at stake relating to agricultural development, trade and food security. Paper for FAO Symposium on Agriculture, Trade and Food Security. Geneva.

Goodman, D., and M. Redclift. 1990. The farm crisis and the food system: Some reflections on the new agenda. In *Political, social and economic perspectives on the international food system*, ed. T. Marsden and J. Little, 18–35. Aldershot, UK: Avebury.

Hayle, C. 2005. Tourism in Jamaica: The impact of the past on the future. In *Caribbean tourism: Visions, missions and challenges*, ed. C. Jayawardena, 119–39. Kingston: Ian Randle.

Jayawardena, C., and D. Ramjee Singh. 2003. Performance of tourism analysis: A Caribbean perspective. *International Journal of Contemporary Hospitality Management* 5 (3): 176–79.

Jamaica Tourist Board (various years). *Jamaica: Annual travel statistics*. Kingston: Jamaica Tourist Board.

Lundgren, J. 1973. Tourist impact/island entrepreneurship in the Caribbean. In *Geographical analysis for development in Latin America and the Caribbean*, ed. R.P. Momsen, 12–19. Chapel Hill, NC: CLAG publications.

Mintz, S. 1985. *From plantations to peasantries in the Caribbean: Caribbean contours*. Baltimore: Johns Hopkins University Press.

Momsen, J.H. 1972. *Report on vegetable production and the tourist industry in St Lucia*. Calgary: Department of Geography, University of Calgary.

———. 1986. Linkages between tourism and agriculture: Problems for the smaller Caribbean economies. Seminar paper no. 45, Department of Geography, University of Newcastle upon Tyne, UK.

———. 1998. Caribbean tourism and agriculture: New linkages in the global era? In *Globalization and neoliberalism: The Caribbean context*, ed. T. Klak, 267–72. Lanham, MD: Rowman and Littlefield.

Pattullo, P. 1996. *Last resorts: The cost of tourism in the Caribbean*. Kingston: Ian Randle.

Pemberton, C.A., L.A. Wilson, G.W. Garcia and A. Khan. 2002. Sustainable development of Caribbean agriculture. In *Natural resource management for sustainable development in the Caribbean*, ed. I. Goodbody and E. Thomas-Hope, 277–306. Kingston: Canoe Press.

Pennicook, P. 2006. The all-inclusive concept: Improving benefits to the Jamaican economy. In *Tourism: The driver of change in the Jamaican economy?* ed. K. Hall and R. Holding, 31–38. Kingston: Ian Randle.

Planning Institute of Jamaica (PIOJ). (various years). *Economic and social survey of Jamaica*. Kingston: Planning Institute of Jamaica.

Rural Agricultural Development Authority (RADA). 2003. Sandals Farmer Programme. Kingston: Rural Agricultural Development Authority. Available at http://www.radajamaica.com.jm/marketing_articles.asp.

Ramjee Singh, D. 2003. Is tourism a developmental tool? The Barbadian challenge. Unpublished paper, Department of Management Studies, University of the West Indies, Mona.

———. 2006. Import content of tourism: Explaining differences among island states. *Tourism Analysis* 11 (1): 33–44.

Sealey, N. 1982. *Tourism in the Caribbean*. London: Hodder and Stoughton.

Taylor, F.F. 1993. *To hell with paradise: A history of the Jamaican tourism industry*. Pittsburgh: University of Pittsburgh Press.

Thomas-Hope, E., and A. Jardine-Comrie. 2007. Caribbean agriculture in the new global environment. In *No island is an island: The impact of globalization on the Commonwealth Caribbean*, ed. G. Baker, 19–43. London: Chatham House.

Timms, B. 2006. Caribbean agriculture-tourism linkages in a neoliberal world: Problems and prospects for St Lucia. *International Development Planning Review* 28 (1): 35–56.

Torres, R. 2003. Linkages between tourism and agriculture in Mexico. *Annals of Tourism Research* 30 (3): 546–66.

Torres, R., and J. Momsen. 2004. Challenges and potential for linking tourism and agriculture to achieve pro-poor tourism objectives. *Progress in Development Studies* 4: 294–318.

Weis, T. 2004. Restructuring and redundancy: The impacts and illogic of neoliberal agricultural reforms in Jamaica. *Journal of Agrarian Change* 4: 461–91.

Witter, M., and G. Beckford. 1980. *Small garden bitter weed: The political economy of struggle and change in Jamaica*. Kingston: Maroon Publishing House.

PART 4

Urban Vulnerability and Urban Change

CHAPTER 12

From Slum to Ghetto

Multiple Deprivation in Kingston, Jamaica

COLIN CLARKE

Introduction

Unlike the burgeoning cities of neighbouring Latin American countries, the major settlements of the Commonwealth Caribbean have expanded since the 1960s without a phase of mass squatting. None of the Commonwealth territories has more than three million inhabitants, and all but Jamaica and Trinidad and Tobago have fewer that one million. Moreover, national population growth has been slow, because out-migration to North America has been an attractive alternative to urbanization for potential migrants with high levels of education. So, even in the largest Commonwealth Caribbean cities, such as Kingston, the capital of Jamaica (with an estimated 2001 population of 870,000) and Port of Spain, the capital of Trinidad and Tobago (with an estimated 2001 population of more than 250,000), the majority of city dwellers are home owners or, more likely, renters rather than squatters, and squatting has been discouraged by vigilant policing. Nevertheless, throughout the cities and towns of the Commonwealth Caribbean there has been, for many decades, a strong socio-spatial association between renting, poor quality housing, high population density, inadequate amenities, unemployment and low class position, and nowhere are these relationships better demonstrated than in the slums of Kingston.

There were signs of the formation of a massive zone of social deprivation in Kingston, Jamaica – notably in West Kingston – dating from the West India Royal Commission Report of 1938–39 (1945) and the Denham Town re-development project of the late 1930s (Central Housing Advisory Board 1936). The development of social deprivation in Kingston may be traced via the report on the Ras Tafari movement (Smith, Augier and Nettleford 1960) and an early paper by Clarke (1966) describing conditions before independence, to the later findings of Clarke (1975 and 2006a) and Eyre (1986a and 1986b) in the 1970s and 1980s. Kingston's late colonial slums were re-designated as "the ghetto" after 1970 (Eyre 1986a and 1986b). More precisely, the ghetto had its origins in the recognized slum areas of West Kingston of 1935 (Clarke 2006a, fig. 25), in the housing areas in poor condition in 1947 (Clarke 2006a, fig. 24), the zones of third-class housing in 1952 (Clarke 2006a, fig. 37), the areas of poor housing in 1960 (Clarke 2006a, fig. 40), and the overcrowded areas of 1960 (Clarke 2006a, fig. 48). Clearly, the slum/ghetto is associated with deprivation, and also with high population density in relation to low social class and poor quality (usually rented) accommodation. Indeed, Eyre pointed to an economic as distinct from an "ethnic" definition of the ghetto in research published over twenty years ago (1986a and 1986b), though this chapter shows that there is a marked racial dimension to the Kingston ghetto, in addition to the class component.

What is the relationship of social deprivation to vulnerability? The term vulnerability is defined by the Oxford English Dictionary as meaning "susceptible to being wounded or injured". Social susceptibility in a developing-world city – such as Kingston – implies liability to deprivation associated with racially black, lower-class characteristics, expressed in marginality in both labour and housing markets, and involving high levels of unemployment and poor-quality housing. Vulnerability has been the hallmark of the occupants of the Kingston slum/ghetto over the last seventy years. In this stagnant urban economy, on the global capitalist periphery, the bulk of the labour force is trapped in menial tasks in the informal sector of the economy, and these individuals experience multiple deprivation on a range of social indicators associated with employment and housing.

The line of argument developed in this chapter is that the Kingston slum in colonial times was the forerunner of the present-day ghetto; both were created by comparable socio-economic processes and have given rise to similar, problematic outcomes. The nomenclatures – slum and ghetto – represent changes in discourse on social deprivation, and an eliding of US and Jamaican urban experience after Jamaica's independence in the 1960s. The colonial slum in Kingston was based on high densities of population, occurring in neighbourhoods with poor quality, rented accommodation. These were also areas in which the highest concentrations of lower-class blacks (the 1960 census used the term "Africans", based on the assessment of enumerators) were recorded, and where the most marginal of the lower class, as measured by unemployment, were located. Massive black populations had built up to create extensive zones of deprivation in West Kingston, and in parts of East Kingston, circumscribed by low status and unemployment. These slum areas were alienated from Jamaican society and dangerous for outsiders, irrespective of their colour, to penetrate (Clarke 2006a). The Kingston ghetto of the period since 1970 – by analogy with the black ghetto in US cities during the late 1960s – reflects an intensification in some (but not all) of these negative characteristics or vulnerabilities. Additionally, the Kingston ghetto has been characterized, since the early 1970s, by gang violence, drugs and warfare, the latter coordinated in downtown Kingston constituencies, at election times, by supporters of Jamaica's two main political parties (Eyre 1986a, 1986b; Clarke 2006b).

What is noteworthy about the present-day Kingston ghetto – as measured by 1991 enumeration district data, the latest small-area data set currently available from the census – is that it is predominantly a black area (more than 92 per cent black by self-assessment), in a city where the black population is 88 per cent of the total (all racial/colour terms and their associated population figures have been drawn from the Jamaican census). So, while the Kingston ghetto conforms to Ward's definition (1982) in that it is racially homogeneous (almost all the remainder of its population is mulatto), it is defined as much by the vulnerability to deprivation of its occupants – and their high density dwelling – as by its exclusive racial characteristics. Moreover, the ghetto has not expanded by flight from white residential heartlands on its periphery, as in the case of Morrill's US ghetto model (1965). Indeed, the middle-class mixed-race (predominantly black-white) districts on the ghetto's northern periphery in Kingston have retained their class status (while becoming noticeably darker) over the last thirty years, and spatial expansion has concentrated in areas that were either vacant (in the west) or have become decayed (in the east). Whereas in 1970, the slum was largely located in West Kingston, the ghetto now extends to East Kingston as well, and the major spatial distinction is between uptown (which is largely upper or middle class) and downtown (which is lower class and houses the core of the ghetto).

Key to understanding the spatial development of the slum into the ghetto is the growth of Kingston's population since 1960 and the spatial expansion of the city from the Kingston Metropolitan Area (KMA) into the Kingston Metropolitan Region (KMR). The population of Kingston increased from 379,980 (KMA) in 1960 to 868,653 (KMR) in 2001 (though in the latter year the KMA recorded a population of only 577,286). The increase in population of almost 490,000 between 1960 and 2001, was, in fact, only just over twice what had been estimated in the early 1960s as possible by 1970 (Roberts 1963). This reduction in anticipated growth was largely due to heavy out-migration to North America starting in the late 1960s, as the United States and Canada revised their immigration policies to select those with high educational credentials (CICRED 1974). Kingston's population increased by a mere 1 per cent per annum in the 1990s – a low figure by urban, third world standards, and way below the figure of almost 4 per cent per annum achieved in the late 1950s, when slum intensification was at its colonial peak.

Nevertheless, the increase in Kingstonians after independence in 1962 could be housed only by the suburbanization of the Ligueanea Plain to the north of the Washington Boulevard (figure 12.1). There was also a subsequent overspill of townspeople from Kingston to the new dormitory suburbs at Portmore, and the suburbs and squatter settlements formed around the former colonial capital, Spanish Town. By 1991, Kingston and these contiguous settlements, collectively making the KMR, had become a single urban zone of 779,016

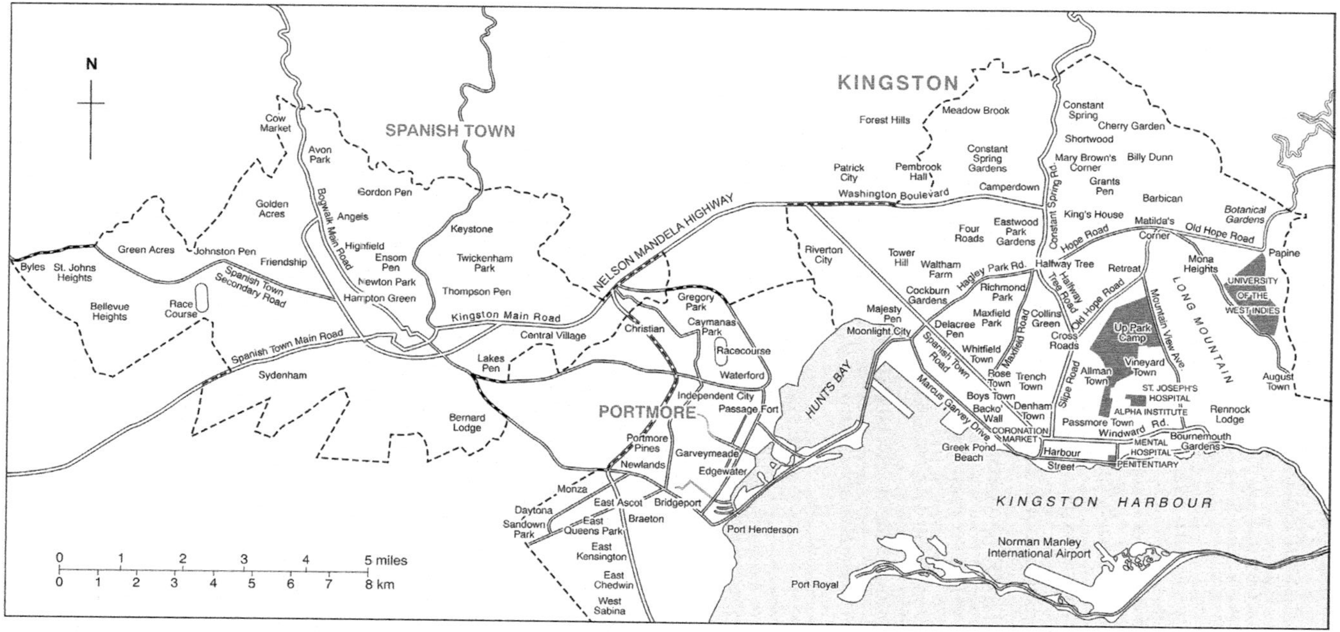

Figure 12.1 Place names in the Kingston Metropolitan Region.

inhabitants. Moreover, at the core of downtown Kingston, a massive zone of citizens vulnerable to deprivation had developed – the ghetto, based on the colonial slum and housing almost half of the KMR's total population.

Population growth was deeply implicated in the expansion of slum conditions (soon to be captured as the ghetto), since 130,000 new residents were added to Kingston between 1960 and 1970. Data from the Sanitary Survey of Kingston and St Andrew show that the tenements, yards and government housing schemes adjacent to Spanish Town Road housed 164,000 people in 1967 (Clarke 2006a), whereas the slum areas of 1935 had recorded just under 20,000 residents, concentrated in ". . . a solid wedge in the rent-yards of West Kingston, in the tenements to the north and east of the central business district, and three areas to the north of Cross Roads" (ibid., 105).

Deprivation in the Slum in the Early 1960s

When the term "slum" was used in Kingston in the late colonial period, it implied a continuous zone of high population density and deprivation that was dangerous for outsiders to penetrate and difficult for insiders to escape. Delimiting the slum has been attempted by mapping variables dealing with deprivation: namely, low socio-economic status, high rates of unemployment, high population density, a preponderance of rental housing and lack of water piped into the dwelling. Slum conditions, derived from poor-quality housing and high population densities, are frequently associated with low employment rates and very low and insecure income, and are themselves deeply influenced by low-class status. In each case, the variable that reflects, say, class or unemployment, has been calibrated to a yardstick determined by map inspection.

Map evidence published in Clarke (2006c), showed that low socio-economic status occurred where the percentage of high-, middle-, and lower-class occupations (multiplied by 1, 2 and 3 respectively and summed) produced a score of more than 260 out of a maximum of 300. Apart from the low-density neighbourhoods adjacent to Long Mountain, the low-class sections of the city were located in a continuous zone

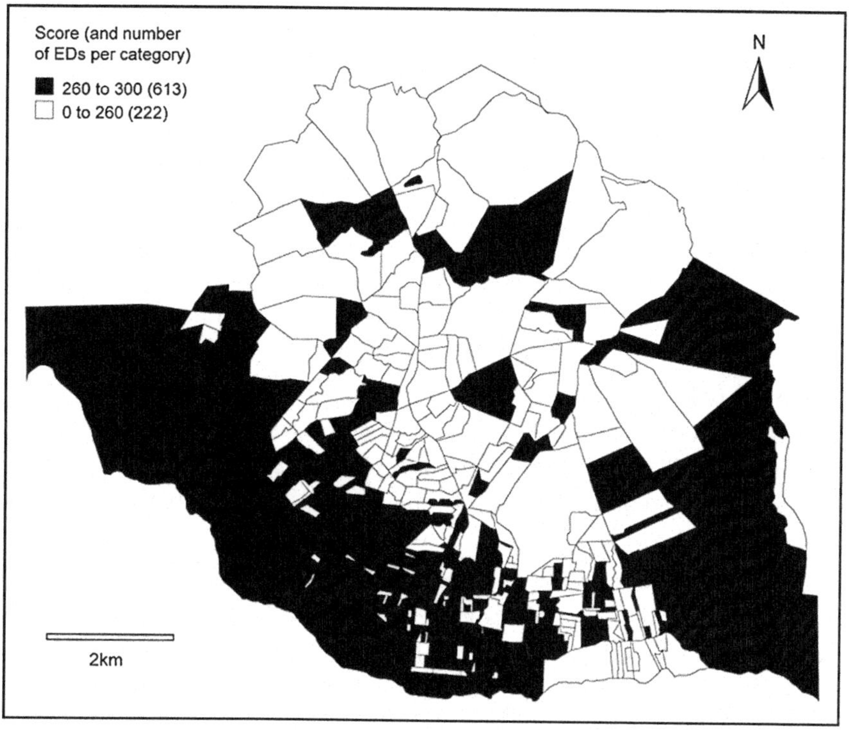

Figure 12.2 Kingston: enumeration districts (EDs) with low socio-economic status, 1960.

throughout much of East and West Kingston and swept both westwards and northwards, across Washington Boulevard, to swing east to the pockets of poverty in the Barbican area (figure 12.2). High unemployment, measured by more than 10 per cent out of work in the year before the census, fitted precisely within the area occupied by low-class neighbourhoods, both downtown and in the suburban shanty towns (Clarke 2006c, fig. 5.2).

High population density, measured by more than twenty-five persons per hectare, concentrated in the downtown enumeration districts of West Kingston and inner East Kingston (Clarke 2006c, fig. 5.3). Tenancy accounted for more than 70 per cent of households and occupied an area of downtown Kingston that was more widespread than that characterized by high densities, but nonetheless conformed to the pattern of lower-class neighbourhoods (Clarke 2006c, fig. 5.4). Lack of piped water in the dwelling, set at fewer than 35 per cent of households, coincided with high rates of tenancy, dense population and lower-class neighbourhoods, but was essentially a West Kingston phenomenon (Clarke 2006c, fig. 5.5).

Probing these circumstances more precisely, it can be shown that 222 out of a total of 829 enumeration districts were of low socio-economic status and accounted for 295,760 of Kingston's population of 380,000 (table 12.1). Moreover, 305 enumeration districts with high unemployment housed 156,160 Kingstonians, while 513 enumeration districts recorded high tenancy rates and housed 208,970 inhabitants. The other measures of deprivation also largely fitted within the spatial pattern of low class and high tenancy. Enumeration districts with high population densities numbered 509 and were occupied by 236,510 residents, while low levels of water supply to dwellings accounted for 220 enumeration districts and 280,020 residents.

In round figures, the lower-class area seems to have accommodated just under 300,000 in 1960 (or 80 per cent of Kingston's population), among whom 100,000 to 200,000 (26 to 52 per cent, depending on the variable selected) would have been vulnerable to some overlapping forms of multiple deprivation. Not everyone in the lower-class areas would, of course, have been lower class, nor would everyone in the areas of multiple deprivation have been vulnerable. Nevertheless, these estimates of vulnerable Kingstonians mesh well with others in Kingston on the eve of independence. In 1960, the town planning department employed its land-use survey in combination with the census to calculate that 120,000 (32 per cent) were living in poor-quality accommodations, the majority in West Kingston, while, using sanitary survey data and the census, it concluded that 80,000 (21 per cent) were living at a density of more than two persons per room, or eight to a hygienic water closet. This population was found mostly in West Kingston or inner East Kingston (Clarke 2006a). Furthermore, Clarke estimated that with 33,000 persons unemployed in Kingston in 1960, they and their dependants numbered about 60,000, or 16 per cent of the population (ibid., 170). But not all the jobless were confined to the slum tracts.

How did these indicants of deprivation relate to race? All the areas of deprivation recorded high proportions of Africans (70 per cent of

Table 12.1 Measures of Deprivation in the Slum and Ghetto Areas in Kingston in 1960 and 1991, Respectively

1960 KMA*	Limit	Entire District	Area in ha	Population Total	Populaton Density	African %	% Afro-European
Socio-economic status score	260+	222	5,600	295,760	52.8	74.2	13.5
Unemployment Population	10%+	305	2,143	156,160	72.9	76.7	12.3
Population density/ha	25/ha+	509	1,069	236,510	221.2	78.9	11.9
House rental	70%+	513	3,033	208,970	68.9	75.7	13.3
Water into dwelling	35%+	220	2,930	280,020	95.6	70.5	15.1
African	80%+	428	2,881	202,450	70.3	93.2	2.9
1991 KMR**							
Socio-economic status score	205+	732	9,293	387,628	41.7	91.6	6.6
Unemployment	10%+	409	5,627	204,500	36.3	90.0	7.2
Population density/ha	120/ha+	558	1,463	286,034	195.5	91.2	6.7
House rental	60%+	494	2,918	198,610	68.1	90.1	8.1
Water into dwelling	30%-	480	7,390	445,144	60.2	86.7	7.0
Shared toilet	30%+	728	6,110	372,067	60.9	90.4	7.6
Black	92%+	607	6,886	340,919	49.5	95.7	3.3

Note: KMA* = Kingston Metropolitan Area. KMR** = Kingston Metropolitan Region.

Kingston's total in 1960) among their residents: 70.5 per cent in areas lacking water supply to the dwelling, 74.2 in low-status areas, 75.7 where there was rental and 78.9 where population densities exceeded 25 persons per hectare. Moreover, Afro-Europeans would have raised the proportion of "blacks" by between 10 and 15 per cent for each aspect of deprivation. For example, a population density of more than 25 persons per hectare was associated with a population that was 90.8 per cent black or mulatto (whereas their combined proportion of the city's total population was 86 per cent).

There were 428 enumeration districts in Kingston where the population was more than 80 per cent African in 1960 – an appropriate evaluation of their likely percentage within the slum zone (Clarke 2006c, fig. 5.6). They coincided closely with areas of low socio-economic status, and more loosely with high vulnerability to multiple deprivation. This African area accounted for 202,000 residents, 93.2 per cent of whom were black, and 3 per cent of whom, additionally, were Afro-European. On the basis of these figures, the total slum population is estimated at 200,000 in 1960 (rather than the figure of 295,000 implied by the number of residents in the low-class areas), or just over 50 per cent of Kingston's total. Within that deprived zone, some 16 to 20 to 30 per cent of Kingston's population (defined respectively as 60,000 unemployed or 80,000 overcrowded or 120,000 in poor housing) were already vulnerable to very low standards of living. Slum characteristics were concentrated on about 3,000 hectares (or 25.4 per cent of the KMA total of 11,804 hectares) – mostly located downtown and more especially concentrated in West Kingston.

Deprivation in the Ghetto in the Early 1990s

As mentioned above, the term "ghetto" has been applied to the slum areas of Kingston since the early 1970s. The violence associated with the electoral process, together with turf wars among the gangs, turned downtown Kingston into an impenetrable area for Kingston's non-slum inhabitants and for slum residents outside their own neighbourhood – hence the substitution of the name "ghetto" for the less exclusionary term "slum" after 1970 (Eyre 1986a, 1986b).

Nevertheless, socio-economic deprivation in 1991 continued to be based on the same variables as in 1960, as the remainder of this section will demonstrate. Structural adjustment policies, for example, curbing state involvement in housing, which had been substantial in downtown Kingston in the 1970s (Klak 1997), certainly intensified ghetto circumstances, but it did not create them. One further point needs clarification: the use of the term "ghetto" in Kingston also alludes to the explosive nature of the black ghettos of the major cities of the US in the late 1960s and early 1970s. However, in Kingston, one was dealing with a near-homogeneous black downtown set in a majority-black city, in contrast to the intensely black zones in predominantly white cities, which was the American norm (Massey and Denton 1993). In Kingston, whites, Chinese and East Indians were largely absent from the 1960 slum, and in the 1991 ghetto they became even more rare, but more because of their upper- and middle-class status than because of their aversion to blacks *per se*.

It seems reasonable to achieve a rough approximation to the size and shape of the Kingston ghetto in 1991 by focusing on the areas where population had built up since independence; namely, where the population density exceeded 120 persons per hectare (bearing in mind that the cut-off indicating high density in 1960 had been only 25 persons per hectare). This overcrowded zone, in turn, coincided approximately with the lower-class areas that achieved a socio-economic status score of 250 and above (260 in 1960), and with enumeration districts where more than 10 per cent of the labour force were unemployed (as in 1960).

Similar zones of deprivation were depicted by enumeration districts where more than 60 per cent of the housing was rented (70 per cent in 1960), more than 30 per cent of households had shared toilets (not in the spatial data set in 1960) and fewer than 30 per cent had water piped into the dwelling (35 per cent in 1960). These changed parameters reflected population build-up in the downtown neighbourhoods, set against the development of a more sophisticated occupational structure in the city after independence, and improved housing and social provision in Kingston after 1960.

Low-class status in Kingston, established with socio-economic scores of between 250 and 300, characterized the whole of downtown, and extended westwards and northwards across Washington Boulevard to pockets of poverty at Barbican (figure 12.3). The threshold for those working for only six to nine months in 1991 was set, by map inspection, at 10 per cent. These highly problematic circumstances fitted, spatially,

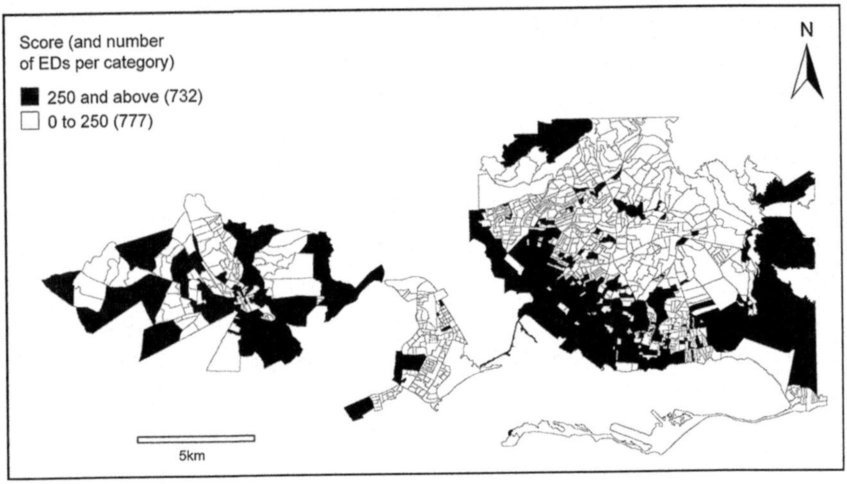

Figure 12.3 Kingston Metropolitan Region: enumeration districts (EDs) with low socio-economic status, 1991.

within the distribution of the low-status zones outlined above, but were less extensive (Clarke 2006c, fig. 5.8).

High population density of more than 120 persons per hectare largely reproduced the low-status downtown pattern in Kingston and at Portmore, but in Spanish Town high density was largely a city-centre characteristic (Clarke 2006c, fig. 5.9). Low-density peripheral areas in Spanish Town were either modern suburbs or squatter settlements. The high-density zone in Kingston could be traced back via the census of 1960 to 1943 (Clarke 2006a, 2006b, 2006c), though it had expanded with the passage of time. Strongly related to the pattern of population density – through the residential connection – was rental, for which the threshold was set at more than 60 per cent tenants (Clarke 2006c, fig. 5.10)

Two further measures of poor residential conditions were selected for 1991: enumeration districts where 30 per cent of households or fewer had water supplied into the dwelling (Clarke 2006c, fig. 5.11) and 30 per cent or more had shared toilets (Clarke 2006c, fig. 5.12) – both of these indicators had been established measures of deprivation in Kingston dating back to the colonial period (Clarke 2006a, 2006b, 2006c). Each of these measures of deprivation in household services was strongly associated with rental tenure.

When the number of enumeration districts falling within the problem areas is aggregated, some interesting commonalities emerge. Out of a total of 1509 enumeration districts in the KMR, 732 were low status and had 387,628 occupants; 558 recorded population densities of more than 120 per hectare with 286,034 inhabitants; and 728 had more than 30 per cent of households with shared toilets and 372,067 inhabitants. Clearly, the areas depicted by these conditions do not overlap completely, nor can we infer that all their inhabitants were vulnerable to the multiple deprivations being investigated. But the problematic enumeration districts coincided quite spatially close to each other, and there are logical reasons for attributing high population density and poor household facilities (as in 1960) to low class (table 12.1).

Within the broader zone of low status and its associated attributes, there is a smaller, core zone of particularly vulnerable circumstances. Within this nucleus, 409 enumeration districts, with a total population of 204,500, recorded more than 10 per cent of the population as working for only six to nine months of the year; there were 494 enumeration districts with 198,610 population where more than 45 per cent of households rented their property; and there were 480 enumeration districts with 445,144 inhabitants in which fewer than 30 per cent had water supplied to their dwelling. The areas depicted by these thresholds/variables do not coincide exactly spatially, but they clearly suggest the existence of core areas of extreme vulnerability within the broader framework of low-status Kingston. However, it is this broader framework, comprising roughly 700 enumeration districts, with a population of more than 350,000 and covering 9,293 hectares or 41 per cent of the KMR (a total of 22,453 hectares), that should be taken as the outer limit of the ghetto in 1991 (table 12.1).

This is corroborated, as most Kingstonians would anticipate, by race. The zone where 92 per cent or more of the population were black in 1991 (in a city where black residents now comprised 88 per cent), coincided closely with areas of lower-class status (Clarke 2006c, fig. 5.13); indeed, 95.7 per cent of the residents of this zone were black (table 12.1). Moreover, it contained 340,000 residents – compared to 387,000 in lower-class neighbourhoods (table 12.1). The boundary of the ghetto, and the conditions that have engendered vulnerability to poverty in its various guises, are clearly as related to class as to race – especially in the post-colonial period. Nevertheless, it is in the ghetto that poor blacks survive their deprivation.

The inner, or smaller, ghetto in 1991 contained just under 500 enumeration districts, and a population of 250,000. The area covered, and population located within it, was similar (but not identical) to the slum area of 1960 (population of 200,000) less Spanish Town. Overall, the ghetto is concentrated in downtown Kingston, in uptown pockets associated with the gully courses, typically at Barbican, and reproduced in the detached zones of Spanish Town's inner-city tenements and squatter camps. Approximately 45 per cent of Kingstonians lived in the greater ghetto in 1991, while 32 per cent concentrated in the nucleus of deprivation.

Kingston's ghetto is similar to US urban ghettos in that it is densely populated, and its inhabitants are vulnerable to deprivation, but it is unlike the US ghettos in that, while over 90 per cent or more of the population in the ghetto area was, in general, black in 1991, the majority of blacks did not live in the ghetto – as they did to the tune of 90 per cent in Chicago as early as the 1930s, according to Philpott's 1978 account. Only 49 per cent of Kingston blacks lived in the zone defined by low-class status, and 44 per cent lived where blacks accounted for more than 92 per cent of the population. The proportion of the total population that was black, in the low-status areas, was 91.6 per cent and 91 per cent where there was high unemployment. In high-density zones, the figure was 91.2 per cent; in high-rental areas, 90.1 per cent; where there was a lack of water supplied to the dwelling, 86.7 per cent; and where there were shared toilets, 90.4 per cent - in a city where blacks accounted for 88 per cent of the entire population. If mixed race (white-black) population is added into the analysis, it accounted for about 7 per cent of the total population of each of the vulnerable areas.

By 1991, the race-class ghetto ran from pockets of deprivation, set among the more affluent neighbourhoods located to the north of the Sandy Gully at Barbican, through the extensive swathes of Cockburn Gardens, Whitfield Town, and Trench Town to Central Kingston, and on through East Kingston to the bottom of Mountain View Avenue (figure 12.1). This is an area vulnerable to multiple deprivation in which low-class position and unemployment have created housing deficiencies and social exclusion on a substantial scale. Similar circumstances of deprivation are repeated in the tenements and squatter settlements at Spanish Town and in impoverished pockets at Portmore.

However, the expansion of the KMA into the KMR, which doubled the built-up area between 1960 and 1991, led to an overall reduction in population densities in the problem zones, taken characteristic by characteristic, largely because the areas they covered expanded spatially at a rate that outstripped the growth of population within them. As we have already seen, 50 per cent of the KMA population were resident in slum areas in 1960, but they covered only 25 per cent of the city's area. By 1991, 45 per cent of the KMR's population was in the expanded ghetto and this ghetto occupied 41 per cent of the area. But, whereas the overall population density in the area with more than 25 persons per hectare was 221.2 in 1960, it dropped to 195.5 where the population density averaged over 120 persons per hectare in 1991. The population densities in all the other areas of deprivation fell (though less precipitously), between the two dates, except for the rental zones, where they remained stationary (table 12.1).

Conclusion

Structural adjustment, since the early 1980s, has constrained government involvement in housing, especially for the poorest, least creditworthy segment of the lower class, and to that extent it has sustained deprivation downtown, but it has certainly not had a universally deleterious effect on housing in Kingston. It must be emphasized that deprived areas in 1991 were measured by lower criteria than in 1960, reflecting improvements in the occupational structure, housing and services in the city. Many of the latter were associated with government housing projects of the 1970s and hurricane rehousing following Hurricane Gilbert in 1988. Among the criteria for ghetto status, only the population density yardstick and the proportion of the population who were black increased between the censuses. It will be interesting to examine the 2001 census data when they become available and extend this ghetto analysis for Kingston in 1991 into the twenty-first century. However, it is unlikely that there will have been big improvements in vulnerability to deprivation over so short a period. Kingston's economy has been flat, and government capacity to intervene has been hampered by national indebtedness.

In 1960, the Kingston slum housed arguably 52 per cent of the Kingston population in 440 enumeration districts, with between 21 and 32 per cent living in the worst affected areas. These areas were lower class, more than 80 per cent black, densely populated at more than 25 to over 100 persons per hectare, and, in more than 35 per cent of households, without a piped water supply. All these slum tracts were in downtown Kingston or in outliers (mostly at low population density) in the Barbican area close to the Sandy Gully. By 1991, the population of the ghetto had increased to about 350,000 (living in 700 enumeration districts) from the slum population of 200,000 in 1960 – an expansion of 150,000 people in thirty years. During the same time-period, the population growth of Kingston on the Liguanea Plain was 183,000, indicating that the bulk of the increase (discounting overspill to Portmore and Spanish Town) was concentrated in the slum belt, thereby intensifying it into what is now called the ghetto.

Not only did the slum/ghetto population increase hugely between 1960 and 1991, but it became proportionately blacker, racially, according to the census, as self-evaluation by individuals replaced classification by colonial enumerators. In 1960, focusing on the five measures that were used to indicate areas of deprivation (low socio-economic status, high population density, high unemployment, high percentage rental, and shortage of piped water), the first three zones recorded percentages of blacks varying from 74 to 82 per cent, the fourth giving a percentage of 70.5. By 1991, six measures (the above five plus the high incidence of shared toilets), picked out areas of vulnerability to deprivation, five of which recorded the percentage of blacks in the low 90s, with only water supply, once more, dropping below (to 81 per cent black) – a clear measure of the "blackening" of the worst areas, as racial self-ascription replaced enumerator categorization.

Acknowledgements

The author is grateful to the Leverhulme Trust, which made a grant to the University of Oxford in 1998 to fund the project entitled Decolonising the Colonial City: Kingston, Jamaica 1940–1991 (Grant Ref. F/773/B). This chapter is based on part of that project, which was directed by Professor Colin Clarke, with Dr David Howard as the principal research assistant. Professor Clarke wishes to express his thanks to Nigel James of the Geography and Map Room of the Bodleian Library, Oxford University, who carried out the analyses using MAPINFO and produced the maps for this chapter. Additional grants to support the research were provided by Oxford University, Jesus College, Oxford and the Oppenheimer Fund, Queen Elizabeth House, Oxford University, all of which are gratefully acknowledged.

References

Central Housing Advisory Board. 1936. *Memorandum dealing with development of Trench Pen township and improvements of Smith Village and surrounding districts.* Kingston: Government Printing Office.

Committee for International Cooperation in National Research in Demography (CICRED). 1974. *Recent population movements in Jamaica.* Kingston: CICRED.

Clarke, C.G. 1966. Population pressure in Kingston: A study of unemployment and overcrowding. *Transactions of the Institute of British Geographers* 38: 165–82.

———. 1975. Ecological aspects of population growth in Kingston, Jamaica. In *Geographical analysis for development in Latin America and the Caribbean*, ed. R.P. Momsen Jr, 42–55. Chapel Hill, North Carolina: CLAG Publications.

———. 2006a. *Kingston: Urban development and social change.* Kingston: Ian Randle.

———. 2006b. From slum to ghetto: Social deprivation in Kingston, Jamaica. *International Development Planning Review* 28 (1): 1–34.

———. 2006c. *Decolonizing the colonial city: Urbanization and stratification in Kingston, Jamaica.* Oxford: Oxford University Press.

Eyre, L.A. 1986a. The effects of political terrorism on the residential location of the poor in the Kingston urban region, Jamaica, West Indies. *Urban Geography* 7: 227–41.

———. 1986b. Party political violence and the struggle for residential space in Jamaica. In *Competition for rural and urban space in Latin America: Its consequences for low-income groups*, ed. J.M.G. Kleinpenning, 125–39.

Netherlands Geographical Studies no. 25. Amsterdam-Nijmegen: Royal Dutch Geographical Society.

Klak, T. 1997. Obstacles to low-income housing assistance in the capitalist periphery. In *Self-help housing, the poor and the state in the Caribbean*, ed. R. Potter and D. Conway, 102–19. Knoxville: University of Tennessee Press.

Massey, D.S., and N.A. Denton. 1993. *American apartheid: Segregation and the making of the underclass*. Cambridge, MA: Harvard University Press.

Morrill, R. 1965. The Negro ghetto: Problems and alternatives. *Geographical Review* 55: 339–61.

Philpott, T.L. 1978. *The slum and the ghetto: Neighbourhood deterioration and middle class reform: Chicago, 1880–1930*. New York: Oxford University Press.

Roberts, G.W. 1963. Provisional assessment of growth in the Kingston-St Andrew area, 1960–70. *Social and Economic Studies* 12: 432–41.

Smith, M.G., R. Augier, and R. Nettleford. 1960. *The Ras Tafari movement in Kingston, Jamaica*. Kingston: Institute of Social and Economic Research, University College of the West Indies.

Ward, D. 1982. The ethnic ghetto in the United States: Past and present. *Transactions of the Institute of British Geographers* 7: 257–75.

West India Royal Commission. 1945. *West India Royal Commission Report*. London: HMSO.

CHAPTER 13

Asset-Based Responses to Urban Vulnerability

An Analysis of Livelihood Strategies in Waterhouse, Jamaica

DAVID DODMAN

Introduction

The title of this collection ties the theme of "Global Change and Caribbean Vulnerability" to the question of the extent to which the environments, economies and societies of the Caribbean are "at risk". In the urban setting, this leads to further questions about the ways in which global and local processes combine to make urban residents "vulnerable" to shocks and stresses. However, despite these challenges, urban residents utilize a variety of assets and livelihood strategies to combat vulnerability and support themselves and their families.

In the preceding chapter, Clarke explores the evolution of Kingston's "ghetto" communities, and documents in detail the aspects of multiple deprivation affecting their inhabitants. This chapter, in contrast, identifies some of the techniques used by residents of inner-city communities to counter the many social, economic and environmental problems that they face, and analyses these through the concepts of vulnerability and resilience. The first part of the chapter identifies the different ways in which local and global processes link together to make urban residents vulnerable to shocks and stresses, and discusses the key challenges facing individuals living in cities in the global south. These processes are illustrated through an examination of the assets utilized by the urban poor in Waterhouse, an economically deprived community in West Kingston. These assets, which include labour, housing and skills, along with social networks, are used in a variety of creative ways to achieve livelihoods and build resilience to vulnerability. In conclusion, the chapter seeks to link the issues of vulnerability and assets by assessing the extent to which the mobilization of different assets as a response to vulnerability can result in the creation of sustainable livelihoods for the urban poor.

Vulnerability and Resistance

Vulnerability is a term that has many uses, and there is no real consensus on its definition. According to Moser (1996), vulnerability is a more useful way of assessing socio-economic well-being than income or expenditure levels. For Moser, vulnerability is

> the insecurity of the well-being of individuals, households, or communities in the face of a changing environment. Environmental changes threatening welfare can be ecological, economic, social, or political, and they can take the form of sudden shocks, long term trends, or seasonal cycles. With these changes often come increasing risk and uncertainty and declining self-respect. Because people move into and out of poverty, the concept of vulnerability better captures processes of change than more static measures of poverty. (Moser 1996, 2)

Pelling (2001, 2003) speaks specifically in relation to natural disasters, and breaks down vulnerability into the components of exposure, resistance, and resilience:

> Exposure is largely a product of physical location and the character of the surrounding built and natural environment . . . Resistance reflects economic, psychological, and physical health, and their systems of

> maintenance . . . Resilience . . . can be thought of as the ability of an individual or collective actor to cope with or adapt to . . . stress . . . Together, exposure, resistance, and resilience are all conditioned by an actor's access to rights, resources, and material assets. (Pelling 2001, 182–83)

However, urban residents face a particular array of opportunities and challenges. Many cities around the world exhibit higher average wages and incomes than the surrounding areas, yet this use of averages is "oversimplified and conceals gross inequalities in incomes and well-being within urban areas" (Rakodi 2002a, 29). Inequality is generally higher in urban areas, and poor urban dwellers are likely to be subject to higher levels of air pollution, crime and violence than their rural counterparts. In addition, factors such as wage freezes, public sector retrenchment, abolition of controls and subsidies on food, and reduced public sector spending during structural adjustment hit urban residents particularly hard.

The implications of unemployment for urban residents are also different: on the one hand, urban life is far more commoditized with an increasing requirement for money; on the other, the city provides opportunities for hustling and making a living in a variety of ways, each providing some form of income for individuals. The particular vulnerabilities associated with urbanization have been identified by several authors. Wratten (1995, 20–26) lists these as follows: urban environmental and health risks; vulnerability arising from commercial exchange; social diversity, fragmentation and crime; and vulnerability arising from the intervention of the state and police. Meikle (2002, 48–49) provides another list including vulnerability related to the following: legal status (particularly in relation to employment and land rights), inadequate services and infrastructure, and the challenges of local physical and social environments. However, she stresses the vulnerability caused by a dependence on the cash economy. Because "free" goods and services, such as common land, water and fuel, are rare in cities, most of the basic living needs of the urban poor must be paid for in cash, leading to a particular vulnerability to debt.

The concept of vulnerability can be further broken down into adverse trends and shocks. Adverse trends that affect many urban dwellers include congestion, land disputes, and safety and health hazards, but they can also be related to increased competition for raw materials and declining markets. Shocks may vary from the general (such as forced displacement, the implementation of structural adjustment programmes, or the rapid devaluation of a currency) to the personal (such as the loss of a partner or sudden household costs, perhaps associated with the need for medical care) (Lyons and Snoxell 2005). The vulnerability faced by urban residents in Jamaica and elsewhere in the Caribbean results from a combination of these factors.

Urban Vulnerability

Poor urban residents in Jamaica are subject to both public and private vulnerability. Public vulnerability refers to the challenges posed by structural problems of limited economic growth and high unemployment rates, while private vulnerability is related to a lack of individual skills and resources.

On a structural level, and utilizing a purely statistical approach, residents of the Kingston Metropolitan Area (KMA) appear to be less vulnerable than their rural counterparts. Since 1989, the official poverty rate in the KMA has been substantially lower than in rural areas, and consistently lower than in other Jamaican towns. In 2003 (the most recent figures available), the poverty rate in the KMA was 9.5 per cent, compared to 15.8 per cent in other towns and 24.2 per cent in rural areas (figure 13.1). However, the measures of poverty used by the Planning Institute of Jamaica to compile these figures are based on a consumption-based, money metric indicator. This type of measure makes little allowance for the cost of non-food essentials such as transport, keeping children at school, and paying for water and health care. It also fails to account for such aspects of poverty as poor quality housing, inadequate access to emergency services and legal protection, and lack of voice within political systems (Satterthwaite 2001). In this regard, the main factors leading to the apparent lowering of poverty rates in the KMA were a growth in income and wealth, a reduction in inflation, and a fall in the relative price of food (PIOJ 2005a).

The differences in unemployment rates between the KMA and rural areas are much less distinct than those in poverty rates, although there does seem to be a higher rate of unemployment in towns other than the KMA. Unemployment in the KMA was running at 11.7 per cent in April 2004 (most recent figures available) – the lowest for the four year period from 2001 to 2004.

The use of statistical generalizations, however, hides the fact that vulnerability is distributed unevenly at a variety of scales, including the city, the community, the household and the individual. Taking just one variable into account – gender – we can note that there is a labour force participation rate of 73.1 per cent for males and only 56.2 per cent for females; an unemployment rate of 8.2 per cent for males and 18.8 per cent for females; and a job-seeking rate of 4.6 per cent for males

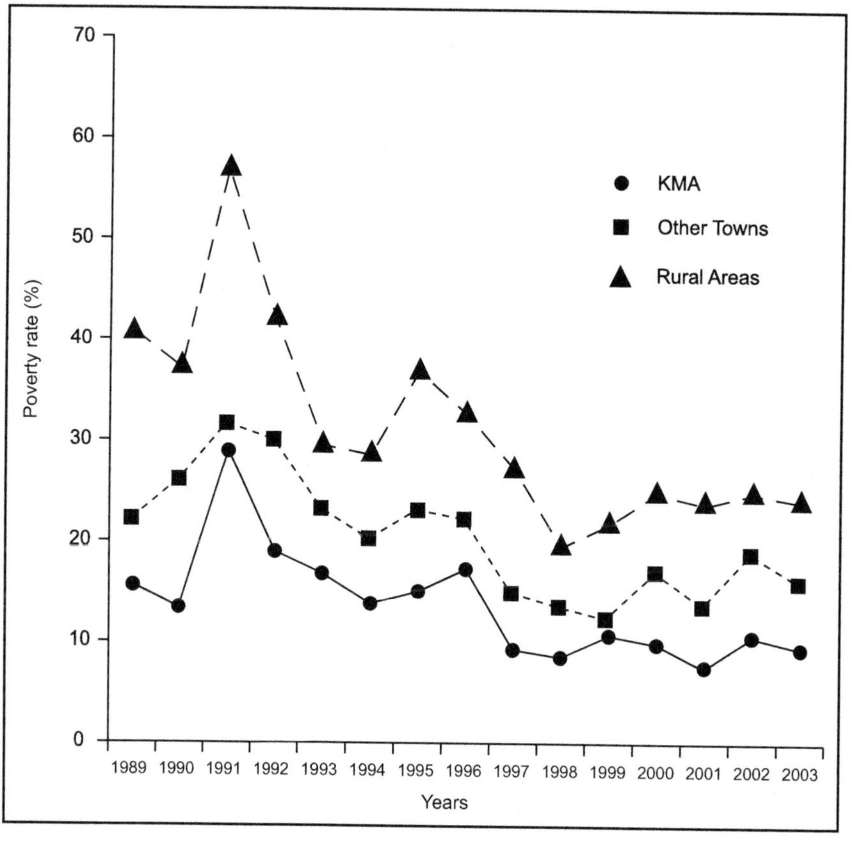

Figure 13.1 Incidence of poverty in Jamaica, 1989–2003.
Source: PIOJ 2000a, 2005a.

and 9.8 per cent for females (all figures are for April 2004, Planning Institute of Jamaica 2005b). In 1999, 67 per cent of "female-headed" households reported unmet basic needs, compared to 58 per cent of "male-headed" households. However, it is not particularly important whether urban residents are "more" or "less" vulnerable than their rural counterparts, but rather to identify that there are specific aspects of vulnerability that affect people living in cities.

Henry-Lee (2005) examined the specific aspects of public and private poverty affecting individuals and communities in the so-called garrison constituencies of the KMA. Her findings suggest that although there has been a reduction in poverty at the macro-level in these areas, the quality of life for residents has not improved significantly and "everyday life is deeply entrenched in crime, violence and political patronage" (Henry-Lee 2005, 83). The areas experiencing these multiple levels of vulnerability are characterized by low-class status, black race, and high and persistent rates of rental and unemployment (Clarke, chapter 12). Some of this vulnerability can be accounted for by the shocks created through the implementation of structural adjustment programmes (SAPs) in Jamaica in the 1980s. Indeed, Clarke describes Kingston as being "comparable in its employment and housing problems to adjacent Latin American cities that have experienced structural adjustment over the last twenty to twenty-five years, and to an even wider range of post-colonial cities that are undergoing rapid political or economic transformation, including globalization" (2006, 247). Although Gordon, Anderson and Robotham (1997) identify the "years of the crisis" in Jamaica as beginning with the OPEC oil price rise of 1974, they stress the harsh experience of the 1980s in which the stabilization and adjustment process was translated into public-sector retrenchment, reduced provision of basic social services and rapid inflation.

Structural adjustment was intended to stabilize the Jamaican economy, but instead reduced the scope of government, opened the economy to foreign goods (but limited capital), and re-produced the colonial version of a non-dynamic, labour-surplus urban economy in Kingston (Clarke and Howard 2006). Most importantly, however, "the process of structural adjustment was one that changed the balance of opportunities for different classes, rather than being simply a process of mass immiseration" (Gordon, Anderson and Robotham 1997, 195). In this

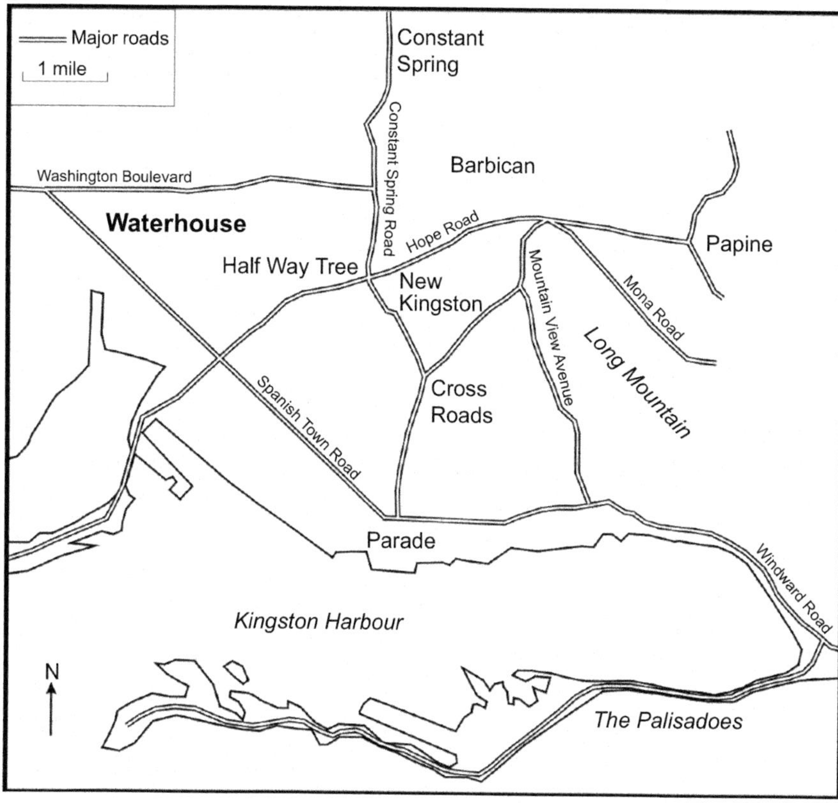

Figure 13.2 Location of Waterhouse in the Kingston Metropolitan Area.

regard, structural adjustment caused trends and shocks that increased the vulnerability of the urban poor, while it reduced the vulnerability of more affluent groups within the Jamaican society.

Urban Vulnerability in Waterhouse

Waterhouse is an inner-city community with a population of approximately 14,000 (Social Development Commission 2004) situated between Spanish Town Road and the Sandy Gully in the KMA (figure 13.2). It is located on the fringes of the broad area of West Kingston, and is bordered to the south and east by other inner-city communities and to the north by Washington Boulevard and the lower-middle-income communities of Maverly and Pembroke Hall. Unemployment is a serious problem, particularly among the young: approximately 65 per cent of residents aged between 18 and 24 are unemployed (Social Development Commission 2004). This problem may be exacerbated through an illiteracy rate of 25 per cent (ibid.) and limited skills training – a recent survey suggested that fewer than 45 per cent of the residents of Waterhouse had a certified skill or trade (van den Idsert 2007). In a letter requesting assistance from the Local Initiative Facility for the Urban Environment (LIFE), a Kingston-based NGO, the residents of Waterhouse described their surroundings in the following words: "Our community is a depressed one and we would like to upgrade same. Many homes does not have lavatory facilities and when people defecate they throw same in the gully. We need to organize clean and healthy lifestyle in our community but without your help we cannot do so."

Participatory research carried out by this NGO provided a list of "Poor Sanitation and Environmental Practices", including an irregular garbage collection system; improper garbage disposal; garbage in gully and gully cleaning; rat infestation; industrial waste from a dairy factory, smelting and other environmentally unfriendly operations; homes without toilets; and the dumping of human waste in the gully. The community faces particular problems with inadequate sanitation facilities – human waste is frequently dealt with by "parachuting", in which plastic bags containing this were swung into the gully. As a result of irregular municipal garbage collection, much household waste is deposited in the gully. The blockage of the gully leads to problems with rats. The *Daily Gleaner* (23 February 2000) reported that the area "needs a Pied Piper". Residents reported "twelve-inch rats, with tails twice as long", who were not even afraid of cats ("Dem nuh 'fraid of the puss as sometimes they are even bigger than the puss and not even we dem 'fraid of"); some had contracted meningitis and leptospirosis as a result of rats contaminating food; and one woman had "nearly lost her fingers and toes after rats climbed into her bed one night".

Many of these characteristics are depressingly familiar to residents of socio-economically deprived urban settlements in the Caribbean and elsewhere in the developing world. However, the residents of Waterhouse have adopted a wide variety of strategies to utilize the assets available to them in the pursuit of sustainable livelihoods for themselves and their households.

Resilience in Waterhouse: Assets and Their Uses

A variety of systems have been proposed to describe the assets that are available to the urban poor. Carney (1998) (cited by Rakodi 2002a) and Meikle (2002) both use frameworks that describe different types of "capital": human capital, social and political capital, physical capital, financial capital, natural capital. Human capital refers to labour resources which have both quantitative (number of household members, amount of time) and qualitative (level of education and skills) dimensions; social and political capital refer to network and group membership, engagement in relationships of trust and reciprocity, and access to wider institutions of society; physical capital includes access to various types of infrastructure and production equipment. The UK government's Department for International Development (DFID) has also used this system of classification to produce a livelihoods framework, which links an "asset pentagon" (of natural capital, human capital, social capital, physical capital and financial capital) with livelihood strategies and livelihood opportunities. These assets, strategies and opportunities are influenced by infrastructure and service provision; policies, institutions and processes; and the vulnerability context (Rakodi 2002b; see also Verrest, chapter 16). Rather than being purely descriptive, therefore, this model aims to provide a basis for identifying appropriate objectives and interventions to support livelihoods.

However, this focus on "capital" of various forms appears to imply at least a semantic bias towards the importance of economic factors in determining vulnerability. Although similar in many respects, the asset-based approaches proposed by Moser (1996) and Pelling (2003) serve to focus attention on both financial and non-financial resources. Moser (1996) identifies and describes the following assets: labour (which she identifies as "the poor's greatest asset" [5]); economic and social infrastructure; housing (which she refers to as "an important productive asset" [9]); household relations (although these are often associated with inequalities – particularly because women often bear a

Table 13.1 Responses and Survival Strategies Used by Jamaicans

Type of Asset / Response	Survival Strategies
Labour/income-generating activities	– Vending – Own-account businesses – Vice – Begging – Theft – Informal savings schemes – Music
Household economies/family relations	– Use of housing/land – Varied household formations – Help from family
Social and economic infrastructure	– Legal and illegal access to state and NGO services
Social capital/establishing socio-economic support groups	– Unofficial community soldiers – Sports – Street dances and shows – Mutual aid societies – "Corner groups"

Source: Modified from Planning Institute of Jamaica 2000b.

disproportionate share of the burden in responding to vulnerability, as external money-earning activities are added to household responsibilities); and social capital (but she recognizes that stresses can both strengthen this through the increased use of reciprocity networks, and weaken it as the ability to cope deteriorates and community trust breaks down). Pelling (2003) proposes a similar list, including labour (which he views as a critical asset in the commodified economy of the city), housing, possessions, "tools of the trade" (58) and social networks.

Which of these assets are available to the residents of Waterhouse, and how do they use these as a response to vulnerability? At the national level, Anderson (2000) identified a variety of what she termed "Coping strategies and social assistance among the poor and needy in Jamaica".

These include foregoing basic needs, increasing or maintaining existing levels of debt, using savings or assets, seeking help from relatives and friends, seeking extra work or hustling, and praying. The Jamaica Human Development Report (PIOJ 2000b) also identifies a set of assets or "survival strategies" used by Jamaicans, including labour/income generating activities, household economies/family relations, social and economic infrastructure, and social capital/establishing socio-economic support groups (table 13.1).

These are self-identified strategies for coping at specific times of financial difficulty, but the residents and community of Waterhouse also adopt a wide variety of responses on a daily basis. These responses are mainly related to the use – in particular ways, at particular times – of different assets that are available to these individuals. The main assets utilized in Waterhouse are labour and employment, skills and tools, access to land, social networks, and financial capital. However, these are inter-related in complex ways: social networks are frequently used to acquire the use of land for productive purposes; connections with friends and relations overseas provide access to financial capital through remittances; and labour power facilitates the learning of new skills which, in turn, enable the more productive use of labour power.

Despite high levels of unemployment, residents of Waterhouse use their labour power in a variety of ways. Some work in a variety of jobs in the formal sector. Examples of this include the women who work as community health workers in the local clinic and young men working in low-skilled jobs such as pumping petrol. A very noticeable group, however, are engaged in informal sector employment of various kinds. This sector has been "a significant social space within the national economy [of Jamaica] and part of the country's economic landscape at least since the eighteenth century" (Gray 2004, 268). However, although much of the literature addressing the informal sector has concentrated on petty trading and service activities, "in the Caribbean context, the informal sector encompasses a range of retailing, production and service activities, including street trading, craft production, smuggling, small-scale farming, food retailing, home-based clothing and electrical production, local household repairs and vehicle maintenance" (Lloyd-Evans and Potter 2000, 159–60).

Various estimates have been made of the size of Jamaica's informal sector. Witter and Kirton (1990) suggest that for the period between 1962 and 1985, the estimated average of the size of the informal economy varied from 14 per cent to 34 per cent of the nation's GDP. A more recent study suggests that around 53 per cent of the employed labour force in Jamaica is informal, and about 40 per cent of the national GDP is produced by the informal sector (GRADE 2002). The statistics on overall employment in Jamaica seem to indicate that the informal sector continues to play a significant role in providing employment. The most recent Labour Force survey suggests that of the 1,040,300 people employed in Jamaica in April 2004, one-third were what is termed "own account workers", who may or not have formal documentation for their businesses (Statistical Institute of Jamaica 2005).

Waterhouse has a high concentration of small-scale, informal sector, metal smelting operations, which are highly visible, provide a substantial amount of employment and represent an important source of income to the community (Dodman 2008). In these workshops, furnaces burning waste engine oil are used to melt down scrap aluminium collected from the Riverton solid waste disposal facility, which is then moulded to produce pots, pans, spoons and other metal objects. Around twenty-five of these workshops exist in the community, varying in size from small sheds run by a single individual to larger workshops with six or seven men working at the same time. These operations are a significant employment opportunity for community members, represent one of very few examples of recycling taking place in Jamaica and provide cheap locally produced utensils to Jamaicans. This final outcome has the additional impact of reducing the need to buy more expensive imported products. Yet at the same time, concerns have been voiced by residents of the community and members of environmental groups that these operations create serious environmental and health problems through the emission of smoke from the furnaces and the dumping of unwanted scrap metal in the adjacent gully.

One adverse trend that is affecting the metal smelters of Waterhouse is the reduced availability of the necessary raw material of scrap aluminium. According to recent fieldwork (September 2006), dealers from Cuba have been purchasing scrap metal in Kingston, thereby increasing the price of this raw material. As a result, many of the metal smelters

Figure 13.3 Tools as a livelihood asset in Waterhouse.

who manufacture cooking pots have been unable to purchase adequate quantities of this resource. These manufacturers have long been susceptible to seasonality in the demand for their products, which has consistently peaked around Christmas. The challenge has been compounded by the limited flexibility in retail prices for pots and pans, particularly related to competition from ever-cheaper imported items. However, the smelters who are producing more specialized items that require smaller amounts of this raw material and higher levels of skill, such as gas stove burners, have a greater level of resistance and resilience to the challenges of material shortages and seasonality.

Workers in the informal metal smelting industry have a variety of skills which they have acquired over time. Obtaining the necessary skills to work in the sector requires appropriate social contacts in the form of established smelters who are willing to give an individual the opportunity to learn the techniques of the trade. In addition to the skills, ownership of (or at least access to) specific tools – such as a furnace and moulds – is required to operate successfully in this mode of production. Increased specialization requires more sophisticated tools, some of which may be self-made. For example, the tools shown (figure 13.3) are used to make burners for gas stoves by drilling and clamping metal pieces together. The more successful operators have also developed strong links in the retail sector, with market traders and with the operators of formal sector retail outlets, both of which have helped to facilitate the sale of their products.

Access to land for productive use – which does not necessarily imply ownership – is usually linked with tenure for housing. All the individuals operating their own smelting workshops in Waterhouse share the characteristic of having access to a piece of land in the community that they can use for this purpose. Those who had been unable to set up their own workshops did not have the same access to land or have fewer direct linkages to individuals with this land space. Land is also used for operating small retail outlets: in some cases, a smelting workshop is operated at the back of the house by a man and a store at the front operated by his common-law wife.

The residents of Waterhouse draw on a variety of social networks, incorporating both local and global linkages, as a means of responding to urban vulnerability. The Building Together Citizens Association is a community-based organization that acts as an advocate on behalf of the community, and its involvement with organizations such as LIFE helped to acquire funds for sanitation projects. It also helps to run an old-age home in the community. In turn, LIFE is a Jamaican NGO that was initially established by the United Nations Development Programme in the wake of the 1992 Rio de Janeiro "Earth Summit". The local health centre was built through international funding (provided by the Dutch government). The community also draws on social networks incorporating the local private sector – the football club and community have benefited from extensive sponsorship by Tank-Weld, one of Jamaica's largest civil engineering firms. These connections operate in a wide range of formats across a variety of scales to position the community in a variety of networks that can facilitate sustainable responses to the challenges faced by its residents.

Even though Waterhouse is an economically poor community, individuals are able to access financial capital in a variety of ways. Anderson (2000, 94) identifies several sources of external support from relatives and friends, including child support, assistance from local relatives or friends, and assistance from overseas relatives or friends. Remittances from overseas are an increasingly important source of income for Jamaican households: in 2003, the volume of registered remittances sent to Jamaica was worth US$1.362 billion (PIOJ 2005c), more than the US$1.350 billion earned by the tourism industry in the same year (PIOJ 2005b). Although no specific data were collected on this topic in Waterhouse, many poor urban residents save through informal "partner" schemes (similar to those known as *su-su* in Trinidad [see Pamuk 2003]) in which deposits are made on a scheduled basis (usually monthly) with members of the scheme taking turns receiving all the payments for a particular month.

The metal smelting industry therefore interacts with other activities in the community of Waterhouse in a variety of ways. Backward linkages are created with suppliers of waste engine oil and scrap metal, while forward linkages exist with middle-men (who transport the end products around the island) and retailers. The financial earnings of the individuals involved in the industry appear to support the petty retail sector in the community, and are described as helping to provide necessities for children's healthcare and education. In addition, engagement in a similar livelihood activity appears to generate a sense of social cohesion, manifested in the community's support for the local football team and in its rousing celebrations when Waterhouse resident Shelly-Ann Fraser won the 100-metre gold medal at the 2008 Olympic Games (*Daily Gleaner*, 17 September 2008).

Conclusion: Towards Sustainable Urban Livelihoods

In this context of vulnerability, therefore, how can the mobilization of different assets result in the creation of sustainable livelihoods for the urban poor? A sustainable livelihood incorporates the capabilities, assets and activities which are available to poor men and women to make up a living; is dynamic and adaptable; is related to poor people's own priorities, interpretations and abilities; is household- and community-centred; and achieves these components without undermining the natural resource base (Meikle, Ramasut and Walker 2001).

Many of the assets utilized by the residents of Waterhouse and similar communities in Kingston exhibit a high level of fungibility. This means that available resources can be converted swiftly, conveniently and without loss from one use to another. The resources that can be distributed in this way consist not only of cash (which can be spent on either domestic or production costs), but also of time (which can be distributed flexibly between domestic and economic tasks) and space (which can be used for a range of activities that change throughout the day or on a seasonal basis) (Kellett and Tipple 2000). The conditions that persist in many households in Waterhouse reflect the fact that "the distinction between reproduction (domestic activities) and production (economic activities) is not clearly drawn in most households in Africa, Asia and Latin America" (Kellett and Tipple 2000, 205). Even though the economic rewards that are gained from domestically based economic activities may be fewer than those that could be obtained in the formal sector, the domestic setting offers flexibility as well as income-generating opportunities for children and other people in the household who might otherwise be excluded from economic activity. This high level of fungibility means that resources can be reallocated readily in response to adverse trends and shocks in the national and household economy.

The concept of the natural resource base is also relevant in the context of Waterhouse. The activities of the metal smelters can be seen as an economically successful response to the problems of solid waste management and the cost of importing metal goods, yet the industry's practices simultaneously result in negative environmental impacts that serve as a point of contention between groups within communities and in the city as a whole. The negative environmental impacts of this particular industry, particularly the thick smoke generated by the oil-burning furnaces, are frequently linked with poor health among children. In addition, the absence of protective equipment creates a generally unsafe working environment for the men engaged in this activity.

For this and other reasons, involvement in informal sector activities should not be viewed as a sustainable livelihood strategy in its own right. Based on a survey of the informal sector in the Pretoria-Witwatersrand-Vereeniging region, the economic hub of South Africa, Rogerson (1996) concluded that "survivalist enterprises" (activities undertaken by individuals unable to obtain regular wage employment) generally "fall short of even a minimum standard of income" (Rogerson 1996, 171). In contrast, many "growth enterprises", which lack the trappings of formality in relation to business licences and operating permits, "have the potential to develop and flourish" (ibid.). Rather than a rigid dichotomy, however, I suggest that there is a scale between these two extremes, representing continuous variation between levels of vulnerability and sustainability. The extent to which the different livelihood strategies adopted by urban residents in Waterhouse can be defined as sustainable is therefore probably linked with their position on this scale.

Much of the literature on "vulnerability" (and even "urban vulnerability") focuses on susceptibility to "natural hazards". However, some of the key concepts in reducing vulnerability to natural hazards are also important in influencing levels of exposure, resistance and resilience to a variety of social, political and economic shocks and stresses. Lewis and Mioch (2005) provide a framework for reducing urban vulnerability to disasters through "good governance". This involves political commitment and integrated policy development; increasing decentralization of responsibilities and resources to municipalities, thereby encouraging the development of integrated and locally applicable measures that reduce vulnerability (to disasters); civil society and multi-stakeholder participation; and urban poverty reduction.

These same approaches are very relevant for reducing vulnerability in a wider sense. Efforts are being made to improve governance in the KMA through the Local Government Reform programme, and the implementation of participatory planning (such as the creation of the Local Sustainable Development Plan for Kingston and St Andrew) (Ministry of Local Government 1993, 1995, 2001). However, much of this is still "donor-driven" and in response to international demands, rather than arising from grassroots civil society action. Greater effort needs to be made to facilitate the "industry" of poor urban residents who have already devised sophisticated strategies to address their own vulnerability.

References

Anderson, P. 2000. Coping strategies and social assistance among the poor and needy in Jamaica. Special module in *Jamaica survey of living conditions 1999*, Kingston: Planning Institute of Jamaica/Statistical Institute of Jamaica.

Clarke, C. 2006. *Decolonizing the colonial city: Urbanization and stratification in Kingston, Jamaica*. Oxford: Oxford University Press.

Clarke, C., and D. Howard. 2006. Contradictory socio-economic consequences of structural adjustment in Kingston, Jamaica. *Geographical Journal* 172 (2): 106–29.

Dodman, D. 2008. Making a living: Small-scale metal smelting as a livelihood strategy in Kingston, Jamaica. In *The Caribbean city*, ed. R. Jaffe, 227–46. Kingston: Ian Randle.

Gordon, D., P. Anderson and D. Robotham. 1997. Jamaica: Urbanization during the years of the crisis. In *The urban Caribbean: Transition to the new global economy*, ed. A. Portes, C. Dore-Cabral and P. Landolt, 190–223. Baltimore: Johns Hopkins University Press.

Gray, O. 2004. *Demeaned but empowered: The social power of the urban poor in Jamaica*. Kingston: University of the West Indies Press.

Group of Analysis for Development (GRADE). 2002. Informal sector study for Jamaica: Final report. Consultancy requested by the Inter American Development Bank, November.

Henry-Lee, A. 2005. The nature of poverty in the garrison constituencies in Jamaica. *Environment and Urbanization* 17 (2): 83–99.

Kellett, P., and A. Tipple. 2000. The home as workplace: A study of income-generating activities within the domestic setting. *Environment and Urbanization* 12 (1): 203–13.

Lewis, D., and J. Mioch. 2005. Urban vulnerability and good governance. *Journal of Contingencies and Crisis Management* 13 (2): 50–53.

Lloyd-Evans, S., and R. Potter. 2000. Environmental impacts of urban development and the urban informal sector in the Caribbean. In *The urban Caribbean in an era of global change*, ed. R. Potter, 153–71. Aldershot, UK: Ashgate.

Lyons, M., and S. Snoxell. 2005. Sustainable urban livelihoods and marketplace social capital: Crisis and strategy in petty trading. *Urban Studies* 42 (8): 1301–20.

Meikle, S. 2002. The urban context and poor people. In *Urban livelihoods: A people-centred approach to reducing poverty*, ed. C. Rakodi with T. Lloyd-Jones, 37–51. London: Earthscan.

Meikle, S., T. Ramasut and J. Walker. 2001. *Sustainable urban livelihoods: Concepts and implications for policy*. Working Paper no. 112. London: Development Planning Unit, University College London.

Ministry of Local Government. 1993. *Ministry paper 8/93: Reform of local government*. Kingston: Government of Jamaica.

———. 1995. *Outline of the local government reform project proposal* (updated version). Kingston: Government of Jamaica.

———. 2001. *Local government reform: A regional framework for local governance and development*. Kingston: Government of Jamaica.

Moser, C. 1996. *Confronting crisis: A comparative study of household responses to poverty and vulnerability in four poor urban communities.*

Environmentally Sustainable Development Studies and Monographs Series, no. 8. Washington, DC: World Bank.

Pamuk, A. 2003. Informal Institutional Arrangements in Credit, Land Markets and Infrastructure Delivery in Trinidad. *International Journal of Urban and Regional Research* 24 (2): 379–96.

Pelling, M. 2001. Natural disasters? In *Social nature: Theory, practice and politics*, ed. N. Castree and B. Braun, 170–83. Oxford: Blackwell.

———. 2003. *The vulnerability of cities: Natural disasters and social resilience*. London: Earthscan.

Planning Institute of Jamaica (2000a). *Jamaica survey of living conditions 1999*. Kingston: PIOJ/SIJ.

———. 2000b. *Jamaica human development report 2000*. Kingston: PIOJ/SIJ.

———. 2005a. *Jamaica survey of living conditions 2003*. Kingston: PIOJ/SIJ.

———. 2005b. *Economic and social survey Jamaica 2004*. Kingston: PIOJ/SIJ.

———. 2005c. *Jamaica human development report 2005*. Kingston: PIOJ/SIJ.

Rakodi, C. 2002a. A livelihoods approach: Conceptual issues and definitions. In Rakodi with Lloyd-Jones 2002, 3–22.

———. 2002b. Economic development, urbanization and poverty. In Rakodi with Lloyd-Jones 2002, 23–34.

Rakodi, C. with T. Lloyd-Jones, eds. 2002. *Urban livelihoods: A people-centred approach to reducing poverty*. London, Earthscan.

Rogerson, C. 1996. Urban poverty and the informal economy in South Africa's economic heartland. *Environment and Urbanization* 8 (1): 167–79.

Satterthwaite, D. 2001. Rural and urban poverty: Understanding the differences. *Economic Perspectives* 6 (3): 33–36. http://usinfo.state.gov/journals/ites/0901/ijee/satterthwaite.htm.

Social Development Commission. 2004. *Community profile: Community of Waterhouse (St Andrew)*. Kingston: Social Development Commission.

Statistical Institute of Jamaica (SIJ). 2005. *The labour force 2004*. Kingston, Jamaica (SIJ).

van den Idsert, H. 2007. Informal entrepreneurs in the midst of neo-liberalism: A Jamaican case study on scrap metal molders in the informal economy. MSc thesis, University of Amsterdam, Netherlands.

Witter, M., and C. Kirton. 1990. The informal economy in Jamaica: Some empirical exercises. Institute for Social and Economic Research (ISER). Working Paper no. 36. Kingston: ISER.

Wratten, E. 1995. Conceptualising urban poverty. *Environment and Urbanization* 7 (1): 11–33.

CHAPTER 14

Urban Violence, Crime and the Threat to "Democratic Security" in the Dominican Republic

DAVID HOWARD

Introduction: The *Barrio Seguro* Project

During the summer of 2006, over half the combined military and police forces of the Dominican Republic were mobilized on thirteen neighbourhoods in the capital of Santo Domingo. The cause of this massive militarized presence on Dominican streets was not immediate civil war or foreign invasion, but the implementation of a recent government policy to protect national "democratic security" from the perceived threat of increasing levels of violence and crime within the city.

The summer experiences of city living in the Dominican Republic would seem to fit squarely with the frequent narratives of pending urban doom – the notion that urban conglomerations across the globe face imposing biophysical, social, economic and political calamities, leaving denizens' livelihoods and futures implicitly susceptible to extreme pressures of densely concentrated urban living (Schneider and Susser 2003; Thrift 2005). Socially produced vulnerabilities are the outcome of social and economic inequalities, and the relative access to supporting services or opportunities that affect the ways in which populations are able to respond to traumatic events. The social vulnerability of urban residents to environmental hazards has been analysed and categorized for a range of cities, but these calculations are assessed largely in the context of biophysical threats (Cutter, Boruff and Shirley 2003). The social vulnerability addressed in this chapter relates to increasing levels of urban violence and governmental response in the form of intensive and place-specific policing strategies.

The following pages concentrate on two main aspects of urban social vulnerability: first, state responses to urban violence and criminality, most notably the introduction of intensive and spatially selective policing, known as the *Barrio Seguro* (meaning "safe neighbourhood") programme; and second, the political rhetoric and logic that has employed the vulnerability of Dominican society to civil disorder and "external threats" to sovereignty. This chapter thus links territory, policing strategies and attempts to control lived environments through imposing expressions of strength. The empirical evidence is derived from four months of fieldwork during 2006 in Santo Domingo, the capital city of the Dominican Republic, and it forms part of a wider research effort focusing on urban policing, territory and violence.

The discussion is presented in the context of the steady rise in reported racist and violent attacks against people presumed to be of Haitian origin in the Dominican Republic over the last five years. In August 2005, 2,000 people were deported from the country, primarily on the basis of "looking Haitian" rather than due to a verified lack of legal documentation (Groupe d'Appui aux Rapatriés et Réfugié 2005). During this period, thirteen racially motivated murders of Haitian-Dominicans were detailed in the media; all were reported as part of a wave of increasing violence and crime, commonly referred to as *la delicuencia* – a generic term for crime, violence and disrespect for others.

Here specifically, I analyse the ways in which the Dominican government has imposed a zero tolerance policing programme in a selected part of Santo Domingo that effectively zones off the so-called dangerous areas from the rest of the city. While the aims of the *Barrio Seguro* programme are to combat urban crime, popular association in the Dominican Republic of civil disorder and crime with Haitian immigration, anti-Haitian sentiment and emphasis on arresting people not in immediate possession of identity papers, all suggest that urban residents of Haitian-origin may be explicitly targeted.

The study mirrors to some extent current research on governmental responses to impoverished urban living conditions which might cause popular unrest and form a potential risk for civic stability. Urban neighbourhoods are "spatially reified [by policy measures] as forms of such 'threats' ", and whereby urban policy has shifted from concerns over the material deprivation of certain areas, to seeing these zones (and by default their residents) as potential threats to the democratic whole (Dikeç 2006, 162).

Drawing on an additional and earlier conceptual linkage, the work reflects Herbert's research on territoriality and policing techniques used by the Los Angeles Police Department. The connection is not so far-fetched, given that one of the "selling points" of the Dominican government's recent revamping of urban policing was that many officers would receive training within the United States. Herbert writes: "Just as social-structural works often neglect the shaping influence of culture, they also regularly overlook the spatial embeddedness of social action … Analysis of everyday police behaviour, in other words, must pay attention not only to its social and cultural construction, but also to its intractable spatiality; in working to uphold socially constructed notions of public order, officers seek to control the spaces they patrol" (Herbert 1997, 20–21). Thus, I am interested in the urban environment in terms of the policing of territory, the "defence" of boundaries and violence, and the concomitant perceived vulnerability of democracy.

The Dominican Context of Vulnerability

Geographical studies of human vulnerability have often been associated with biophysical environmental risk, "natural" hazards or aspects of sudden, often disastrous, "calamity" (Hewitt 1983; Wisner et al. 2004). Authors working within the social sciences, however, have also consistently portrayed cities to be "at risk". In addition, they occasionally confer simultaneous narratives to infrastructure and people, thereby labelling cities as "wounded" (Schneider and Susser 2003) or as sites that are embedded in globally "splintering" urbanisms (Graham and Marvin 2001).

The current concern over levels of urban violence in the Dominican Republic, and across the Caribbean region as a whole, have reinvigorated a focus on the human and societal vulnerabilities of cities. The threats posed and damage caused are not biophysical but societally induced, and the context of the urban environment can perhaps similarly be defined in terms of "hazards". In response to growing public fears for personal safety, the political response outlined below has been to defend not only the individual well-being of citizens from levels of violence, but to "protect" the very basis of national sovereignty from implied external threats. Sovereignty and the exercise of power by a sovereign state, however, does not imply personal freedom for its citizens and may indeed be the antithesis to personal liberty (Hobbes 2005). In the context of the Dominican Republic, the intensified policing programme, in place since 2006, has been the government's main response to ongoing portrayals on behalf of media and politicians of a Dominican society exposed to uncontrolled Haitian immigration and increasing levels of criminals deported from the United States.

The Dominican example fits squarely with the growing concern across the Caribbean about levels of crime and violence in the region. In April 2008, a special meeting of the Caribbean Community (CARICOM) was held in Trinidad at which the respective heads of government pledged to act collectively in a region which has been highlighted as having the highest murder rate per capita in the world (United Nations Office on Drugs and Crime and the World Bank 2007; CARICOM 2008). Combined responses to individual state vulnerabilities, notably revising shared information systems for intra-regional flights, deportation schedules and maritime security were proposed. The collective states' intent was to reduce regional vulnerability to organized crime and challenge increasing levels of violence in individual countries.

The Dominican Republic is not a member of CARICOM, but via a series of trade and political initiatives, as well as joint strategies, Dominican governments have been increasingly associated with pan-Caribbean programmes since the late 1990s. Tackling issues of violence and crime is likely to be of increasing importance in terms of the regional agenda. Since the early 1990s, verbal and media reports of increased violence and robbery have matched a steady rise in the number of gun murders to over 1,000 per year in the Dominican Republic (Bobea 2004).

Between 2001 and 2005, 9,300 murders were reported in a population of nine million, of which 790 victims were women. In 2005, 2,382 murders were documented – double that of 2001 – and the annual level of violent deaths has remained at over 2,000 per year, making per capita murder rates comparable to the Jamaican and Trinidadian contexts as the highest in the region (United Nations Development Programme 2007; United Nations Office on Drugs and Crime and the World Bank 2007). Increased reported crime rates have been matched by intensified police and military responses. Since 2005, sixteen out of every one hundred reported violent deaths in the Dominican Republic have involved "death at the hands of the police in the course of their duty" (United States Department of State 2008).

Despite the general heightened awareness of, and evidence for, increased rates of crime, the core concern remains spatially, gender- and age-specific. Public fear and police intervention have concentrated on the young male population of low-income urban neighbourhoods in the two main cities of Santo Domingo and Santiago de los Caballeros. This chapter will focus on the problematic issues of violence and public order control faced by one neighbourhood in the former city that has been targeted as part of the government's recent anti-crime initiative.

Fear of violent crime has become the leading social issue for many urban residents across all income groups and social classes (Pérez and Milán 2005). The most recent national household survey suggests that delinquency rates only second to employment as the principal problem in the country (Oficina Nacional De Estadística 2005). Just over 6 per cent of adult respondents in Santo Domingo said that they were victims of household burglary and 7 per cent were subjected to personal robbery or attack during the preceding year (this figure rises to 14 per cent when the last five years are taken into account). Only one in four reported these personal incidents to the police; just under half believed that the police would not, or could not, do anything to assist them.

Dominican media accounts and everyday conversations, with noticeably more frequency than during the author's previous research visits during the 1990s, have recurrently focused on *la delincuencia*. Such reports have propelled crime-related concerns to epidemic status over the last five years, often in areas relatively unscathed by serious criminal acts. To tackle this "crime wave", the government instituted a programme to protect "democratic security" during 2005.

Defending Society? The Plan for Democratic Security

The Plan for Democratic Security has been the cornerstone of the current Dominican government's social policy since first announced in 2004. The programme refers directly to a Colombian security policy implemented during the administration of President Álvaro Uribe (2002–2006), which aimed to confront the violent effects of political violence and the illegal drug trade. The Dominican campaign sought to build on the rhetorical ideas of a government defending sovereignty, by bringing together notions of national, household and personal security, responsible citizenship, and the fear of a "threat" to Dominican society. At his inaugural presidential address, on 16 August 2004, Dr Leonel Fernández rhetorically linked all concerns together: "The Dominican Republic cannot continue the way it is going. Citizens' insecurity cannot persist . . . With the influx of [bad] influences. With clientelism. With illegal profiteering. With the abuse of power. With disrespect. Above all, without serious intent in everything . . . Let us construct a new society. More supportive. More just. More prosperous. More human. More democratic. More transparent. More participatory" (author's translation). As this quote illustrates, the government has seized upon the notion that the Dominican "nation" is vulnerable to unqualified, and often externalized, threats, particularly the urban concentrations of population.

The wider context of the speech, and subsequent government policy, suggest that the negative "influences" have arrived from non-Dominican sources in the form of illicit trade and illegal immigration, principally referring to the longstanding immigration of Haitian labour. The government's role is therefore to defend consensual ideas of democracy and national security. As Hobbes (2005) noted, however, such a "social contract" for the whole will necessarily restrict individual freedom of action for all citizens. It is argued below that the government's anti-crime initiative and concomitant call for a collective defence of democracy, by default the protection of sovereign power over citizenry and territory, impinges on a wide range of everyday legal as well as illegal practices.

The *Barrio Seguro* Project

In June 2005, the attorney general, the national police and the Ministry of the Interior and Police formally announced the launch of the Plan for Democratic Security (Procuraduría General de la República, Secretaría del Estado de Interior y Policia y Policía Nacional 2005). The plan is a mixture of strong rhetoric, highly visible media campaigns and policing initiatives, all meant to underpin President Fernández's election pledge to "secure" Dominican society. Most evidently the scheme relies on intense military and police surveillance of selected low-income neighbourhoods in the main cities of Santo Domingo and Santiago. This chapter explores the effects of this plan, and concomitant project, to produce "safe neighbourhoods" in the country's capital, a city with a census population of two million people (Oficina Nacional de Estadística 2004).

At the start of 2006, President Leonel Fernández proclaimed the introduction of *Barrio Seguro*, the flagship project of the Plan for Democratic Security. The planned mobilization, and subsequent recruitment drive, of sixteen thousand military and police offers, was announced to eradicate and prevent, primarily, criminal and anti-social acts. A secondary motive was to promote the recruitment into the police of existing government employees, and also "sportsmen and women, doctors, musicians, and private security workers" (*Hoy*, 8 January 2006, 1). Controversially, the secretary of the interior and police, Franklin Almeyda Rancier, also announced that twenty thousand police officers were not being deployed correctly for the public good, but were informally working as private security officers. This was strongly denied by his colleague, Francisco Domínguez Brito, the chief of police (*Hoy*, 9 January 2006, 3).

A pilot programme had been running in the Santo Domingo neighbourhood of Capotillo since August 2005. This neighbourhood has long been seen as one of the "hottest" – politically and in terms of "civil disobedience" – in the capital city. The president suggested that Dominicans would soon "breathe a more secure, more hopeful and more optimistic atmosphere" (*Hoy*, 8 January 2006, 1). Rhetorically and practically, the project sought to cleanse the urban environment and those who lived there.

The cornerstone of the *Barrio Seguro* project in Santo Domingo is to create a *cordon sanitaire*, as described by a government employee involved in implementing the project, thus making thirteen contiguous neighbourhoods in the northern part of the city "safe" (figure 14.1).

The Dominican government selected these neighbourhoods as key areas of high criminal activity, but detailed area-specific census or survey data as evidence to support this assessment remains lacking. All are located in the north east of the central city, bordering the River Ozama. The most recent household survey and national census indicate that the residents of these neighbourhoods share economic and health indicators that are similar to the national average, or at least are not among the poorest areas of the country, which tend to be in rural areas. The average percentage of people who are under- or unemployed in the thirteen neighbourhoods is 40 per cent for men and 60 per cent for women, compared to the national average of 39 per cent and 59 per cent respectively. The national average for literacy is 39 per cent for men and 40 per cent for women, compared to 42 per cent for men and 45 per cent for women in the thirteen neighbourhoods (Oficina Nacional de Estadística 2004, 2005, 2008). The thirteen neighbourhoods may be considered as among the poorest areas of the city, but their inclusion into a contiguous zone for the *Barrio Seguro* project has labelled them as "dangerous" and poor.

Much of the data for this chapter is based on police reports and interviews by the author with police officers and community activists directly involved or affected by the *Barrio Seguro* project. Thirty-seven residents in the neighbourhood of Villas Agrícolas were interviewed by the author between February and May 2006 (see figure 14.1). Villas Agrícolas is one of thirteen low-income, contiguous urban neighbourhoods in Santo Domingo initially selected for the intensified policing scheme. Interviewees were selected initially at random by the author, and, subsequently, through a process of recommendation and introduction among neighbourhood residents. A conscious attempt was made to avoid overt political, age or gender bias among the overall range of interviewees. The interviews were semi-structured and all were recorded with the permission of the interviewee. Given the sensitivity of the topics surrounding policing practice and the effects of crime on individuals

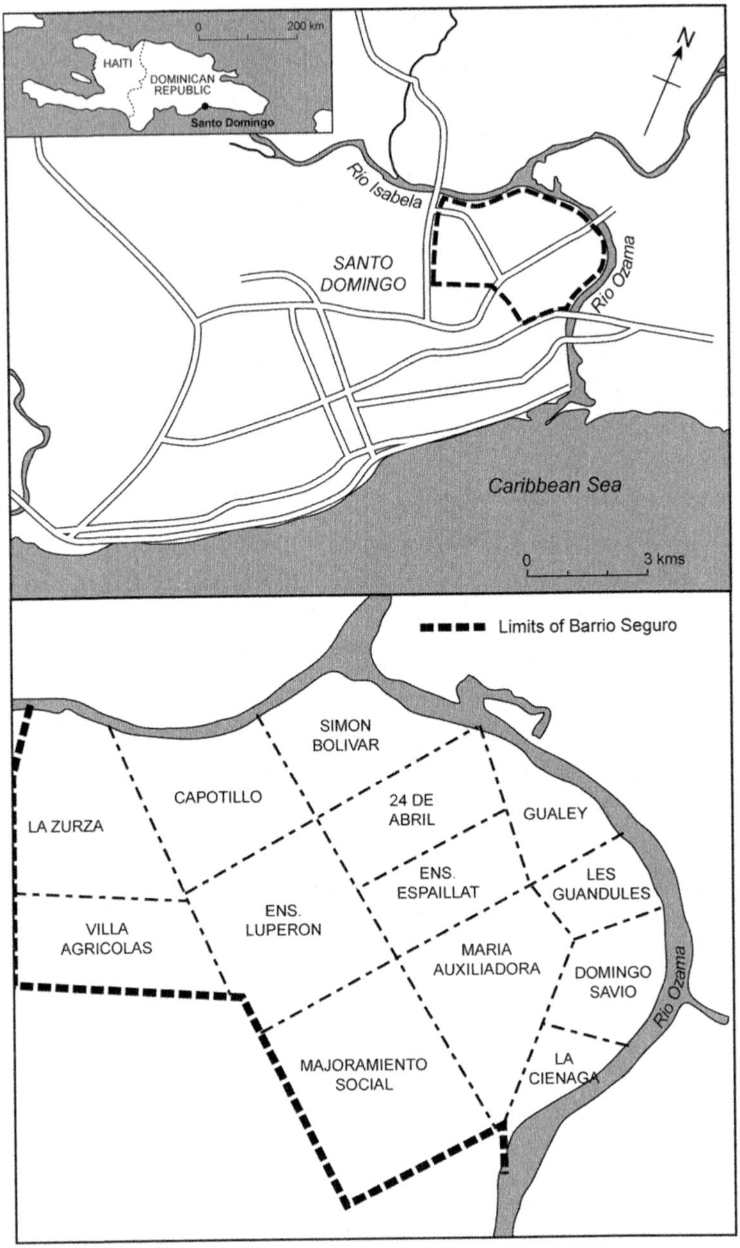

Figure 14.1 The thirteen neighbourhoods that initially constituted the *Barrio Seguro* project in Santo Domingo.

Source: Mapas Gaar 2005.

and the neighbourhood, surprisingly, none of the interviewees chose to remain anonymous. The names of all interviewees, however, have been omitted from this chapter due to the nature of the statements given and the ease by which individuals might be identified within the neighbourhood, police force or government offices. Among the residents interviewed in Villas Agrícolas, seven were re-interviewed at the end of the research period to reflect on the changes that had occurred. In addition, twelve residents in the more affluent central urban neighbourhood of Gazcue, located outside the targeted zone of the project, were interviewed to gain their reflections on the new policing measures.

Villas Agrícolas is a mixed residential and industrial neighbourhood that consists of approximately fifteen thousand residents, and it corresponds to the national urban average, whereby one third of households are female-headed (Oficina Nacional de Estadística 2004). It should be pointed out, however, that the *Barrio Seguro* icon focuses on a nuclear "family" (figure 14.2 [from Procuraduría General de la

Figure 14.2 Logo for *Barrio Seguro*.
Source: Procuraduría General de La República 2005, 66.

Figure 14.3 *Barrio Seguro*: "A triumph for all".
Source: *Hoy*, 16 December 2005.

República 2005, 66]). This nuclear household structure is a relatively rare occurrence for many Dominican households, particularly lower income, in which extended networks are the norm. Among the households interviewed, none reflected this image of two adults/parents and two children under one roof. The use of such an icon exemplifies the branding and myth behind the campaign – the attempt to regain the sense of a traditional, secure past of a solid Dominican nation and safe homestead.

At this point it is important to note also that despite the image of the sheltered nuclear household, the *Barrio Seguro* project makes no specific reference to confronting domestic violence or abuse. By and large, government policy has made a sharp separation between public and private spaces. Challenging domestic abuse has been left to the realm of non-government agencies, such as Profamilia. Indeed, the first government-initiated conference on household violence and HIV/AIDS only took place in June 2006.

The wider Plan for Democratic Security is meant to address a broad range of issues: health, welfare, sport and leisure. A basketball court was renovated in Villas Agrícolas, but the overriding media and practical message is of an anti-crime, intensive policing agenda. The aesthetics of the campaign have been crucial to deliver this theme (figure 14.3).

Policing Measures and Resident Responses in "Safe Neighbourhoods"

Arendt's (1969) argument that the loss of power is often the motive to substitute violence for power is exemplified by the "zero tolerance" policy which is current in Santo Domingo since early 2006, and has since been extended to the second city of Santiago de los Caballeros. Clearly, a contradiction exists in the use of aggression or even violence to quell violence; here it corresponds to a popular idea that the government has "lost control" of parts of the city, namely the poor or "marginal" *barrios* that border the river. Once borders or boundaries are located in terms of physical or emotional realms, then these become markers of identity, and thus entities to be "defended" in certain circumstances and by certain actors (Arendt 1969). The Dominican government's policing policy covered by the *Barrio Seguro* project is all about the defence of these boundaries. The zoning of this area has created the territorial basis for the policing of violence and crime. The practical components of the initiative are fourfold.

First, the emphasis on highlighting transnational, specifically US-based, training links, is an attempt to enhance the professional image of the police force. A substantial proportion of the police officers involved have received, or are due to receive (fifteen hundred being the estimated final number) specific training in the United States, at police facilities in New York and Florida. Residents quite rightly have challenged the feasibility and ethical basis for expecting (often widely accepted) corrupt police officers, who were directly implicated in much of the criminal activity, to gain the respect of the local population. The implied acceptance of "poacher turned gamekeeper" is a critical weakness of the campaign.

Second, also aiming to present a clean, new image of the policing regime, was the purchase of a new look. The newly trained teams have been equipped with new jeeps, Harley Davidson motorcycles and uniforms to mark them out as specific members of the *Barrio Seguro* task force, although they remain members of the National Police. Much was made of the fact that new jeeps had lap-top computers with GPS navigation systems – which does not tally well with the rhetorical notion of community policing, emphasized by the president, where upstanding residents and police share neighbourhood knowledge. One police officer emphasized the new, hi-tech edge of the police force, but admitted that no relevant database exists which could be accessed by remote lap-tops.

A neighbourhood that forms part of the project is placed under intense supervision by the police, and more recently the army, although the intensity of this surveillance has varied between areas. In July 2006, between 15,000 police and military (half the combined human resources of the police and military) were mustered at once to patrol the streets as part of the project – this was the largest military mobilization since the "April Revolution" of 1965. Three thousand further police officers were to be recruited during the ensuing six months.

As part of the same measures to control social behaviour, all alcohol sales were prohibited after midnight from Sunday to Thursday, with extensions to 2 a.m. permitted on Friday and Saturday. The Dominican government has recently invested $25 million in tourism promotion – a nationwide curfew would seem to counteract this publicity drive. All bars and restaurants have been affected, including tourist resorts. The National Hotel and Restaurant responded by saying: "It is not fair to blame these establishments for the upswing in crime, or paint hundreds of thousands of people as potential criminals or bums if they go to restaurants at night for business, a social event or just to have fun" (Inter Press Service News Agency, 3 September 2006).

Third, intensive, twenty-four-hour surveillance by police was highlighted as a key measure to eradicate criminal activity. The follow-up phase of the project relies on residents to volunteer as lookouts. The incentive to volunteer, beyond that of reducing crime and strengthening the idea of a moral community, was the provision of a mobile telephone to assist rapid contact with the police. A further issue, not fully addressed here, is the party political aspect of the campaign. While the majority of residents interviewed were sceptical about the campaign; supporters of the government were more likely to praise the aims of the project, and more likely to receive roles and largesse. The *junta de vecinos* ("neighbourhood committee") in Villas Agrícolas seemed primarily a political, rather than communal, organization.

The fourth aspect of the programme has been the intensified use of random arrest on "suspicion" of criminal behaviour. The failure to present identity papers on demand is one of the most common grounds for arrest and falls most heavily on the poorest members of a neighbourhood who may not have the necessary resources to obtain a full set of identity documents, for example, a birth certificate and residential details. In particular, given the large number of legal but undocumented residents, or workers of Haitian origin, in Dominican cities, the patterns of arrest are likely to take on a racialized context. The enforced round-ups of Haitians "without papers" has been a regular aspect of Dominican police and military intervention in Dominican cities during the last two decades, as the formerly rural migrant labour force has extensified to urban construction and service sectors (Báez Evertsz 1999).

Direct connections were made by one quarter of interviewees (nine out of thirty-seven residents interviewed by the author) between the increasing experiences of crime and the Haitian urban presence. One respondent argued that theft and robbery tended to occur largely in areas where "Haitians" (which might include anyone perceived to have darker skin colour) were involved in new building construction. Many of the new phase of arrests, which are group-focused round-ups initiated as part of the *Barrio Seguro* project, may also merge into *de facto* responses to the popular and government-cited concern about Haitian immigration to the Dominican Republic, particularly to urban areas.

Capotillo, seen by the government and National Police as a flagship neighbourhood for the project, has received the most attention and publicized police activity. General crime statistics exist for the country, although the full results of recent police intervention, as part of the new programme, have yet to be released. The minister coordinating the campaign made the bold exclamation that, after five months, crime had been reduced to zero in the capital's thirteen neighbourhoods. This fantasy was largely treated as such.

Not long after this statement, the president had visited one of these "saved" and now "safe" neighbourhoods via helicopter, heavily guarded by swarms of military and police. Police presence had been further intensified during the preceding three days in this neighbourhood of Gualey before the president's arrival. Such a militarized airlift into an urban community, which lies 5 kilometres from the presidential palace, to proclaim the success of an intensely aggressive campaign, ultimately mocked its own veracity. Clearly a mixed message exists: militarization of urban space brings safety for the common good, but this alleged communal security rests on the presence of aggressive policing and the manipulation of fear. The use of extreme surveillance for the presidential visit evidently showed the state's might and means, but it of course raises contradictory questions over the need for such protective measures.

Concluding Comments on New Policing Measures

At this stage, the results of the *Barrio Seguro* programme are not clear. Informal evidence from interviews suggests that crime has decreased in the areas that are being intensively policed, but that criminal activities are merely displaced to "quieter areas" (author's interview with *Barrio Seguro* police officer, 12 May 2006; author's interviews with residents in Villas Agrícolas, February–May 2006). The responses from residents were largely ambivalent in many cases, and showed ongoing mistrust of the police force and the government's political motives. As one resident said: "*Barrio Seguro* arrived two weeks ago, stayed two days, and finished."

Alternatively, one resident reflected a common distrust of government intervention and highlighted the open acknowledgement on all sides of police and military involvement in criminal activities: "Robberies are by men in military clothes, driving military cars with military guns . . . what do you make of that? Who are the delinquents?"

The criminals continue to be largely perceived by interviewees in the current study to be young, male and occasionally either of Haitian-origin or Dominican deportees from the United States.

Pedro de Jesús Candelier, the chief of police between 1999 and 2002, and an unsuccessful presidential candidate during 2008, has summed up the security policy doctrine now enforced through *Barrio Seguro*: "Let's confront the situation with the means that we have, and in the criminal's own neighbourhood, but note that nothing will stop us" (cited in Bobea 2004, 175).

A similar hard-line approach was evident among current policy makers in the office of the Plan for Democratic Security. One interviewee notably made explicit links between cleansing a "dangerous" urban environment and thus "sanitizing" residential space: "We are talking of *el mundo salvaje* here [in the thirteen neighbourhoods] . . . A lumpen mass of problems: violence, extreme violence, delinquency" (interview with author, 5 May 2006).

Perhaps the most significant impact of *Barrio Seguro* has been to provide a clear basis for the mobilization of a militarized security forces and an enhanced spatial focus on the fear of crime. Interviews with residents outside the thirteen neighbourhoods under the project tended to disassociate themselves from people living in "those" *barrios*, *barrios* that are now segregated politically and practically from the rest of the city. The branding of *Barrio Seguro* thus reflects the importance of territory in defining forms of policing and governance. *La zone norte* of the city is more commonly referred to as a dangerous, ghettoized space since the introduction of the project. Compared with interviews undertaken by the author across the city towards the end of the 1990s, this enhanced social distancing of low-income riverside neighbourhoods indicates an explicit "marginalization" of people and place in the context of economic and material conditions, but also markedly in emotional terms.

The government has set out to "cleanse" thirteen northern neighbourhoods in the capital; the residents for their part (speaking from the evidence in one neighbourhood) have further removed themselves from the police and this form of urban governance and policing. Thus, a reversal of community policing and involvement, opposite to that intended, has occurred, despite being a key, rhetorical component of the "safe neighbourhood" policy (CONARE 2005a, 2005b). Police surveillance has served to heighten tensions in several instances, since all residents have become "watched". The wider mission of defending state sovereignty from external "influences", as outlined by the president, has come at the cost of the freedom of the individual, to some extent. Interviewees in Villas Agrícolas suggested that, while the streets were empty during the daily evening curfew, they noticed more that their personal mobility seemed to suffer greater restrictions rather than their neighbourhood becoming "safer". To venture outside without personal identity papers further risked arrest.

The discursive emphasis placed on the branding of the *Barrio Seguro* project as a fight for democracy and national security underpins Rancière's (1999) and Dikeç's (2002, 2005) focus on first, the spatial context of the political (*le politique*); and second, if we look at the emphasis on visual aesthetics and media involvement, on the spectacle of the political. Dikeç (2005) has argued that a place becomes "political" when it becomes a site through which wrongs and inequalities are being addressed. The *Barrio Seguro* project aims to use a spatial framework to coordinate an anti-crime and social justice initiative, but initial results suggest that the outcome may be producing new restrictions to everyday routines without addressing directly the personal injuries of largely individually focused and practiced criminal activity or violent behaviour. The project itself continues to be promoted intensively across a range of media, and it remained as the main "flagship" policy for President Fernández's successful re-election campaign in May 2008. The practice and spectacle of the *Barrio Seguro* project seems thus to have been supported by the majority of Dominican voters. The minority of the electorate living under the new regime of intensive policing in the selected "safe neighbourhoods" appear, however, to share mixed opinions about the impact of a national security agenda on their daily lives.

References

Arendt, H. 1969. *On violence*. San Diego, CA: Harvest/HBJ.

Báez Evertsz, F. 1999. Migrantes haitianos en un barrio de Santo Domingo. *Estudios Sociales* 22, no. 118: 45–83.

Bobea, L. 2004. Economía política de la inseguridad y deasfíos a las políticas de seguridad ciudadana en la República Dominciana: Cero tolerancia bajo la mirilla. In *Entre el crimen y el castigo: Seguridad ciudana y control democrática en América Latina y el Caribe*, ed. L. Bobea, 175–227. Santo Domingo: FLACSO.

CARICOM. 2008. Statement issued by the Conference of Heads of Government of the Caribbean Community at its Thirteenth Special Meeting, 4–5 April, Trinidad and Tobago. http://www.caricom.org/jsp/communications/meetings_statements/13sphgc_statement.jsp.

Consejo Nacional de Reforma del Estado (CONARE). 2005a. *¿Qué es la participación?* Santo Domingo: Consejo Nacional de Reforma del Estado.

———. 2005b. *¿Qué es el presupuesto participativo? Para el desarrollo del municipio*. Santo Domingo: Consejo Nacional de Reforma del Estado.

Cutter, S.L., B.J. Boruff and W.L. Shirley. 2003. Social vulnerability to environmental hazards. *Social Science Quarterly* 84 (2): 242–61.

Dikeç, M. 2002. Police, politics and rights to the city. *Geojournal* 58: 91–98.

———. 2005. Space, politics and the political. *Environment and Planning D: Society and Space* 23: 171–88.

———. 2006. Badlands of the republic? Revolts, the French state and the question of *banlieues*. *Environment and Planning D: Society and Space* 24: 159–63.

Graham, S., and S. Marvin. 2001. *Splintering urbanism: Networked infrastructures. Technological mobilities and the urban condition*. London: Routledge.

Groupe d'Appui aux Rapatriés et Réfugié. *2005*. R. Dominicaine: Neuf Haitiens tués en 8 jours. http://www.garr-haiti.org/actu/index.html.

Herbert, S. 1997. *Policing space: Territoriality and the Los Angeles Police Department*. Minneapolis: University of Minnesota Press.

Hewitt, K., ed. 1983. *Interpretations of calamity*. London: Allen and Unwin.

Hobbes, T. 2005. *Of man*. London: Penguin.

Oficina Nacional de Estadística. 2004. *VIII Censo nacional: Población y vivienda*. Santo Domingo: Oficina Nacional de Estadística.

———. 2005. *Encuesta nacional de hogares de propósitos múltiples*. Santo Domingo: Oficina Nacional de Estadística.

———. 2008. *Encuesta nacional de hogares de propósitos múltiples*. Santo Domingo: Oficina Nacional de Estadística.

Pérez, C., and G. Milán. 2005. Los niveles de la percepción de la criminalidad en la República Dominciana: Caso de las provincias de la Romana, Santo Domingo y el Distrito Nacional. *Ciencia y Sociedad* 3 (3): 510–39.

Procuraduría General de La República, Secretaría del Estado de Interior Policia y Policía Nacional. 2005. *Plan de seguridad democrática República Dominicana*. Santo Domingo: Secretaría del Estado de Interior y Policia.

Rancière, J. 1999. *Dis-agreement: Politics and philosophy*. Minneapolis: University of Minnesota Press.

Schneider, J., and I. Susser, eds. 2003. *Wounded cities: Destruction and reconstruction in a globalized world*. Oxford: Berg.

Thrift, N. 2005. But malice aforethought: Cities and the natural history of hatred. *Transactions of the Institute of British Geographers* 30 (2): 133–50.

United Nations Development Programme (UNDP). 2007. *Human Development Report 2006: Beyond scarcity – Power, poverty and the global water crisis*. New York: UNDP. http//hdr.undp.org/hdr2006/.

United Nations Office on Drugs and Crime (UNODC) and the World Bank. 2007. *Crime, violence, and development: Trends, costs, and policy options in the Caribbean*. Report no. 37820. Vienna, Austria: UNODC.

United States Department of State. 2008. *Country reports on human rights practices 2007*. Washington, DC: United States Government. http//www.state.gov/g/drl/rls/hrrpt/2007/100637.htm.

Wisner B., P. Blaikie, T. Cannon and I. Davis. 2004. *At risk: Natural hazards, people's vulnerability and disasters*. London: Routledge.

CHAPTER 15

Conflicting Environments

Negotiating Social and Ecological Vulnerabilities in Urban Jamaica and Curaçao

RIVKE JAFFE

Introduction

Caribbean islands, including Jamaica and Curaçao, form one of the "biodiversity hotspots" selected by transnational environmental organizations such as Conservation International. Maintaining global biodiversity hinges on the protection of the exceptionally diverse ecosystems, and dozens of endangered and often endemic species the region harbours. Apart from the intrinsic value of Caribbean ecologies, many of the region's tourism-based economies are strongly dependent on (the idea of) unspoiled natural landscapes and an image of the region as paradise. The coastal zones which encompass much of Caribbean nature are especially vulnerable to both natural and human-made hazards. In Jamaica and Curaçao, local environmental non-governmental organizations, and to a lesser extent governments, connect with global environmental discourse and lobby for the protection of ecosystems and biodiversity, with an emphasis on the marine environment.

In a number of ways, the discourse employed by these environmental professionals is a continuation of colonial ideas of an Edenic Caribbean. To many citizens of these urbanized islands, who live in polluted city neighbourhoods, this form of environmentalism has limited appeal. This chapter examines the clash between a discourse of ecological vulnerability conveyed by non-governmental and governmental stakeholders, and the contradictory experience and communication of social vulnerability by residents in blighted urban areas. While supralocal stakeholders may tend to portray the natural environment as threatened by the uneducated masses, stakeholders at the community level are preoccupied by a different range of environmental problems which they associate with their socially vulnerable positions.

This chapter analyses how constructions of Caribbean environments are associated with class, ethnicity and scalar orientation, while possibilities are explored for reconciling social and ecological vulnerabilities. It is based on twelve months of fieldwork in Kingston, Jamaica and Willemstad, Curaçao with residents of low-income areas and with policy makers, NGOs and academics (see also Jaffe 2006).

Caribbean Vulnerabilities

Both Jamaica and Curaçao are considered to be small island developing states (SIDS). SIDS share certain economic, social and ecological traits that are all related to vulnerability, a key term in speaking of such territories. Vulnerability is defined as "exposure to risk and an inability to avoid or absorb potential harm", and its twin concept is found in resilience, "the capacity to adjust to threats and mitigate or avoid harm" (Pelling 2003, 5). Vulnerability and resilience are seen as the result of differential access to economic, political, social, environmental and geographical assets, while the distribution of these assets is determined by both human and physical forces (Pelling and Uitto 2001, 51).

The ecological vulnerability of SIDS relates in part to their high exposure to natural hazards. The active plate boundary nature of the Caribbean's geological setting leads to earthquake and volcanic hazard, as witnessed, for example, by the 1995–1997 eruptions in Montserrat which obliterated two-thirds of the island. Other common hazards throughout the centuries have been hurricanes, flooding, droughts and mudslides. Recent examples are heavy floods in Suriname in 2006, Guyana in 2005 (Williams and Johnson-Bhola, chapter 4) and Hurricane Ivan, which devastated Grenada and parts of Jamaica, in September 2004. Hurricanes appear to be increasing in frequency and intensity in recent decades, presumably as the result of global climate change (for discussion, see Gamble, chapter 2). Likewise the rising of sea level connected to global warming will have devastating effects on low-lying islands like those in the Caribbean, and tsunamis, such as the one that devastated large parts of Asia in December 2004, are not unthinkable (Zahibo and Pelinovsky 2001).

One of the properties of SIDS is that a relatively high proportion of their land consists of coastal zone, recognized as especially vulnerable to a range of natural hazards. This is especially urgent in the Caribbean, as the majority of human life and property is concentrated along the coastline. Mangrove forests, dunes and coral reefs shield shorelines and beaches, but, as coastal development threatens these ecosystems, the coast itself becomes subject to erosion. Conservation of the natural environment is essential in protecting cities and villages from environmental hazards including flooding and storms. This relation is all the more critical when population pressure on resources, lack of regulatory authority and poverty lead to inadequately planned construction in inappropriate areas, illustrated, for instance, by extensive damage to improperly or illegally constructed dwellings in the Jamaican parish of Clarendon following Hurricane Ivan in September 2004. Urban areas are especially vulnerable to this combination of anthropogenic and natural hazards, as poorly regulated urban expansion equates to social and spatial concentrations of risk. Informal settlements that have been developed without land tenure, basic services and infrastructure aggravate this vulnerability.

It is in this area that ecological vulnerability merges with social vulnerability. Vulnerability to all types of environmental hazards is connected to poverty, social isolation and political marginalization. Environmental risk tends to be increased for groups and individuals with limited financial, physical, social and political assets (Pelling 2003, 67). The urban poor, stakeholders with limited access to various assets, are most likely to both suffer the ill effects of environmental problems and be blamed for their prevalence. Poverty and urban environmental issues are related in various, not necessarily straightforward, ways. While urban environmental degradation and hazards do contribute to poverty, activities and lifestyles of the urban poor do not directly cause environmental problems. Rather, macroeconomic conditions impede environmental management, as municipal and national governments suffer a shortage of resources. Yet inadequate and inefficient governance is an important mediator in this regard. This is apparent in the environmental degradation associated with rapid, unplanned urban expansion, and the environmental injustice produced by environmental policies which disregard equity and inequity issues.

As Satterthwaite (2003, 76) argues, "the environmental problems that low-income groups face are often more related to inadequate provision of infrastructure and services, lack of any rule of law, discrimination, and lack of political influence than to a lack of income". He reasons that many studies have neglected the contribution of middle- and upper-class urban residents to environmental degradation, resulting in inappropriate policies that tend to blame the poor, such as the clearing of slum settlements or the obstruction of informal livelihood strategies. The removal of informal sidewalk vendors from downtown Kingston or the eviction of squatters could be seen as Jamaican examples.

In linking vulnerability and Caribbean environments, the distinction between ecological and social vulnerability is in large part the difference between "green agenda" and "brown agenda" approaches to environmental problems. There is no rigid dividing line between the two approaches, but they can be differentiated along spatial, temporal and political dimensions. The green agenda refers to dispersed and delayed problems that affect future generations. The issues prioritized by the green agenda impact mainly on ecosystem health, and the scale at which they operate is global or regional. Characteristic attitudes include the following: nature should be protected and worked with, people should be educated, and environmental services should be used less. The brown agenda, in contrast, addresses local, more immediate, problems that predominantly affect the poor. Brown agenda issues relate mainly to impacts on human health. Nature is seen as a domain that should be manipulated to serve human needs, people are to be worked with and more environmental services should be provided (McGranahan and Satterthwaite 2002).

Another emerging idea is the "blue agenda", which refers to problems associated with water supplies, watershed management and coastal and marine resource management (IIED 2001). In the context of the Caribbean, this blue agenda, to a large extent, overlaps with the green agenda. While the green and brown agendas appear to be conflicting, they can also be framed as complementary, "meeting the needs of the poor is not a major threat to sustainability except when it allows environmental abuse by all sectors of society. Similarly, pursuing environmental sustainability is not a major threat to the environmental health of the poor except when it is used to justify maintaining the most deprived residents' already inadequate access to environmental resources" (IIED 2001, 3).

Ecological Vulnerabilties: From Colonial to Elite Environmentalism

Colonial Proto-environmentalism

Colonial reports on the Caribbean reveal the wonder Europeans experienced on being confronted with the abundant and supposedly pristine natural landscapes they came across. In 1494, when Columbus and his crew first encountered *Xaymaca*, as Jamaica was called by its original Taino inhabitants, they expressed their delight at its natural beauty: "there silhouetted against the evening sky, arose sheer and darkly green Xaymaca. It is the fairest island that eyes have beheld: mountainous and the land seems to touch the sky; very large, bigger then Sicily, and full of valleys and fields and plains" (Floyd 1981, 25).

The roots of the environmental movement can be traced to such colonial encounters with what they saw as a tropical paradise. Various authors (Grove 1995; Barton 2002) point to European expansion as the context in which environmental awareness first materialized and the role of islands within these developments. Environmental degradation resulting from colonial practices, combined with the image of the colonies as so-called tropical Edens, gave rise to the emergence of European proto-environmentalists from the seventeenth century onwards. Insular ecosystems display environmental degradation rapidly and natural resource depletion, deforestation, soil erosion and localized climate change were particularly evident in the Caribbean island colonies. Watts (1987) traces the earliest evidence of environmental degradation to the introduction of plantation agriculture in the early sixteenth century. Through the sixteenth century and into the seventeenth century, deforestation for plantation agriculture progressed until, as Watts (1987) notes, very little natural forest remained in islands such as Barbados by the latter half of the seventeenth century. By the mid eighteenth century, soil exhaustion on Barbadian sugar plantations was of such grave concern that efforts were made to import more fertile soil from the Guyanas (DeLoughrey 2004, 299). The visibility of such degradation contributed to the sensitization to, and understanding of, human-environment interactions. A basic sort of environmental awareness developed following interactions with natural environments as the demise of natural landscapes provoked the image of a "paradise lost". Explorers, naturalists and scientists, including the staff of botanical gardens, recognized the detrimental effects of certain policies and practices, and took colonial governments to task, with varying levels of success.

Apart from direct confrontation with human-induced environmental change, the environmentalism of these first "activists" was also influenced by philosophical currents such as Romanticism (Tomalin 2004). The notion of discovering the Garden of Eden and the fear of losing it again is a recurrent theme in fuelling conservation, and it extends from colonial times to the present (Grove 1995). Such a narrative, however, focuses on the relationship between European – or later Euro-American – outsiders and Caribbean nature, to the exclusion of both population groups and Caribbean urbanization. In colonial accounts, the "natives", who are mainly African slaves and their descendants, are generally depicted as childlike creatures, incapable of managing their natural surroundings sensibly. Connections between ecological and social vulnerabilities are seldom made. In a similar manner, urban areas, present from the early days of colonialism, are elided from celebratory descriptions of lush, tropical settings and supposedly pristine islands.

Though the concerns expressed by these proto-environmentalists were predominantly related to "green" problems of resource depletion, Sachs (2003) argues that colonial figures such as Alexander von Humboldt did link the exploitation of natural resources to the exploitation of specific social groups, resembling the connections made by the environmental justice movement centuries later.

Current Professional Environmentalism

Certain parallels can be observed between the colonial proto-environmentalists and present-day environmental professionals in Jamaica and Curaçao. Like their predecessors, current environmentalists tend to focus predominantly on ecological vulnerabilities, to the exclusion of social vulnerabilities. Most efforts at the level of environmental non-governmental organizations (ENGOs) and governments go towards "green" issues, including biodiversity, marine environment and coral reefs, and nature conservation within a framework of sustainable tourism. Supralocal environmentalism tends to neglect or ignore the urban, which figures only as the cause of environmental degradation. Similar to the situation under colonialism, a majority of the population is often excluded. The "green" environmental discourse encountered at the professional level is an elite phenomenon, interwoven with specific socio-economic and ethnic constellations, and it is indicative of globally hegemonic forms of environmentalism.

A focus on ecological vulnerabilities is evident in the supralocal environmental discourse in Jamaica and Curaçao, which is overwhelmingly "green". This disposition towards this type of environmental problems is clear in governmental policy documents and ENGO campaigns and websites. For relevant policy documents see Vomil/Mina 1998, 1999, 2000, 2001, 2004; NRCA 1998; NEPA 1999, 2002.

The island governments do attempt to address "brown" issues, but financial interests in utilities, industry and environmental services complicate decisive action. Most ENGOs tend to focus on less "dirty", more attractive, issues. Prominent Curaçaoan environmental organizations include Reefcare, Uniek Curaçao and Carmabi. The first, which was founded in 1992 by a group of dive operators, marine biologists and educators, is dedicated to the "worldwide protection and preservation of the Coral Reef" (see www.uniek-curacao.an). The organization's website states Reefcare's motivation as the fact that the reef is a unique and beautiful natural phenomenon, which generates income for Curaçao through dive tourism. Their activities include research and monitoring of the reef, and environmental education and awareness raising projects, including snorkelling lessons for "underprivileged children".

Uniek Curaçao is a foundation with a "creative, educational and recreational character" that organizes nature tours and is involved in both environmental education and the raising of awareness. Its goal was stated, in an interview with one of its (Dutch) interns, as "getting the local population to realize that nature should be respected and not regarded as a trash can". Carmabi, the Caribbean Research and Management of Biodiversity Foundation, started as a marine biology institute but merged with another organization into a foundation responsible for managing Curaçao's protected areas, acting as a consultant for the government and establishing environmental education programmes. Their initial focus on the marine environment has broadened to nature in general.

In Jamaica, leading ENGOs include the Jamaica Conservation and Development Trust (JCDT), the Jamaica Environment Trust (JET) and the Environmental Foundation of Jamaica (EFJ). JCDT was established in 1988 and has been charged with co-managing Jamaica's Blue and John Crow Mountains National Park, with two governmental agencies. It is also involved in environmental education and communication, as well as advocacy related to protected areas management (see www.greenjamaica.org.jm). Attempts at working with local communities and building the capacity of community-based organizations bordering the park has been, by their own admission, challenging (Otuokon 2007).

JET started off in 1991 as a group concerned about the fact that Jamaica's unique natural environment was being destroyed. Funded through international NGOs and development agencies, the organization specializes in environmental education, especially through school programmes, and advocacy. Its director has been a prominent media presence. The EFJ is a foundation that promotes and sponsors sustainable development activities through the distribution of funds from a debt-for-nature-swap with the US. Some of the largest project grants disbursed have gone to conservation-oriented organizations such as JCDT, JET, Friends of the Sea, Dolphin Head Trust and Negril Coral Reef Preservation Society (EFJ 2005).

The prioritization of conservation by Jamaican and Curaçaoan organizations appears to be typical of the Caribbean (cf. Jácome 2006). At this level of environmental professionals, a stress is placed on diminishing biodiversity, deforestation and degradation of the marine environment. In part, this focus on ecological vulnerability reflects the importance of tourism to Caribbean governments and business elites. Perceived as the motor for development in most islands, Caribbean tourism is connected to longing for tropical abundance and supposedly "pure", unspoiled nature. Such a tourist vision generally ignores or erases the existence of "impure" cities in the region. Additionally, global and local factors explain the near-exclusive orientation towards green agenda problems: global, in that this environmental agenda is pushed by international stakeholders, ideas and funding; and local, as green concern – apart from reflecting economic interests in the tourism sector – becomes a symbolic marker of distinction within society.

Globally, environmental thought and discourse is disseminated in particular ways through international, but Western-dominated, bilateral and multilateral fora. Green discourse and practice is diffused and imposed through donor countries, international financial institutions and international ENGOs. As Hartwick and Peet (2003, 289) argue, through an increase in global governance mechanisms, "environmental concern has been ideologically and institutionally incorporated into the global neoliberal hegemony". In the discourse that accompanies this sphere, "global environmental problems are presented as being *a priori* of a different order, and thus marginalize many other environmental concerns that might affect many people or eco-systems much more directly" (Hajer 1995, 11). Perhaps unwittingly, Caribbean governments and NGOs – under external financial and ideological pressure – reinforce the hegemony of this discourse, one which prioritizes ecological vulnerability and environmental problems with a global dimension. Environmental organizations in Curaçao are, to a large extent, dependent on the Dutch government for funding, while the Jamaican situation finds that the best-funded organizations are those with access to international funding, whether bilateral, multilateral or non-governmental. The implications are similar on both islands: "the interests of these international organizations determine which of the NGO activities are funded, and by extension which NGOs survive at an operational level" (Witter 2002, 24).

The globally oriented green discourse that dominates Jamaican and Curaçaoan environmentalism remains at the level of elites and segments of the middle class. Bilateral and multilateral aid agencies increasingly direct funding from governmental to non-governmental organizations, based in part on the assumption that NGOs work more effectively through participation and are more representative of the communities the funds are supposed to assist. Lundy (1999) argues that, in the case of Jamaica, these are mistaken assumptions, as ENGO membership is overwhelmingly well educated and middle class, and ENGO activities reflect the concerns and priorities of local elites, thus inadvertently reinforcing inequality in social relations. A similar situation is found in Curaçao. Specific environmental prioritizations are associated with the socio-economic and ethnic backgrounds of stakeholders. At the ENGO level, differences between their members and the majority of the urban population are evident, most apparently in class, racial and ethnic identities. In Jamaica, many environmental organizations are headed by either foreigners, often expatriate Americans, or by upper-middle-class, often light-skinned, Jamaicans (cf. Carrier 2003). In Curaçao, the number of Afro-Curaçaoans in the environmental movement is fairly limited, certainly as the heads of organizations; white Curaçaoans and European Dutch émigrés tend to dominate as members and leaders. On both islands, the movement is not a grassroots phenomenon; it is plausible that displaying an interest in and commitment to conservation and other green issues has become a local form of elite and middle-class distinction. ENGO membership, nature hikes and support of national parks can all be means of acquiring symbolic capital (cf. Bourdieu 1984).

Despite the fact that ENGO leaders and members are usually extremely well-meaning, hard-working, idealistic individuals, it is not hard to discern something of a condescending attitude towards "the community". The president of one ENGO describes how "many in Jamaica do not have an appreciation of nature – they do not comprehend the relevance of lizards, insects and plants (except in the light of economic gain) and *have to be assisted to 'see'* the wealth and beauty of our island" (Levy 1996, 25, emphasis added). In Curaçao, one of the more successful ENGOs, led by a Dutchman, has as its slogan *konosé bo isla*, "know your island". Another NGO leader, also Dutch, stated, when interviewed, that environmental problems on Curaçao were a cultural problem, because the people did not learn to love themselves or their environment. The portrayal of members of the local population as insufficiently knowledgeable – or appreciative – of their own island reflects a similar sentiment to that expressed by the Jamaican ENGO leader quoted.

A general assumption made by decision makers is that environmental problems result largely from a lack of environmental awareness on the part of the population. The majority of Jamaicans and Curaçaoans are depicted as unaware or problematic. These ideas are evident in the number of governmental and non-governmental campaigns and programmes aimed at raising awareness and environmental education. The "local" population is often seen as incapable of sustainable resource management. As a result, there has been a strong propensity towards top-down conservationist strategies in which nature is, or becomes, "unpeopled".

People, and especially urban people, are seen primarily as the cause of environmental problems, rather than the victims. However, while both governmental and non-governmental environmental organizations are largely based in urban areas, their green focus incorporates an image of Caribbean islands which does not depict urban areas or their problems. This environmentalist vision merges with the tourism industry's representations of paradise-like, non-urban landscapes. These "natural" landscapes are implied to be representative of the island, without due note of the fact that many of these are endangered. Cities and their residents are either ignored or depicted as aberrations, disturbing the natural order.

Social Vulnerabilities: Blighted Cityscapes

There is a persistent image of the Caribbean region as a range of lush, verdant natural landscapes that must be conserved, both as intrinsically valuable, high-biodiversity ecosystems and as crucial resources for tourism (Sheller 2003). In contrast with this "natural" representation, Caribbean islands are largely urbanized. In Curaçao and Jamaica, 90 and 50 per cent of the population, respectively, lives in urban areas, and most of those urban citizens reside in the capitals of Willemstad and Kingston. At odds with the tourist image of the islands, there is considerable poverty, and a significant proportion of the population lives in poor-quality housing in low-income urban areas (Clarke, chapter 12; Weeber 2004), rather than in a picturesque shack on a beach. These urban residents do not consider their surroundings Edenic. In Willemstad, the centrally located oil refinery – previously owned by Shell, now leased by the Venezuelan oil company PDVSA – is responsible for noxious emissions in the downwind "marginal" neighbourhoods or *barios*. In addition, the consequences of inadequate solid waste management are visible in the piles of litter that are found throughout the entire island, from construction materials and garden waste to plastic bags, beer cans and dead dogs. In Kingston, residents live in areas adjacent to the garbage dump, heavy industry or the polluted harbour. Open sewage and irregular garbage collection contribute to polluted surroundings that are hazardous to health.

These types of brown agenda problems receive highest priority among urban residents, in stark contrast with the focus of ENGOs and the government. In the perception of residents of low-income areas in Willemstad and Kingston, the islands are characterized not by threatened natural landscapes and ecological vulnerability, but by blighted cityscapes and social vulnerabilities. In these urban settings, brown environmental problems are linked to an inequitable social context. The state of the urban environment as perceived by residents can be conceptualized as "urban blight": the combined and interrelated effects of environmental and infrastructure degradation, poverty, violence and crime, and social disintegration (Jaffe 2006).

Jamaican and Curaçaoan residents do not feel that the environmental problems in their communities are isolated from other problems affecting the broader urban environment. In what are sometimes high-crime settings, violence, for instance, is integral to understandings of environment. A qualitative study conducted among Jamaican lower-income youth, in the context of an environmental education campaign, found that any interpretation of environment includes social aspects; all definitions of "good environment" referred to peace and unity. To the majority, the word "environment" was defined as "what they saw in their surroundings: their community, people and their behaviour (including 'war'), the state of the infrastructure, living conditions, flowering plants, painted and decorated corners, buildings, trees and animals" (Hope Enterprises 1999, 20). Similarly, focus groups in Kingston in a study by Dodman (2004) identified the ten top "environmental problems" in their city as unemployment, water, toilets, housing, garbage, lack of education, pollution, war/violence, electricity and sewage. This study also found a distinct division in environmental priorities between suburban middle-class and inner-city, low-income respondents. Brown environmental problems, including air pollution, inadequate sanitation and solid waste management, are associated with social and economic concerns such as health, poverty, social equity, social disintegration, and violence and crime. These social and economic problems receive priority over environmental problems as presented by supralocal stakeholders: the environmental NGOs and government. For the most part, the professional portrayal of environment and sustainable development remains distant from local, specifically urban or low-income, priorities.

Residents of environmentally degraded and hazardous urban areas connect environmental problems to their status as low-income, politically marginalized communities. Environmental issues are seen in the light of socio-economic inequities at the urban and national levels, which are evident in a lack of political concern for their living conditions and an inability on their part to mobilize effectively against or influence policy that affects them. Low-income urban residents lack economic and political assets to solve or prevent brown environmental problems. Multiple protest marches in low-income Curaçaoan *barios* do nothing to diminish the toxic refinery emissions. Yet when a twist in the island's trade winds causes the fumes to affect wealthier areas, the anger of more powerful citizens makes the papers. Community clean-up initiatives in Jamaican neighbourhoods fail when government trucks neglect to collect the accumulated garbage. High-income neighbourhoods need not depend solely on the government's irregular services and can charter private waste collection companies to keep their streets clean. A reduced exposure to brown environmental problems significantly diminishes the urgency of such issues to elites.

Issues of equity and urban environmental justice demand attention. The spatial distribution of urban environmental "bads" is skewed towards the homes, workplaces and communities of the urban poor who are disproportionately affected by brown problems. Urban blight, or the nexus of violence, poverty and environmental degradation, is, not coincidentally, concentrated in certain areas. Privatization of basic services and security, following neo-liberal restructuring, results in cities where a safe and healthy environment is available at a price that not everyone can pay. The ensuing concentration of urban blight ultimately reinforces existing patterns of socio-spatial fragmentation through fear, repulsion and isolation.

The spatial concentration of poverty and urban pollution can result in what Drackner (2005) refers to as "social contagion": polluted streets or neighbourhoods reflect on their residents who may be seen as "dirty" or "nasty" people on account of their surroundings. In these historically divided Caribbean cities, elites may unconsciously associate this dirty reputation with the ethnicity or skin colour of "downtown" or "marginal" residents. This often entails some causal confusion: to many, poor and polluted neighbourhoods are dirty because of the polluting, "unhygienic" or "asocial" poor people who live there. In such causal constructions, ghetto dwellers are seen as "nasty" – why else are their areas of residence so polluted? The dirt must be caused by the environmentally unaware, unconscious state of being of the poor, hence the professional, elite emphasis on environmental education and raising awareness. This sometimes morphs into discourse and policies that blame the poor. As noted previously, such understandings of poverty-environment interactions neglect the mediating effects of vulnerability, resource access (to assets, labour, credit or markets) and institutions (including governance and tenure systems). As often as not, the "nasty" neighbourhoods are polluted because poor people do not receive environmental services and infrastructure, such as solid waste collection and adequate sewage systems.

The "downtown", "ghetto" or "marginal" landscapes of urban blight are the reality in which many residents of Kingston and Willemstad live. These cityscapes, narrated powerfully in Caribbean popular culture, offer a different view from the landscapes related by past and present environmentalists. Well-known examples of Caribbean popular music and film that place a strong emphasis on urban issues are the decidedly urban genres of dancehall and reggaeton, and movies such as *The Harder They Come* or *Rockers*. This urban emphasis within popular culture entails a, perhaps inadvertent, mission of putting the city on the map, in contrast to tourist or conservationist images that limit the functions of Caribbean islands to tropical décor and biodiversity zones.

Where environmental professionals are influenced by global environmental thought, and focus on problems with a global scale, low-income residents display an environmentalism that is constructed in an explicitly local manner. They connect environmental degradation in their immediate surroundings to other localized urban problems. They draw attention to the socio-spatial patterns of pollution, poverty, violence and exclusion within their city, using explanatory frameworks that incorporate local power relations. In their interweaving of environmental, social, political and economic concerns, this form of environmentalism avoids the people-nature dichotomy implicit in professional discourse, which tends to preclude tackling social and ecological vulnerabilities in tandem.

Conflicting Environments

Within Jamaica and Curaçao, environments and environmental problems are constructed and explained in drastically different and apparently conflicting manners. Professional environmentalism emphasizes and draws on a global scale; its elite stakeholders employ environmental discourse that is non-urban, excludes local participation and emphasizes ecological vulnerability. The "lay" environmentalism encountered in urban ghettos and marginal *barios* is explicitly local in scale, and its urban discourse stresses the centrality of social relations and social vulnerability. Despite widespread global-institutional support for participatory approaches, and the incorporation of "local knowledge" in environmental and natural resource management, in practice, strong hierarchical divisions still exist between different types of environmental knowing. Conflicts of knowledge and power, reinforced by the social and spatial situatedness of the stakeholders involved, have been documented in the cases of protected area management and conservation projects in Trinidad, Grenada, Belize, Guyana and Suriname (Rosenberg and Korsmo 2001; Heemskerk 2002; Brown 2003; Colchester 2005; Sletto 2005). Such conflicts tend to be overlooked in urban areas, as if local knowledge were an exclusively rural phenomenon.

Within a structure of discursively maintained power and an expert-oriented epistemology of development (Ramphall 1997), environmental professionals have the authority to portray residents as uneducated, polluting urban dwellers, who must be led to conscious patterns of thought and behaviour by way of environmental education programmes. This mandate of "educating the masses" is not always far from a subtle form of environmental neo-colonialism, as the colonial situation of outsiders and cosmopolitan elites organizing around the theme of ecological vulnerability is mirrored in the contemporary Caribbean. This association of a category of problems with specific groups of people can result in ecological vulnerability being perceived as an exclusionary sphere, the domain of the light-skinned elite and foreigners.

While not necessarily an intentional result, the emphases and tone employed by Caribbean environmental professionals elide and exclude the importance of local, urban issues and social vulnerabilities. They have also tended to favour conservationist solution strategies that are not based on participation of low-income populations. The professional environmentalists are themselves influenced and sometimes pressured by external bilateral and multilateral stakeholders; they generally operate against many odds with the wholehearted intent of achieving certain ends, such as maintaining biodiversity or promoting sustainable development. However, to attain these goals, the most efficient strategy is to adapt government and non-governmental policies to acknowledge and utilize local priorities and perspectives, so as to gain the support, or at least co-operation, of a majority of the (predominantly urban) population.

As noted earlier, the green and the brown agendas need not be mutually exclusive, nor conflicting. Neither does combating ecological vulnerability necessitate the exclusion or exacerbation of social vulnerabilities. The environmental issues that preoccupy low-income urban residents – sewage, garbage and air pollution – are connected with the professionally prioritized issues of biodiversity and the protection of coastal and marine ecosystems. Marine pollution is intrinsically related to land-based, urban problems such as wastewater and solid waste management; much of the uncollected garbage and unrefined sewage polluting poor neighbourhoods ends up in the sea and washes up on beaches, degrading ecosystems. The urban and industrial pollution that endangers human health, specifically in low-income areas, has similar harmful effects on non-human species. Ecological vulnerability is aggravated when causes of social vulnerability, such as limited access to basic services, infrastructure and tenure, are left unchecked. Conversely, including the socially vulnerable in environmental protection and conservation programmes, both in and outside urban areas, can provide a financial and social boost to disenfranchised citizens.

Given the urgency of both brown and green environmental problems, environmental organizations should avoid environmental policy and awareness-raising campaigns that do not connect to the reality of large segments of the population. Shifting the emphasis towards participatory urban environmental action can enhance ownership of environmental programmes, while furthering environmental improvements through an inclusive, learning-by-doing process. What is needed is an increased recognition of the interconnectedness of first, various environmental problems, and second, the relation between environmental and socio-political issues. Such an understanding could serve to establish environmentalism as a field in which disparate social groups find common ground, rather than having environmentalism function as a divided and divisive space, in which the connectivities between social and environmental issues and inequalities remain obscured.

References

Barton, G.A. 2002. *Empire forestry and the origins of environmentalism.* Cambridge: Cambridge University Press.

Bourdieu, P. 1984. *Distinction: A social critique of the judgement of taste.* London: Routledge.

Brown, D. 2003. A view from the ground: Competing discourses on impoverishment and the coastal environment in Belize. Paper presented at the Society for Caribbean Studies Annual Conference, 7–9 July. Bristol, UK.

Carrier, J.G. 2003. Mind, gaze and engagement: Understanding the environment. *Journal of Material Culture* 8 (1): 5–23.

Colchester, M. 2005. Maps, power and the defense of territory: The upper Mazaruni land claim in Guyana. In *Communities and conservation: Histories and politics of community-based natural resource management*, ed. J.P. Brosius, A.L. Tsing and C. Zerner, 271–304. Walnut Creek: Altamira.

DeLoughrey, E. 2004. Island ecologies and Caribbean literatures. *Tijdschrift voor Economische en Sociale Geografie* 95: 298–310.

Dodman, D.R. 2004. Community perspectives on urban environmental problems in Kingston, Jamaica. *Social and Economic Studies* 53 (3): 31–59.

Drackner, M. 2005. What is waste? To whom? An anthropological perspective on garbage. *Waste Management and Research* 23: 175–81.

Environmental Foundation of Jamaica (EFJ). 2005. Financial statements, 31 July 2005. http://www.efj.org.jm/pdf/efj_2005_audit_fin_stmt.pdf

Floyd, B. 1981. *Jamaica: An island microcosm.* London: Macmillan.

Grove, R.H. 1995. *Green imperialism: Colonial expansion, tropical island Edens and the origins of environmentalism, 1600–1860.* Cambridge: Cambridge University Press.

Hajer, Maarten A. 1995. *The politics of environmental discourse: Ecological modernization and the policy process.* Oxford, UK: Oxford University Press.

Hartwick, E., and R. Peet. 2003. Neoliberalism and nature: The case of the WTO. *Annals of the AAPSS* 590: 188–211.

Heemskerk, M. 2002. Livelihood decision making and environmental degradation: Small-scale gold mining in the Suriname Amazon. *Society and Natural Resources* 15 (4): 327–44.

Hope Enterprises. 1999. *NEEC Campaign Pretest Report: A Qualitative Assessment.* Kingston: NEEC and CGR Communications.

International Institute for Environment and Development (IIED). 2001. *Reconciling the "brown" and the "green" agendas for urban environmental improvement.* Urban Environmental Improvement and Poverty Reduction Briefing Paper no. 6. London: IIED.

Jácome, F. 2006. Environmental movements in the Caribbean. In *Beyond sun and sand: Caribbean environmentalisms*, ed. S.L Baver and B.D. Lynch, 17–31. New Brunswick, NJ: Rutgers University Press.

Jaffe, R. 2006. Urban blight in the Caribbean: City, environment and culture in Curaçao and Jamaica. PhD diss., Leiden University.

Levy, C. 1996. The environmental NGO movement in Jamaica. *Jamaica Journal* 26: 22–25.

Lundy, P. 1999. Fragmented community action or new social movement? A study of environmentalism in Jamaica. *International Sociology* 14: 83–102.

McGranahan, G., and D. Satterthwaite. 2002. Environmental health or ecological sustainability? Reconciling the brown and green agendas in urban development. In *Planning in cities: Sustainability and growth in the developing world*, ed. R. Zetter and R. White, 43–57. London: ITDG Publishing.

National Environment and Planning Agency (NEPA). 1999. *JANEAP (Jamaica National Environmental Action Plan) 1999–2002.* Kingston: NEPA.

———. 2002. *JANEAP (Jamaica National Environmental Action Plan): 2002 status report.* Kingston: NEPA.

National Resource Conservation Agency (NRCA). 1998. *Jamaica state of the environment report.* Kingston: NRCA.

Otuokon, S. 2007. Jamaica Conservation and Development Trust: Building capacity for natural and cultural heritage conservation in the Blue and John Crow Mountains National Park. Presentation, 30 March.

Pelling, M. 2003. *The vulnerability of cities: Natural disasters and social resilience*. London: Earthscan.

Pelling, M., and J.I. Uitto. 2001. Small island developing states: Natural disaster vulnerability and global change. *Environment Hazards* 3: 49–62.

Ramphall, D. 1997. Postmodernism and the rewriting of Caribbean radical development thinking. *Social and Economic Studies* 46: 1–30.

Rosenberg, J., and F.L. Korsmo. 2001. Local participation, international politics, and the environment: The World Bank and the Grenada dove. *Journal of Environmental Management* 62: 283–300.

Sachs, A. 2003. The ultimate "other": Post-colonialism and Alexander von Humboldt's ecological relationship with nature. *History and Theory* 42 (4): 111–35.

Satterthwaite, D. 2003. The links between poverty and the environment in urban areas of Africa, Asia and Latin America. *Annals of the AAPSS* 590: 73–92.

Sheller, M. 2003. *Consuming the Caribbean: From Arawaks to Zombies*. London: Routledge.

Sletto, B. 2005. A swamp and its subjects: Conservation politics, surveillance and resistance in Trinidad, the West Indies. *Geoforum* 36 (1): 77–93.

Tomalin, E. 2004. Bio-divinity and biodiversity: Perspectives on religion and environmental conservation in India. Numen: *International Review for the History of Religions* 51:265–95.

Vomil/Mina (Ministry of Public Health and Social Development, Department of Environment and Nature). 1998. Nota Duurzaam Toerisme voor de Nederlandse Antillen. Curaçao: Vomil/Mina.

———. 1999. *Voortgangsrapportage van de Contourennota van het Natuur-en Milieubeleid van de Nederlandse Antillen 1996–2000*. Curaçao: Vomil/Mina.

———. 2000. *Natuurbeleid van de Nederlandse Antillen: Aan de Dageraad van een Nieuw Millennium 2000–2005*. Curaçao: Vomil/Mina.

———. 2001. *Meerjarenplan Milieu-en Natuurbeleid Nederlandse Antillen 2001–2005*. Curaçao: Vomil/Mina.

———. 2004. *Nature and environment policy plan Netherlands Antilles, 2004–2007*. Curaçao: Vomil/Mina.

Watts, D. 1987. *The West Indies: Patterns of development, culture, and environmental change since 1492*. Cambridge: Cambridge University Press.

Weeber, L. 2004. *Leefsituatie in enkele achterstandsbuurten van Curaçao: Historische achtergronden en statistische feiten*. Curaçao: CBS, Willemstad.

Witter, M. 2002. *Report on a Survey of NGOs in Jamaica*. St Augustine, Trinidad: Caribbean Sustainable Development Network.

Zahibo, N., and E. N. Pelinovsky. 2001. Evaluation of tsunami risk in the Lesser Antilles. *Natural Hazards and Earth System Sciences* 1: 221–31.

CHAPTER 16

Caribbean Urban Livelihoods and Policies

Targeting Assets, Vulnerability and Diversity

HEBE VERREST

Introduction

This chapter focuses on household livelihoods in four low-income neighbourhoods in Paramaribo (Suriname) and Port of Spain (Trinidad and Tobago) (figure 16.1). It discusses how different households make a living and which households are better and worse off in terms of assets and vulnerabilities. The chapter examines the extent to which households operating home-based economic activities (HBEAs) can be distinguished from households without such activities, and concludes with a discussion on the meaning of the findings for poverty reduction policies.

The analysis builds on Caroline Moser's asset-vulnerability framework (1998). Her framework categorizes assets of urban households, and shows how obstacles and opportunities in asset accumulation affect household vulnerabilities. This study adds to Moser's framework by systematically analysing the impact of location and household characteristics on assets, livelihood activities and vulnerability. In addition to commonly used Caribbean household characteristics, such as ethnicity, household head and household size, the impact of household dynamics is assessed.

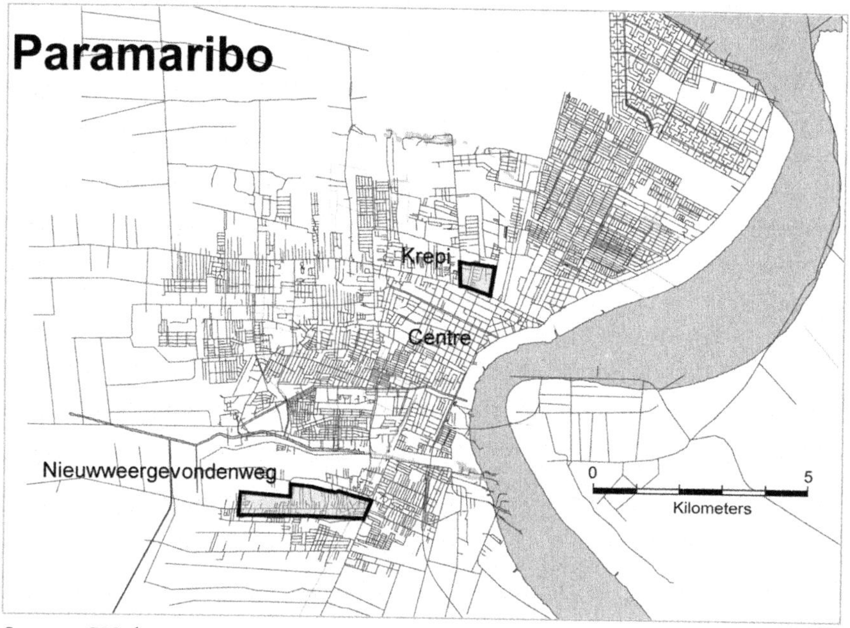

Source: GIS data

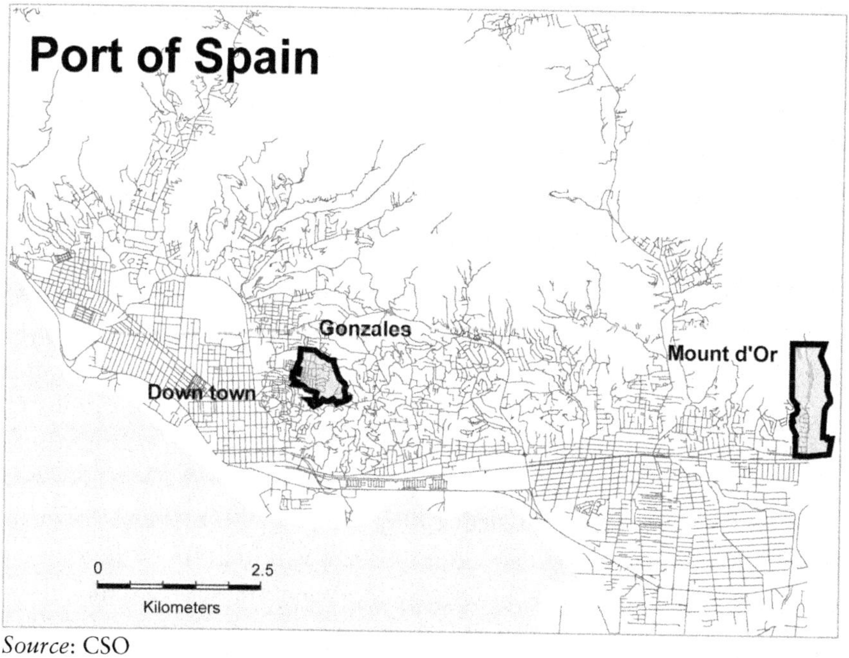

Source: CSO

Figure 16.1 Location of study neighbourhoods: Paramaribo and Port of Spain.

Studies on Caribbean urban livelihoods are relatively scarce, and a large proportion concentrates on Jamaica, for example, Moser and Holland (1997), Moser and McIlwaine (1997), Henry-Lee (2005), Jaffe (2006) and Dodman (2008). This chapter expands the scope of this available research by including two other locations. For Suriname, research by Kromhout (2000) and Verrest (2007) are the most recent comprehensive studies. For studies on urban poverty in Trinidad and Tobago, I refer to Kairi Consultants (2004), Ryan, Cree and St Bernard (1997), Lloyd-Evans and Potter (2002), Janssen (2004) and Verrest (2007, 2008). These studies show how social, environmental and economic vulnerabilities shape daily lives of the Caribbean urban poor. Moreover, they emphasize how households develop a range of strategies to spread risks and counter insecurities.

Suriname and Trinidad and Tobago share a history of slavery, indentureship, (de)colonization and dependence on the export of natural resources. These features impact on current social-economic structures and the multi ethnic composition of the populations. In Trinidad and Tobago, Afro- and Indo-Trinidadians are the main ethnic groups. In Suriname, Hindustani, *Creolen* (descendents of African slaves emancipated in 1863), *Marrons* (descendents from African slaves who were able to free themselves before emancipation) and Javanese are the four major groups.

Recent political and economic developments in the areas are distinct. Suriname's economy was in crisis from 1982, and the country only recently acquired stabilization and economic growth (Dijck 2001; ABS 2005a). Today, an estimated 60 per cent of its population lives below the poverty line (ABS 2001). Trinidad and Tobago's economy was in crisis between 1983 and 1993, but improved from then onwards and has been buoyant from the turn of the century. The oil and gas industry contribute largely to this growth (CSO 2006). However, despite these accomplishments, a substantial proportion of its population is considered poor (Henry 2004 [issued 2006]; UNDP 2004). Unemployment figures in both countries were about 10 per cent in 2003 at the time of the study (CSO 2004; ABS 2005b). In addition to these economic differences, the economic and poverty policies of each country differ. Trinidad and Tobago has a strong neo-liberal policy, emphasizing entrepreneurship, and various programmes to realize this have been established.

Suriname's economic policies are less explicit, but few support programmes are available for entrepreneurial activities. These macroeconomic statistics and policies, however, say little about the everyday conditions faced by individuals who are trying to make a living, and the remainder of this chapter attempts to examine this issue in more detail.

Urban Livelihoods and Vulnerability

Much current research on deprivation and vulnerability employs the livelihoods approach. Ellis (2000, 10) defines livelihoods as "comprising the assets, activities and the access to these (mediated by institutional and social relations) that together determine the living gained by an individual or household" (see also Chambers and Conway 1991; de Haan and Zoomers 2005). Households, individuals and communities develop livelihood strategies based on assets and activities available to them. Most develop a mixture of activities adjusted to their own circumstances and preferences (Farrington, Ramasut and Walker 2002; Rakodi 2002). The strategies that households develop are designed to recover from stresses or shocks, or to maintain or enhance assets or capabilities. Households and individuals do not operate in a vacuum but their access to assets and activities is shaped by interactions with institutions, organizations and social relations.

At such micro-scales, vulnerability refers to the ability of households, individuals and communities to bend and bounce back when confronted with adverse situations (Moser 1998). Shocks and gradual trends jeopardize the well being and security within households. Their resilience (the ease and rapidity of recovery) and sensitivity (the impact of an external event on a system) determine whether and how they can cope with these crises (Moser 1998). An overarching strategy of low-income households is diversification (Farrington, Ramasut and Walker 2002). It reduces dependence on one source of income and enables households to deal with irregularities. As Ellis (2000, 406) explains, diversification is not new or short term but pervasive and widespread. For some groups, notably low-income urban women, it is a way of life (Beall and Kanji 1999; Moser and McIlwaine 1997).

Urban Living

Initially, the livelihoods approach analysed the lives of rural households and communities (Bebbington 1999). On the list of "classic" works on livelihoods (for example, Chambers and Conway 1991; Moser 1998; Bebbington 1999; Leach et al. 1999; Haan 2000; Ellis 2000) only Moser's study of 1998 has an urban focus. That and Rakodi and Lloyd-Jones's edited volume (2002) form the main body of (theoretical) work on urban livelihoods. Despite its rural origins, the livelihoods approach is suitable for analysing urban poverty. It needs however to be adapted to include relevant urban features. First of all, urban life is highly commoditized. Urban residents must pay for most of their needs and can hardly rely on their own production. Therefore, their demand for and dependence on cash income is higher than of their rural counterparts (Farrington, Ramasut and Walker 2002; Meikle 2002). Second, social structures in urban areas differ. Networks are diverse, and go beyond the family and immediate environment of the neighbourhood to include, for example, work-related networks. Third, the urban poor experience specific "urban" forms of vulnerability (Moser 1995, 147). They often live in informal settlements and lack tenure security. This frequently prevents their participation in decision making and makes them vulnerable to harassment and eviction. They also rely on insecure incomes. Their vulnerability is further exacerbated by the violent and environmentally hazardous situation to which their living and working conditions expose them (Satterthwaite and Tacoli 2002, 57; Jaffe 2006). Finally, relations between urban households and the government, organizations, and institutions are complex, and these relations strongly affect the livelihoods of urban households. The ability to exercise citizenship is therefore very important to them. Table 16.1 shows the framework for analysing urban livelihoods.

Caribbean Households: Diversity and Change

People-centred approaches to poverty, such as the livelihoods approach, acknowledge the importance of household diversity. Household characteristics affect the assets and livelihood opportunities both accessible and preferred by households (Gonzáles de la Rocha 1994; Chant 2002;

Table 16.1 Urban Livelihoods Framework

Livelihoods Platform	Access Modified by	In Context of	Resulting in	Composed of	With Effects on
Assets Natural Physical Human Financial Social	**Social relations** Gender Class Age Ethnicity **Institutions** Rules & laws Tenure Market practice **Organizations** Associations NGOs Government Private sectors	**Trends** Migration National economic trends Globalization Labour market Education Crime **Shocks** Flooding Illness Unemployment Social unrest	**Livelihood strategies**	Labour Self-employed Remittances Transfers Savings Credits Agriculture	Livelihood security Size income Regular income Seasonality Risks

Source: Adapted from Ellis's (2000) framework on analysing marine livelihoods.

Rakodi and Lloyd-Jones 2002). Furthermore, gender and age relations within households determine whether households can mobilize additional labour, and are adaptive institutions for pooling income and sharing consumption (Kabeer 1994). This study considers the impact of the following characteristics of Southern Caribbean households: ethnicity (though it should be noted that the ethnic variety of the Surinamese sample is large, but the Trinidad and Tobagonian sample is predominantly of African descent, which constrains examining ethnic differentiation in this study), headship (couple [2P – two person], female [FHH – female headed household] or male [MHH – male headed household]), size and composition (nuclear, extended or other). Households headed by a single female are common throughout the Caribbean and generally considered more vulnerable than two-parent households. However, for example, Chant and Campling (1997) show that FHHs are not necessarily poorer or more vulnerable. Size impacts on livelihoods, as larger households have chances to share costs of living and time for care. However, they require more income and space to fulfil their needs. Finally, Caribbean households are characterized by extension through the inclusion of a third or fourth generation or other laterally related relatives (Safa 1995; Chant 2002).

In addition to these characteristics, I consider household dynamics. Households change over time because of people's capabilities and decisions (for example, on fertility or migration), and generational composition (Rakodi 2002). Gonzáles de la Rocha (1994) distinguishes a household life cycle consisting of three stages. In each stage, households are confronted with specific demands and opportunities to fulfil these needs. As such, this concept is useful for livelihoods analysis. However, it needs to "be considered in relation to the particular socio-economic and cultural characteristics of given societies" (Chant and Campling 1997, 3). The original model applies to nuclear households but has difficulty capturing the development of extended households. Furthermore, the inherent assumption in the model is that households develop according to a cyclic pattern, yet reality shows that some do, but others move back and forth between various phases and never go through all of them. Finally, by seeing the life cycle as a "natural process", we are blind to developments and changes within households that result from either external shocks or intended actions.

Therefore, I do not believe the development of households occurs as a cycle with stages in a predetermined order. Rather, I focus on the modalities of household composition related to dependency ratios and age. Households may experience a specific modality at one or more moments of their existence. I refer to this as the *age-dependency structure* and distinguish four possible situations. Three situations are adjusted from the stages of expansion, consolidation and dispersion of the original life-cycle model (Gonzáles de la Rocha 1994).

Expanding households consist of two generations: children (regarded here as "not able-bodied") and (able-bodied) adults. Such households often experience an uneven balance between able-bodied and not able-bodied members, and between income producers and income consumers. Moreover, they require relatively large amounts of time and capital to enable education, care and housing needs. Expanding households are often considered vulnerable.

Consolidated households consist of two or three generations of which the eldest is below sixty-five and at least one of the youngest above fifteen. Consolidated households have a higher proportion of able-bodied members and demand less time and capital for care and education. They are generally less vulnerable and better equipped with opportunities.

In *dispersed* households, adult children have left the house, and the parents (or parent) are on their own. In the household life cycle approach, this phase is the last stage. Income generating opportunities in such households shrink, and demands for care grow. As a result, livelihood demands and opportunities become unbalanced, and vulnerability grows.

Finally, I distinguish a fourth condition: a state of *transition*. It applies to households of two or more generations in which the eldest generation is older than sixty-five. The relationship between able-bodied and non able-bodied members can be rather unbalanced in these households, and demands for care are possibly high. This can result in large claims on the able-bodied members of the household. Table 16.2 displays the distribution of positions in the age-dependency structure in the selected locations. Consolidated households dominate, and dispersed households are very few. This means that in more than half of households the age-dependency structure is advantageous.

Table 16.2 Age-Dependency Structure Household by Country and Neighbourhood (in Percentage)

	Expansion	Consolidation	Transition	Dispersion	Total
Krepi	24	52	16	8	23
Nieuwweergevondenweg	23	53	18	6	25
Suriname	*24*	*52*	*17*	*7*	*49 (n = 191)*
Gonzales	11	69	14	6	26
Mount d'Or	19	57	18	6	26
Trinidad & Tobago	*15*	*63*	*16*	*6*	*51 (n = 202)*
Total	19 (n = 75)	58 (n = 228)	17 (n = 65)	6 (n = 25)	100 (n =393)

Note: Association countries: Cramer's V 0.13; significant at 0.10; association neighbourhoods: Cramer's V 0.09; significant at 0.36.

Methodology

For the livelihoods analysis, an asset and vulnerability index has been developed. Subsequently, correlations are measured between assets and vulnerabilities on the one hand and household characteristics on the other. The applied variance analysis results create a value for eta^2 (η^2), indicating the fraction of variation in scores on the index attributable to one or more of the household and locational variables. Such quantitative approaches to assess livelihoods are rare (cf. Baud, Sridharan and Pfeffer 2008) and are open to criticism, not least because their development needs validation through fieldwork. However, an advantage of a quantitative approach is the opportunity to detect linkages between household, locational and individual characteristics on the one hand, and assets and vulnerabilities on the other. The selection of indicators is based on (Caribbean) livelihoods literature and has been further refined in various focus group discussions held in the research locations. The data are derived from a survey that was conducted in 2003 among 393 randomly selected households in three neighbourhoods (the sample sizes were approximately one hundred each). The survey gathered data pertaining to household composition, livelihood activities, HBEAs and the quality of the living environment. In addition, two focus groups in each neighbourhood discussed definitions of concepts such as vulnerability and well being.

The Asset Index

The asset index consists of the quality and size of households' human, social, financial and productive assets. For each asset, two or more indicators are selected (see figure 16.2). The value of an asset equals the average score of a household on the indicators for that asset. The possible scores on each indicator have been standardized (with zero and one as the respective minimum and maximum scores). Literature on urban livelihoods indicates that *human assets* are most important. Therefore, human assets are weighted double in the final asset index.

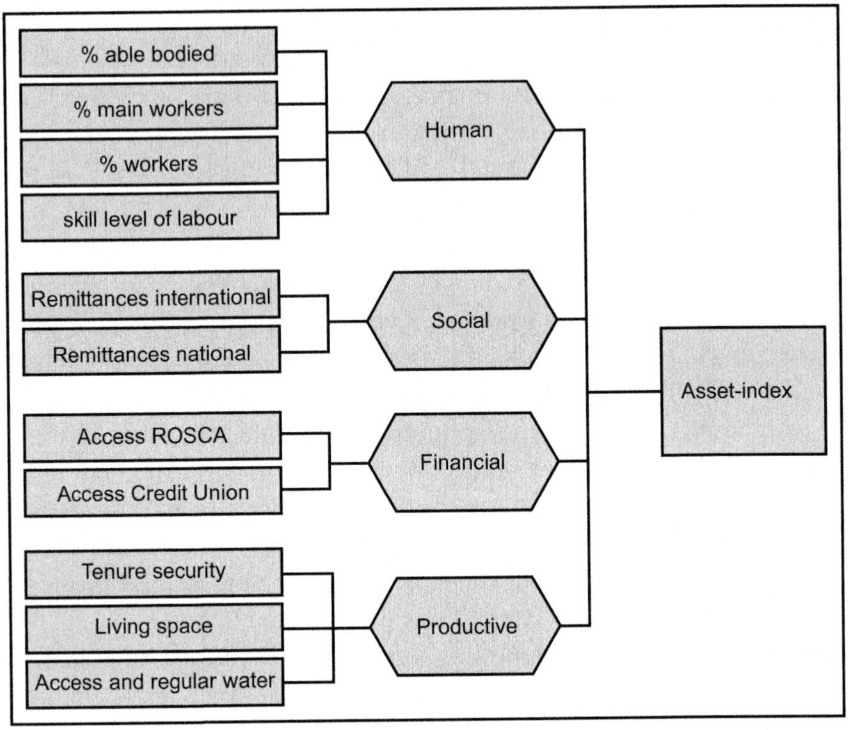

Figure 16.2 The asset index.

Scores for human assets are derived from the following indicators:

- highest level of education of any member: none (0), primary (0.25), secondary 4 (0.5), secondary 6 (0.75) and tertiary (1)
- percentage of household members with labour as main activity
- percentage of household members with paid labour
- skill level of the job (based on *Dictionary of occupation in Trinidad and Tobago*): low skilled (0), clerks and service-workers (0.5), and legislators, managers, associate professionals and technicians (1).

Financial assets are represented through membership of credit unions (CUs) and rotating saving and credit associations (ROSCAs, known as *kasmoni* in Suriname and *sousou* in Trinidad and Tobago): no member (0) or member (1).

Indicators of *social assets* are transnational family networks that result in remittances and local social networks that contribute to household income: having no contributing network (0), having a network that contributes irregularly (0.5), or having a network that contributes regularly (1).

For *productive assets*, the indicators are as follows:

- access and regularity of water: no access to pipe-born water (0), access to pipe-borne water but irregular provision (0.5) and access to pipe-borne water with continuous provision (1)
- the number of household members per bedroom: 3 or more (0), between 1.5 and 3 (0.5), and less than 1.5 (1)
- tenure: squatting (0), renting from private party (0.33), ownership/leasehold by family (0.67), private property/leasehold (1).

The Vulnerability Index

The vulnerability index has four components related to the main dimensions of vulnerability: resilience, diversification and income regularity (see figure 16.3). The possible scores on each indicator have been standardized (with zero and one as the respective minimum and maximum score). The vulnerability index represents households' mean scores on these indicators. Higher values indicate higher resilience and lower vulnerability. Technically, therefore, the vulnerability index is actually a "non-vulnerability" index.

The *percentage of able-bodied* household members indicates a household's ability to mobilize labour, and the share of members that are dependent, financially and otherwise. Small percentages of able-bodied members indicate high vulnerability. The next two indicators capture *diversification* of income sources: number of jobs per household member and the total number of income sources in the household. The final vulnerability indicator is *regularity* of income. Only regularity of labour income is considered. A permanent job provides more security but should by no means be seen as a job for life, since contracts are broken easily and retrenchment operations affect many people with

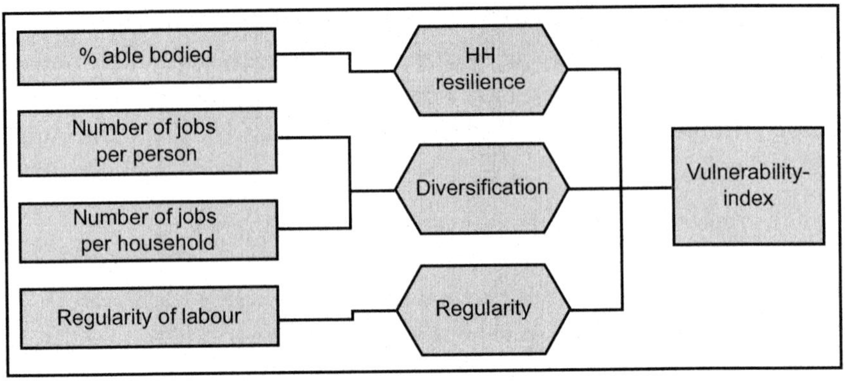

Figure 16.3 The vulnerability index.

"permanent" jobs. At an aggregate level, the total average regularity of jobs within the household is used.

Assets and Low-Income Groups in Paramaribo and Port of Spain

Human Assets

"Je diploma is je eerste man", which is best translated as "your diploma is your first husband", is an often used expression in Suriname. It emphasizes the importance of education in acquiring a living, particularly for girls. In both countries, education is highly valued. For example, during Secondary Entrance Assessment, CXC and A-level examinations in Trinidad and Tobago, local newspapers report extensively on the examination proceedings. Furthermore, in Paramaribo, many people beyond the age of compulsory education (numbering almost 20 per cent in this study) continue to pursue their education. Finally, in both Suriname and Trinidad and Tobago, the perceived diminishing quality of education, as well as high drop-out rates, are subjects of intense discussions in society and the media. In the majority of studied households (54 per cent), at least one member had attained a secondary education. In one-third (31 per cent), the highest level reached was primary school, and, in 13 per cent, at least one member had completed tertiary education. In today's Caribbean, girls perform better at school performance than

Table 16.3 Human Assets of Households by Age-Dependency Structure

		Skill Level Job		% Main Workers		% Workers		Highest Level of Education		Productive Assets	
		μ	η^2	μ	η^2	μ	η^2	μ	η^2	μ	η^2
Life stage	Expansion	.46	.12**	.37	.18**	.41	.14**	.47	.14**	.43	.21**
	Consolidation	.48		.51		.54		.49		.50	
	Transition	.47		.38		.42		.46		.44	
	Dispersion	.1		.02		.1		.25		.12	

Note: **= significant at 0.01 level; *= significant at 0.05 level; μ= mean score; $\acute{\eta}^2$ (eta squared) is the proportion of total variability attributable to a factor.

boys. However, the level of education at the individual level in our sample does not show large differences between men and women.[1]

A mean of 46 per cent of members in each household claimed to be working, despite the fact that employment opportunities in Paramaribo and Port of Spain are limited. For many people, to be employed means to be employed more or less regularly. Self-employment, on irregular bases and generating too little income to maintain oneself, is not considered as having a job. People refer to such labour as “hosselen” (Suriname) or “hustling” (Trinidad and Tobago). The majority (54 per cent) of people aged fifteen or above were working, and 9 per cent were unemployed. A similar percentage of people in each city had a paid job, but in Paramaribo more people were reported to be students or housewives, and fewer were unemployed than in Port of Spain, reflecting that the labour force was smaller in Paramaribo.

A gender-specific analysis shows that more women than men remain outside the labour force (56 per cent versus 35 per cent respectively); while 17 per cent of women are unemployed, compared to 10 per cent of men. In comparison with the overall picture of the working population in each country, highly skilled occupational groups and agricultural groups are under represented. A gender analysis shows that women are more often found working as service workers, clerks and associate professionals, as well as in elementary occupations, whereas males dominate crafts and machine operation.[2]

The age-dependency structure is the major variable explaining variation in the bundle of human assets as well as its individual components (see table 16.3). Each shows the same trend. Consolidated households are better off, and dispersed households have least human assets. Expanding and transitional households have comparable scores. This can be explained from the fact that one of the distinguishing characteristics of age-dependency conditions is the ability to work.

Financial Assets

Differences between Suriname and Trinidad and Tobago exist in financial assets, with Trinidad and Tobagonians attaining higher scores. In particular, membership of CUs was higher. In Suriname, 14 per cent of households claimed CU membership and 16 per cent ROSCA membership as compared to 47 per cent and 31 per cent respectively in Trinidad and Tobago. Suriname's economic crisis was characterized by strong devaluation of the Suriname guilder, in which saving money in local currency became an unattractive proposition. Instead, people tried to buy foreign currency, durable consumer goods or gold instead, by-passing local financial institutions which then became paralysed. Moreover, the devaluation caused steep increases in consumer prices, poverty and interest rates, as well as a reduction in savings. At the time of the survey (2003), Suriname's economy had just started to stabilize. Already, in late 2004 (when a second round of interviews with financial institutions was conducted), banks, CUs and local department stores reported an increase in supply of and demand for products targeting lower-income groups. The impact of both further stabilization and local economic growth are likely to further increase participation in CUs and ROSCAs.

Social Assets

Half of the households receiving foreign remittances receive them regularly. The duration of migration and the closeness to household members, in terms of generational relationships, explain the intensity of remittances. Partners send remittances most frequently, followed by parents and children. Other relatives contribute more irregularly. Variations in local social networks are best explained by household type and headship (7 and 8 per cent respectively). Contributions from parents, partners and children have been described as an important strategy for

FHHs (Wekker 1994; Kromhout 2000). This research confirms this, as 37 per cent of FHHs report having such networks compared to 12 and 20 per cent of 2P and MHHs respectively.

Productive Assets

Home ownership is common in the areas researched in both countries (more than 50 per cent). As a result, a large majority (91 per cent) of households perceive their tenure as secure. Even the majority (76 per cent) of households without home ownership consider their tenure secure. Access to water is severely constrained in each city, with over a quarter (28 per cent) of households having no pipe-borne water on their plots. Fewer than half have a continuous supply of water either directly (39 per cent) or through use of a water pump. All others have water for one or two periods of the day or even the week. Paramaribo households more often have piped water. The hilly areas in Trinidad and Tobago – particularly Mount d'Or and Upper Gonzales – are badly serviced. Krepi, on the other hand, is relatively well equipped. Yet, the regularity of water is better guaranteed in Trinidad and Tobago, specifically in Port of Spain's inner city areas. The age-dependency structure of households and their size each explain 6 per cent of the scores on productive assets. But, as opposed to the trend in other assets, dispersed households do better. They more often own their home, and have more living space and better access to regular water at home.

Total Asset Index

Scores on the asset index show a large variety, which is best explained by country, dominant ethnic group and household age-dependency structure (see table 16.4). Together these variables explain 17 per cent of variation but the age-dependency condition dominates. Other factors that probably affect variations in household asset bases are related to characteristics of individual household members (including gender and level of ambition) and the institutional context. The relationship between assets and age-dependency can be explained from the demographic characteristics of households at each stage of age dependency.

Table 16.4 Household Scores on Total and Individual Asset Index by Various Characteristics

		Asset Index		Human Assets		Financial Assets		Social Assets		Productive Assets	
		μ	ή2	μ	ή2	μ	ή2	μ	ή2	μ	ή2
Country	Suriname	.38	.05**	.43	.01*	.14	.14**	.21	.01	.69	.04
	Trinidad & Tobago	.44		.48		.39		.26		.60	
Age/dependency	Expansion	.37	.08**	.43	.21**	.28	.03**	.16	.03**	.57	.06*
	Consolidation	.43		.50		.28		.24		.64	
	Transition	.42		.44		.30		.23		.71	
	Dispersion	.29		.12		.06		.38		.79	
Ethnicity	Creole/African	.44	.08**	.48	.02	.37	.27**	.25	.03*	.64	.01
	Hindustani/Indo-Trinidadian	.35		.43		.1		.15		.67	
	Marron/Maroon	.35		.40		.1		.29		.55	
	Mixed	.42		.43		.11		.28		.64	
	Other	.36		.45		.31		.15		.69	

Note: **= significant at 0.01 level; *= significant at 0.05 level; μ = mean score; $ή^2$ (eta squared) = the proportion of total variability attributable to a factor.

For example, an expanding household generally only has a few able-bodied members, meaning that few people can take up labour activities. This in turn affects the sum of human assets the household has at its disposal.

Vulnerability and Low-Income Households in Paramaribo and Port of Spain

Diversification

Most households surveyed use more than one source of income. Only 15 per cent depend on one source, and almost a fifth mobilize five or more sources. Others distribute their risks over two (25 per cent), three (26 per cent) or four (15 per cent) of the following sources of income: paid labour, home-based economic activities, remittances and social welfare. These are detailed next.

Paid Labour

Labour is by far the most important livelihood activity. Eighty per cent of households contain at least one person with a job. In three-quarters of households it is the most important, and in 22 per cent it is the only source of income. The mean number of jobs in a household is 1.7. ssessing incomes earned through labour is often difficult and inaccurate, and this study is no exception. The quarter of households that gave accurate income-data earned a mean of US$574 monthly with labour activities.

Home-Based Economic Activities

The second most often reported sources of income for households related to home-based economic activities (HBEAs). Roughly 60 per cent of households operate HBEAs, and 39 per cent earn income from these. HBEAs generate an average income of US$260 monthly, but again these figures are based on few respondents. For 8 per cent of all households, income from HBEAs is the most important, and for 36 per cent, it is the second most important source of income. For households operating HBEAs, these income-generating activities are the most important sources of income for 20 per cent of these households, and the second most important for a further 54 per cent.

Remittances

A third (35 per cent) of households receives foreign remittances, of which half do so regularly. Foreign remittances are more often and more regularly collected in Port of Spain than in Paramaribo. It may be that this is the result of the relation between the senders and receiving households. Close family members (for example, parents, children or partners) send remittances regularly, and family members at further distance do so irregularly. In Port of Spain, several households were encountered where one partner had migrated while the other remained behind or where migrating parent(s) had left young children behind (so-called barrel children because of the barrels of foreign consumer goods sent, particularly at Christmas). The Paramaribo sample did not contain such households. This difference might explain why fewer Paramaribo households received regular remittances. In addition to foreign remittances, 22 per cent of households receive local remittances, that is, from people outside the household but within the country.

Fewer than a quarter were able to report the income derived from local remittances, therefore only their relative importance is discussed here. Foreign remittances were the most important source of income for 3 per cent of all households and the second most important for close to a quarter. For 21 per cent of households with such remittances, these were most important, and for 55 per cent, these were the second most important source. Local remittances represented the most important source for 4 per cent of households. From households receiving support, 23 per cent stated this was their most important, and for 46 per cent the second most important source of income.

Social Welfare

A third of all households receive social welfare, mainly old age pension. Other forms of financial social welfare mentioned are child support and poverty relief. However, the size and prevalence of these forms were marginal. In addition, poor Surinamese have an *onvermogende* or *minvermogende* (less or insufficiently wealthy) card which provides them with (basic) health care. Basic health care through health clinics is also free in Trinidad and Tobago. For 10 per cent of all households, social welfare was the most important source of income.

Vulnerability and Household Diversity

The vulnerability situation of households is, similarly to asset portfolios, most strongly influenced by their age-dependency situation. This factor explains 21 per cent of variation in vulnerability and also affects individual vulnerability components (except for regularity of income). Dispersed households are most vulnerable because they experience disadvantageous dependency ratios, have limited diversification, and rely on social networks and social welfare for income. Expanding households have few sources of income; hence, they rely heavily on their labour. They have few opportunities to expand their participation on the labour market, thus their inability to perform additional labour makes them vulnerable (see table 16.5).

The other characteristics do not influence the vulnerability index significantly. Household size affects the percentages of able-bodied members (larger households have lower percentages of able-bodied members) and total sources of income (larger households have more sources of income). Interestingly, the type of household and whether the household head is male or female does not make much difference. Despite lower access to labour and dependence on others for income, FHHs are able to reduce vulnerability by diversifying their income and investing in skills and education.

Household vulnerability needs to be analysed from the perspective of household asset portfolios. Assets explain 31 per cent of variation in vulnerability, with larger asset stocks resulting in higher levels of resilience. Both human assets and financial assets are relevant, explaining 51 and 9 per cent respectively. The total asset index explains variations in each vulnerability indicator, except regularity of labour income. Interestingly, financial assets impact most strongly on the number of jobs in the household, yet having more jobs stimulates the use of financial

Table 16.5 Household Scores on Vulnerability Index by Various Characteristics

		Total (non)-vulnerability		% Able Bodied		Regularity of Jobs		Total Number of Sources		Jobs per Hh-member	
		μ	η^2	μ	η^2	μ	η^2	μ	η^2	μ	η^2
		.54		.72		.59		.41		.32	
Country	Suriname	.51	.01*	.70	.00	.60	.00	.39	.01	.46	.02**
	Trinidad & Tobago	.56		.73		.58		.43		.55	
Age-dependency	Expansion	.46	.21**	.57	.26**	.58	.00	.25	.09**	.48	.13**
	Consolidation	.60		.82		.58		.45		.57	
	Transition	.55		.68		.59		.51		.45	
	Dispers.	.21		.32		.00		.30		.1	
Ethnicity	African	.56	.03**	.70	.01	.61	.00	.45	.03**	.56	.02*
	East Indian	.51		.76		.58		.32		.46	
	Marron	.52		.64		.60		.46		.45	
	Mixed	.52		.71		.54		.44		.47	
	Other	.52		.75		.61		.33		.50	
Neighbourhood	Krepi	.52	.01	.74	.01	.57	.00	.37	.01	.47	.0**
	Nieuwweergev.	.51		.67		.63		.41		.44	
	Gonzales	.55		.71		.57		.44		.52	
	Mt d'Or	.57		.75		.59		.43		.59	
Size	1–3	.52	.02	.78	.05**	.59	.01	.29	.23**	.55	.02**
	4–6	.65		.68		.55		.44		.49	
	7–9	.60		.62		.67		.67		.42	
	10+	.61		.60		.67		.76		.41	
Household type	Nuclear	.52	.05**	.73	.00	.56	.00	.34	.23**	.51	.00
	Extended	.59		.79		.59		.58		.51	
	Other	.54		.75		.74		.19		.48	
Household head	2-parent	.55	.01	.71	.02**	.59	.00	.44	.04**	.50	.01**
	FHH	.52		.69		.57		.43		.47	
	MHH	.55		.83		.65		.26		.60	
		r^2		r^2		r^2		r^2		r^2	
Assets	Asset profile	.31**		.17**		.00		.20**		.34**	
	Social	.01		.11**		.00		.07**		.04**	
	Financial	.09**		.06**		.00		.06**		.12**	
	Productive	.01		.00		.00		.00		.01**	
	Human	.51**		.07**		.00		.13**		.72**	

Note: **= significant at 0.01 level; *= significant at 0.05 level; μ = mean score; ή² (eta squared) is the proportion of total variability attributable to a factor.

assets. In other words, financial assets are the result, rather than prerequisite, of vulnerability.

Home Based Economic Activities (HBEAs), Assets and Vulnerability

Moser (1998) pointed to the importance of the home in the livelihoods of the urban poor. Through HBEAs, households generate subsistence goods and/or income, which contribute to assets, well-being and vulnerability reduction. As mentioned above, after paid labour, HBEAs are the most often mobilized source of income. Table 16.6 shows the scores of households with and without income-generating HBEAs on the asset and vulnerability indices. Households with HBEAs score better on both indices. Having an income-generating HBEA explains 22 per cent variation in vulnerability. HBEA households are better prepared against shocks and stresses because of better access to financial institutions, higher levels of education and more sources of income. Households

Table 16.6 Household Scores on Asset Index and Vulnerability Index by Productive Use Habitat

		Asset Index		Human Assets		Financial Assets		Social Assets		Productive Assets	
		μ	η^2	μ	η^2	μ	η^2	μ	η^2	μ	η^2
		.41		.45		.27		.23		.65	
Income-generating HBEA	Yes	.46	08**	.52	.08**	.25	.03**	.25	.00	.65	.00
	No	.38		.41		.22		.23		.64	

		Total (non)-vulnerability		% Able Bodied		Regularity of Jobs		Total no. of Sources		Jobs/ Hh-member	
		μ	η^2	μ	η^2	μ	η^2	μ	η^2	μ	η^2
		.58		.72		.59		.54		.32	
Income-generating HBEA	Yes	.67		.74		.73		.56		.65	
	No	.46	.22**	.70	.00	.48	.10**	.32	.23**	.41	.13**

Note: ** = significant at 0.01 level; * = significant at 0.05 level; μ = mean score; $\acute{\eta}^2$ (eta squared) is the proportion of total variability attributable to a factor.

with and without HBEAs show only marginal differences in the types of their income sources, although households with HBEAs tend to have more income generating activities (a mean of 4.0, compared to 2.6 for households without these). This shows that HBEAs tend not to replace an alternative money-generating activity, but rather represent an additional source of household income.

Conclusions

This chapter portrays the assets and vulnerabilities of households in four low-income areas in Port of Spain and Paramaribo, and relates these to location and household characteristics. Despite the distinct economic differences between the two countries, pictures of household assets and vulnerabilities barely differ, suggesting that recent positive macro-economic developments in Trinidad and Tobago have not had a positive impact on the assets and vulnerabilities of households in low-income areas. The structural characteristics of the economy, for example, a lack of low-skilled employment, and of poverty, for example, poor housing, have not changed as result of recent economic developments.

Second, the age-dependency structure best explains difference in scores on asset and vulnerability indices. Consolidated households have the best position, followed by transitional households. Expanding but dispersed households, especially, have fewer assets and are more vulnerable. However, it is necessary to look beyond overall scores to unravel specific strengths and weaknesses of each group.

A final conclusion is that households with HBEAs are less vulnerable and have larger asset bases than households without such activities. HBEAs contribute to diversification and the involved households have more human and financial assets. Yet, the question is whether HBEAs are a cause or consequence of more assets and resilience. The quantitative data used here do not give conclusive evidence, but point in both directions. Qualitative data from interviews with HBEA operators do not clarify this ambiguity. The higher participation of HBEA-operating households in ROSCAs and CUs for example, is the result of HBEAs, rather than a prerequisite for these. On the other hand, building an

HBEA is facilitated and secured by the back-up of other income sources. Hence households with higher levels of assets and security are starting HBEAs more often and are more successful in maintaining them over time.

What Do These Findings Mean for Policy?

Expanding and dispersed households need special attention. Expanding households frequently lack adequate housing and depend heavily on labour. They, and women in these households especially, could be supported by facilities to combine productive and reproductive labour (for example, day care or HBEAs) and better access to labour. Dispersed households lack variety in income sources and depend on social networks. Policies facilitating multi-generational households could well support them. Use could be made of their large number of productive assets. Furthermore, affordable and accessible health facilities reduce the financial impact of illness for all, regardless of their assets.

The findings on the importance of HBEAs have implications for policy as well. Nowadays, entrepreneurship is viewed as the solution to poverty. Several projects, particularly in Trinidad and Tobago, exist to stimulate micro-entrepreneurship, mainly through providing business skills and micro-credit. However, this study shows that more assets and lower vulnerability lead to HBEAs. HBEAs, in turn, induce asset growth and vulnerability reduction. Policies that aim at improvement of human assets or labour opportunities may therefore be much more effective in stimulating entrepreneurship. These policies could entail adult education and stimulation of labour-intensive middle-sized enterprises.

Second, an important question is the applicability of the findings to other Caribbean settings. The similarities in findings between the two areas of study, despite the existing differences in their economic and policy context, allow us to consider painting a more general Caribbean picture of urban livelihoods, assets and vulnerabilities, particularly where it concerns low-income households in medium-sized cities. Such a picture would contribute positively to the development of relevant policies throughout the Caribbean.

Notes

1. Gender: Cramer's V is 0.11; significant at 0.02.
2. Relation between gender and occupational group Cramer's V is 0.40; significant at 0.00.

References

Algemeen Bureau voor de Statistiek (ABS). 2001. Poverty lines and poverty in Suriname. *Suriname in Cijfers* (no. 191-2001/02). Paramaribo: Algemeen Bureau voor de Statistiek.

———. 2005a. Bergen en dalen in de Surinaamse economie: De ontwikkelingen van het Bruto Binnenlands Product 1956–2004. *Statistical Papers 3* and *Statistical Papers 4*. Paramaribo: Algemeen Bureau voor de Statistiek.

———. 2005b. *Suriname census 2004*. Volume 2: *Werkgelegenheids-en Onderwijskarakteristieken*. Paramaribo: Algemeen Bureau voor de Statistiek.

Baud, I.S.A., N. Sridharan, and K. Pfeffer. 2008. Mapping urban poverty for local governance in an Indian mega-city: The case of Delhi. *Urban Studies* 45 (7): 1385–412.

Beall, J., and N. Kanji. 1999. Households, livelihoods and urban poverty. Urban Governance, Partnership and Poverty theme paper no. 3. Department of Social Policy and Administration/London School of Economics.

Bebbington, A. 1999. Capitals and capabilities: A framework for analyzing peasant viability, rural livelihoods and poverty. *World Development* 27 (12): 2021–44.

Chambers, R., and G.R. Conway. 1991. *Sustainable rural livelihoods: Practical concepts for the twenty-first century*. IDS Discussion Paper no. 296. Brighton, UK: IDS.

Chant, S.H. 2002. Men, women and household diversity. In *Challenges and change in middle America: Perspectives on development in Mexico*, Central America and the Caribbean, ed. C. McIlwaine and K. Willis, 26–50. Harlow, UK: Pearson Education.

Chant, S.H., and J. Campling. 1997. *Women-headed households: Diversity and dynamics in the developing world*. New York: St Martin's Press.

Central Statistical Office (CSO). 2004. *Labour force bulletin, fourth quarter, 2003* (vol. 16, no. 69). Port of Spain: Ministry of Planning and Development/Central Statistical Office.

———. 2006. *Gross domestic product, 2002–2006*. Port of Spain: Ministry of Planning and Development/Central Statistical Office.

Dijck, P. v., ed. 2001. *Suriname, the economy: Prospects for sustainable development*. Kingston: Ian Randle.

Dodman, D.R. 2008. Making a living: Small-scale metal smelting as a livelihood strategy in Kingston, Jamaica. In *The Caribbean city*, ed. R. Jaffe, 227–46. Kingston: Ian Randle.

Ellis, F. 2000. *Rural livelihoods and diversity in developing countries*. Oxford: Oxford University Press.

Farrington, J., T. Ramasut and J. Walker. 2002. *Sustainable livelihoods approaches in urban areas: General lessons, with illustrations from Indian cases*. London: ODI.

Gonzáles de la Rocha, M. 1994. *The resources of poverty: Women and survival in a Mexican city*. Oxford: Blackwell.

Haan, L. de. 2000. Globalization, localization and sustainable livelihood. *Sociologia Ruralis* 40 (3): 339–65.

Haan, L. de and A. Zoomers. 2005. Exploring the frontier of livelihoods research. *Development and Change* 36 (1): 27–47.

Henry, R. 2004 (issued in 2006). Going behind the statistics on poverty in Trinidad and Tobago. *Caribbean* Dialogue 9 (4): 45–58.

Henry-Lee, A. 2005. The nature of poverty in the garrison constituencies in Jamaica. *Environment and Urbanization* 17 (2): 83–99.

Jaffe, R.K. 2006. Urban blight in the Caribbean: City, environment and culture in Curaçao and Jamaica. PhD diss., Leiden University.

Janssen, T. 2004. It's the way the system set: An analysis of youth livelihoods in the Beetham neighbourhood, East Port of Spain, Trinidad. MA diss., University of Amsterdam.

Kabeer, N. 1994. *Reversed realities: gender hierarchies in development thought*. London: Verso.

Kairi Consultants. 2004. *Poverty reduction and social development in Trinidad and Tobago*. Washington, DC: Inter American Development Bank.

Kromhout, M.Y. 2000. Gedeelde Smart is Halve Smart: Hoe Vrouwen in Paramaribo Hun Bestaan Organiseren. PhD diss., Universiteit van Amsterdam/FMG.

Lloyd-Evans, S., and R.B. Potter. 2002. *Gender, ethnicity, and the informal sector in Trinidad*. Aldershot, UK: Ashgate.

Meikle, S. 2002. The urban context and poor people. In *Urban livelihoods: A people-centred approach to reducing poverty*, ed. C. Rakodi and T. Lloyd-Jones, 37–51. London: Earthscan.

Moser, C.O.N. 1995. Urban social policy and poverty reduction. *Environment and Urbanization* 7 (1): 159–72.

———. 1998. The asset vulnerability framework: Reassessing urban poverty reduction strategies. *World Development* 26 (1): 1–19.

Moser, C.O.N., and J. Holland. 1997. *Urban poverty and violence in Jamaica.* Washington, DC: World Bank.

Moser, C.O.N., and C. McIlwaine. 1997. *Household responses to poverty and vulnerability.* Washington, DC: World Bank.

Rakodi, C. 2002. A livelihoods approach: Conceptual issues and definitions. In Rakodi and Jones 2002, 3–22.

Rakodi, C., and T. Lloyd-Jones, eds. 2002. *Urban livelihoods: A people-centred approach to reducing poverty.* London: Earthscan.

Ryan, S., R.M. Cree and G. St Bernard. 1997. *Behind the bridge: Poverty, politics and patronage in Laventille, Trinidad.* St Augustine, Trinidad: Institute of Social and Economic Research, University of the West Indies.

Safa, H.I. 1995. *The myth of the male breadwinner: Women and industrialization in the Caribbean.* Boulder: Westview Press.

Satterthwaite, D., and C. Tacoli. 2002. Seeking an understanding of poverty that recognizes rural-urban differences and rural-urban linkages. In Rakodi and Lloyd-Jones 2002, 52–70.

United Nations Development Programme (UNDP). 2004. *Human development report 2004: Cultural liberty in today's diverse world.* New York: UNDP.

Verrest, H. 2007. Home-based economic activities and Caribbean urban livelihoods: Vulnerability, ambition and impact in Paramaribo and Port of Spain. PhD diss., University of Amsterdam.

———. 2008. Working from home, habitat as a place of production: Urban livelihoods in low-income Surinamese and Trinidadian households. In Jaffe 2008, 247–74.

Wekker, G. 1994. *Ik ben een gouden Munt, ik ga door vele Handen maar Verlies mijn Waarde niet: Subjectiviteit en Seksualiteit van Creoolse volksklasse Vrouwen in Paramaribo.* Amsterdam: Vita.

PART 5

Synthesis

CHAPTER 17

Beyond Caribbean Vulnerability

Towards Resistance and Resilience

DAVID DODMAN, DUNCAN McGREGOR AND DAVID BARKER

Introduction

This chapter seeks to identify and draw out the key themes that have driven individual authors and recur in the preceding chapters of this volume. It is not intended as a comprehensive review of these chapters, nor will it be a definitive statement on Caribbean vulnerability and global change. However, we will move beyond the focus on the different aspects of vulnerability addressed by the contributing authors, and attempt to draw on their insights to assess ways in which the lessons learned from the range of research approaches represented here can be applied to decrease vulnerability and increase resilience. It is freely acknowledged here that a majority of our contributed chapters focus on the Jamaican situation, but, as individual contributors have discussed, the vulnerabilities illustrated by particular island situations are very largely generic to the region as a whole.

The previous sections in this volume have addressed several of the key components of global change and Caribbean vulnerability. Perhaps the single most pressing issue of global change, or at least the one with the highest public awareness at present, involves current and projected climate change. Douglas Gamble's chapter provides a substantial, yet accessible, discussion of the implications of climate change for the region that will inform academic researchers and policy makers alike.

The main aspects of vulnerability addressed in the next section involve the management of vulnerable rural environments, with the chapters providing theoretical frameworks for examining this (Kemi George's contribution on the nature of scientific knowledge claims); the effects of global change on particular landscapes (Deborah Rowe et al.'s chapter on coastal management, Patrick Williams and Linda Johnson-Bhola's empirical research into coastal flooding in Guyana, and Mick Day and Sean Chenoweth's chapter on karst landscapes); the influences of tourism on carbon sequestration (Elizabeth Thomas-Hope and Adonna Jardine-Comrie's chapter); and locally based interventions for addressing these challenges (Susan Otuokon and Shauna-Lee Chai's examination of the Blue and John Crow Mountains National Park).

The three chapters that link vulnerability with food supply bridge the gap between the global and the local, as well as the rural and the urban, in the case of Jamaica, by critically examining the relationships between hazards, food security and vulnerability, either through empirical research (McGregor, Barker and Campbell; Clinton Beckford and Steve Bailey), or proposing new mechanisms for enhancing the agricultural sector (Kevon Rhiney, Beckford and Bailey). At a time when, regionally, domestic agriculture is under the pressures not only of climate change, but also trade liberalization and increasing dependence on food imports, food security is an increasing regional concern.

The Caribbean is one of the most highly urbanized regions in the world, with 65 per cent of the region's population living in urban areas, a figure that is predicted to rise to 75 per cent by the year 2025 (Potter 2000). The third section of this volume therefore assesses the particular nature of urban vulnerability. Colin Clarke examines the history of multiple deprivation in Kingston; Rivke Jaffe integrates environmental and social vulnerability in Jamaica and Curaçao; David Howard draws attention to urban violence and crime as a specific form of vulnerability; and David Dodman and Hebe Verrest both address the ways in which the urban poor utilize a variety of assets in order to create livelihoods for themselves and their families. These region-wide examples highlight different aspects of urban vulnerabilities, and therefore, together with Willams and Johnson-Bhola's examination of flooding in urban coastal Guyana, cogently illustrate the complexity of the issues facing the Caribbean urban environment.

In the remainder of this concluding chapter, we revisit the key theme of Caribbean vulnerability and move the discussion in the direction of Caribbean resilience. For each of these themes, we draw on the insights provided by the contributors to this volume. This allows for an assessment of the ways in which their work both reflects and illuminates issues of scale and political ecology, and can be used to identify some key areas for future research and policy. Finally, and in response to the issues eloquently raised by the Honourable Omar Davies in his opening statement to this book, we seek to bridge the gap between academic research and the development of policy by proposing practical interventions to improve resistance and resilience in the Caribbean.

Towards a Revised Understanding of Caribbean Vulnerability

One of the key insights from the contributors to this volume is the complex and varied nature of vulnerability in the Caribbean region (and, indeed, elsewhere). There is a need to reflect on the typologies of vulnerability and not simply to look at vulnerability due to extreme natural events. Vulnerability takes on many different guises in this context, as shown, for example, by the contrasting vulnerability focuses of urban research and rural research.

The Scale of Vulnerability

The causes of Caribbean vulnerability originate at a variety of scales: international, national, communal and individual. However, rather than viewing these as independent and discrete, it is more appropriate to assess the ways in which these factors are mutually constituting and co-dependent (Swyngedouw and Heynen 2003). In other words, many of the key challenges of globalization arise from the ways in which ostensibly "global" processes both influence and are affected by "local" events, and vice versa. It is important, therefore, to work through how regional and local processes both contribute to vulnerability in any given situation. Scale issues in urban systems are equally important drivers of vulnerability, where historic patterns of poverty and insecurity of livelihoods may become entrenched (Clarke, chapter 12) but also where local initiatives may present islands of relative (if not absolute) prosperity (Dodman, chapter 13).

Rather than simply stating the existence of cross-scale linkages, it is necessary to interrogate the nature of these. It is important to recognize the possibility of complex response within the system, where either the same stressor acting in two different situations may cause different outputs, or different underlying processes in different circumstances may lead to the same output. Detailed empirical research is required to resolve cross-scale linkages, and Verrest's (chapter 16) analysis of different types of low-income household assets in two different settings, Paramaribo and Port of Spain, shows both the complexity of linkages and how one variable, in this case age-dependency structure, may help to explain differences in vulnerability between households.

This is relevant for both natural and human systems. Researchers need to look at focusing more specifically than from the global and regional scales, especially in terms of climatic parameters, in order to define marginality of agricultural and urban systems more effectively. What might be seen as an issue at the regional scale might not be so at the local scale, and vice versa. To take climate, Gamble (chapter 2) makes the pertinent point that at present, Global Circulation Models do not have the spatial resolution to deal with the very significant range of climatic conditions experienced across mountainous Caribbean islands such as Dominica (for example, see Walsh 1980) and Jamaica (as shown in McGregor, Barker and Campbell, chapter 9). It follows that the current, and likely continuing, wide variations of rainfall seen across larger (for example, Jamaica) and smaller (for example, Dominica) islands will inevitably lead to differential effects of climate change, on agricultural systems in particular, at different spatial scales. Despite this, generalizations with regard to the effects of climate change on key Caribbean physiographies are still valuable. One such example is illustrated by Day and Chenoweth (chapter 5), who explain how changing climates will modify karstic hydrological behaviour throughout the region, influencing community vulnerability to variables as different as flooding and health.

Similarly, the changing nature of global economic processes, such as neo-liberalism, also have unique and particular effects in different societies and economies, and differentially within both more marginal and less marginal sectors of individual nation states. Indeed, a focus on neo-liberalization as a process, rather than on neo-liberalism as an end-state (Peck and Tickell 2002), helps raise awareness of the changes in systems and patterns, and the unique outcomes that these changes can provide. It has been suggested that historically the Caribbean "is perhaps the most globalized of world regions" (Klak 1998, 6), yet neo-liberal globalization has ushered in a new form and intensity of integration with a global economy, as well as significant shifts in the way external relations have been managed and organized. The liberalization of trade through the reduction of tariff and non-tariff barriers has intensified through compliance to new rules-based trade regimes. These are enacted and enforced through a range of different actors including the WTO, US multinationals, and European Union agreements (Dodman and Newstead 2008).

One of the most important contributions made by the authors in this volume is showing that the global and the local, the physical and the social, are all inter-related. This is perhaps most evident in the chapters examining agricultural production (9 to 11), which, although focused on Jamaica, depict the ways in which climate change, land degradation, global economic changes and the rise of the local tourism industry all contribute to the continued viability, or lack thereof, for agricultural livelihoods. As the concluding discussion in chapter 9 indicates, the broad vulnerabilities of Jamaica's smallholder farmers are mirrored throughout the region.

The Political Ecology of Vulnerability

We also need to take into account the heterogeneous nature of Caribbean societies, and recognize that within the region, the nation and even the community there will be groups with different degrees of vulnerability to the same stressor. There is also a dislocation between the land resources that are present in the region and their availability to different groups of people. Issues of tenure are just one factor here that may have a significant impact on vulnerability, particularly for smallholder farmers (see Beckford and Bailey, chapter 10).

A political ecology approach, which examines the struggle over resources and the ways in which the costs and benefits of environmental change are distributed unevenly, may help to identify populations which are particularly at risk. Political ecology recognizes that the transformation of the physical environment is not independent from specific historical, social, cultural, political or economic conditions (for example, see Mistry, Berardi and McGregor 2009), and that the process of socio-environmental change "is never socially or ecologically neutral" (Swyngedouw 2004, 23). This approach can enhance our understanding of vulnerability by encouraging a deeper examination of the inequitable social distribution of the risks of vulnerability (to both physical and social phenomena), and by elucidating and challenging the complex power relations at the root of this reality.

In a politicized environment which is characterized by marginality and vulnerability (as is frequently encountered in the Caribbean situation), environmental vulnerability is most often encountered at the grassroots level. This is traditionally associated with rural livelihoods, but in the increasingly urbanized Caribbean, fuelled by rural-urban migration, vulnerability persists in urban areas (Clarke, chapter 12; Howard, chapter 14). As Howard shows, for Santo Domingo, self-help is one way forward for poor, urban neighbourhoods. As urbanization and periurbanization proceed, social and ecological vulnerabilities are exacerbated by differences of class and ethnicity, as Jaffe (chapter 15) shows notably in her research.

Bryant and Bailey (1997) explore the relationships between NGOs, grassroots actors and adaptation to vulnerability. Their call for local empowerment is echoed here by Jaffe (chapter 15). She points out that the globally oriented, green discourse that dominates Jamaican and Curaçaoan environmentalism promotes top-down conservationist strategies which do not resolve the problems of the rural and urban poor. Jaffe argues that a more enlightened approach to combating ecological and social vulnerability is required, adapting governance to acknowledge and utilize local priorities and perspectives. Only thus will "joined-up" solutions to vulnerability succeed, she argues. It is encouraging to note that international agencies are now increasingly recognizing this need, as exemplified by the GEF small grants approach mentioned below.

"Empowerment" undoubtedly requires more efficient communication between grassroots advocacy and governance than has been extant hitherto in the insular Caribbean. George (chapter 7) illustrates some of the issues faced by local people, his "epistemic communities", in attempting to interface with decision-makers. Otuokon and Chai (chapter 8) provide an example of the structures required for successful national park management, including a significant input from those living in and around the protected area. Both interventions present lessons for the wider Caribbean. An excellent example of what can be achieved is the regionally based International Development Research Centre (IDRC) programme to foster better community-based resource management structures (Breton et al. 2006). This programme has ranged over several territories and identified common factors of vulnerability and resilience in coastal communities throughout the region. Among their recommendations for more efficient coastal zone management are a request for more interdisciplinary research and more local autonomy (while without isolating communities), a desire for less negative attitude from state agencies to collaborative work, a need for better training of social scientists to enable more holistic solutions, and a plea for necessary attention to the gender gap which denies women's potential to innovate and contribute to decision-making.

The Global Context

Global environmental change is undoubtedly a critical issue for the Caribbean region, and the way in which this may affect the magnitude and frequency of extreme events could become an even more critical issue, particularly in the coastal zone, as witnessed by Rowe et al.'s analysis (chapter 3) of boulder fields around Jamaica's coastline. As Rowe et al. discuss, many of the Caribbean's coastlines have boulder fields, and share the vulnerability to storm surge and tsunami events.

Further, most authorities now consider that it is likely that global warming will lead to more frequent tropical storms and a greater probability that some of these will develop into hurricanes (Emanuel 2005; Angeles et al. 2007; Saunders and Lea 2008), though conflicting interpretations of the data remain to be reconciled (Shepherd and Knutson 2007). Further, although again data are not clear-cut across the region, declining rainfall has been seen in recent decades in many parts of the Basin. Medium-term rainfall figures for Jamaica are discussed in chapter 9, while the deepening nature of the Caribbean early season drought is considered by Gamble in chapter 2. When allied to the likelihood of rising sea levels in parts of the Basin (Mimura et al. 2007), and with the increased use of irrigation marginalizing water resources in drier parts of the Basin, these climate trends point towards increasing marginality of agricultural livelihoods.

Many Caribbean livelihoods are also at risk as a result of "human" factors such as globalization. Rural dwellers who are engaged in small-scale agriculture are particularly at risk, as they rely on imported inputs whose prices fluctuate according to foreign exchange rates, while simultaneously facing competition from mass-produced and cheaper imported food items. Although urban dwellers may have benefited from competition and cheaper imported food products, they have also been affected by globalization, particularly through the imposition of structural adjustment policies that have reduced the ability of the state to provide certain services. Clarke and Howard (2006, 106) have explained how this has led to a re-production of the "colonial version of a non-dynamic, labour-surplus urban economy", with public-sector layoffs, declining provision of basic social services and rapid price inflation; although the process changed the balance of opportunities for different classes rather than being simply a process of mass immiseration (Gordon, Anderson and Robotham 1997).

The Regional and Institutional Context

Vulnerability has been a factor of Caribbean life throughout recorded history. Progressive deforestation started with the plantation economies of the seventeenth century and was accompanied by progressive land degradation (Watts 1987). The occupation by freed slaves, following emancipation, of more marginal land exacerbated this trend (for example, see Richardson 1992). High-magnitude climatic events (such as tropical storms and hurricanes) and tectonic-related hazards (such as volcanic eruptions, earthquakes and tsunami) add to the physical hazards throughout the insular Caribbean (for example, see Ahmad 1992, 1997; Lashley and Bandara 2001). Relatively high population densities, particularly in the coastal zone, where most of the region's primate cities are found, add to vulnerability. Economic dependence on a limited natural resource base as well as limited human and financial capacities, along with small size, are typical features of these small island developing states (SIDS).

Nor has the advent of mass tourism resolved coastal vulnerabilities. Conflicts abound between infrastructure development and environmental sustainability, as discussed for Jamaica by Thomas-Hope and Jardine-Comrie (chapter 6); while the expansion of the tourism industry has also failed to alleviate the social problems faced by many members of Jamaican society (Dodman 2009). These situations throughout the Caribbean are amply illustrated by Pattullo (2005).

Individual states have long recognized their vulnerability and that of their citizens. Integrated action has, however, been of more recent date. This has undoubtedly followed from the recognition of the particular problems of SIDS (formalized by the Barbados SIDS meeting of 1994) and the recognition of the consequences of contemporary global warming – indeed, the Fourth Assessment Report of the Intergovernmental Panel on Climate Change (IPCC) specifically noted that SIDS are likely to be among the countries most affected adversely by climate change (Mimura et al. 2007). The high economic and social vulnerability of the Caribbean region was further recognized by the Commonwealth Secretariat (2000).

Key international focus for regional action following the 1994 meeting in Barbados has been on the Caribbean Programme for Adaptation to Climate Change (CPACC), which ran from 1997 to 2001, and its successor, the Mainstreaming Adaptation to Climate Change (MACC) project (from 2003 to 2008). These were linked (from 2001 to 2003) by the Adaptation to Climate Change in the Caribbean (ACCC) project. These internationally funded initiatives have undoubtedly done much to raise awareness of climate change issues in the region at the political level and have contributed significantly to a focus on vulnerability reduction. This remains a work in progress, though a significant outcome of the international effort has been the setting up of the Caribbean Community Climate Change Centre (CCCCC) established in Belize in 2005. Also, the focus on setting up finer-scale climate models (the PRECIS Project, run from the Institute of Meteorology in Cuba) is already leading to more detailed prediction of regional and national climate trends for the rest of this century (Taylor 2007). At the grassroots scale, the recent approach of the Global Environment Facility (GEF) has been to fund small-scale and locally run projects. This represents a participatory approach to adaptation, and evaluation of the relative sustainability (in both environmental and socio-economic terms) of this approach is keenly awaited.

In a wider political and academic sense, issues of vulnerability have moved since the 1990s and were perhaps encapsulated by SIDS Barbados 1994. There has been a change from issues of mitigation of extreme events, traditionally the focal point of governments and of institutions such as the Caribbean Disaster Emergency Response Agency (CDERA) and its predecessors, towards a deeper appreciation of the added and interlinked contexts of globalization and climate change, and an increasing focus on proactive adaptation to climate change.

The problems of forecasting and adapting to climate change have been well rehearsed (for example, see Huq et al. 2003; Witter 2007; McDonald 2007; UNFCCC 2007), often amid much rhetoric, but there remain significant dislocations between theory and practice, academic research and policy output, and assessment of vulnerability and the building of capacity to reduce risk. As yet, however, the relatively coarse scale of current global climate models reduces their effectiveness at the island scale, and therefore prejudices their impact on national planning authorities. This is in many ways a vulnerability in itself, specifically in the reluctance by decision-makers to commit resources to issues that remain, at present, inadequately defined for detailed, proactive planning. In this respect, the Caribbean mirrors the wider context of developing areas, and a recent review of the linkages between adaptation to climate change and development (Schipper 2007) points to the undeniable fact that much remains to be done.

Future Directions for Vulnerability Research

In assessing levels of vulnerability, a holistic approach to determining community vulnerability is very important. As witnessed by many chapters in this collection, a key factor in community vulnerability is social capital. This is implicitly, if not explicitly, recognized in much current donor agency thinking. For example, in respect to climate change, mainstreaming strategies are focused on building capacity to assess and reduce vulnerability at the institutional level. But access to resources and public education are also supported by the programme, recognizing the importance of accessibility to adaptation and mitigation.

This is, of course, a two-way process, and it is critical that "political ecological" structures be set up to ensure that adaptation to, and mitigation of, vulnerability recognizes and builds on existing social capital. As Pelling and High (2005) point out, there is, at present, an imperfect understanding of how social capital can best be used in assessing the practicalities of adaptation to vulnerability. This requires, they argue, urgent research into how social capital is formed, how it operates and how it may be utilized. Much of the empirical research presented earlier in this volume examines the interface between people, their environment and vulnerability, notably in rural (chapters 9 to 11) and urban (chapters 12 to 16) contexts. Social capital has frequently been pointed to as an existing resource. We suggest that this is not only exploitable in its own right, but represents a fundamental linkage between the essentially regional-scale and institutional mainstreaming approach and the actual amelioration of vulnerability at the grassroots level. Further research into social capital is thus, we argue, a necessary direction for future vulnerability research in the Caribbean context.

However, one obvious problem relating to a holistic approach to assessing levels of vulnerability is the issue surrounding the homologation of physical and human data. The use of Geographical Information Systems (GIS) presents one possible approach with the ability to bridge this divide, particularly through the use of participatory GIS (Sieber 2006) that involves the insights of local residents as well as macro-scale remote sensing data.

In our call for more integrated research that brings together the global and the local, alongside the environmental, social and economic, we are conscious that there is a strong need for more empirical data that isolate the different components of vulnerability in order to determine the relative levels of importance. However, as the earlier volumes in this series have demonstrated, there is an abundance of individual research approaches and outputs. Almost all of this research underlines the vulnerability of Caribbean livelihoods, even if this is not invariably stated. It would be instructive to revisit these and similar collections in order to synthesize approaches to vulnerability-related research undertaken by geographers and those in cognate disciplines so as to better inform future directions for vulnerability research. However, this is outwith the remit of this chapter, and we conclude this section by pointing to a further issue which this research has raised.

There is clearly a need for more longitudinal datasets. These do exist for climatic parameters, but generally not for human factors. The work of Janet Momsen (formerly Henshall) (for examples, see Henshall 1966; Momsen 1998, 2005) illustrates what can be achieved by repeated surveying of eastern Caribbean stakeholders. This does, however, require lengthy commitment on behalf of individual researchers, which is difficult under current research funding regimes. But some dedicated school teachers in the region have been collecting longitudinal datasets on aspects of human vulnerability, and a means of funding production of outputs from this research would handsomely repay investment.

Towards Resistance and Resilience

Whereas a better understanding of vulnerability helps us to recognize the dimensions of the problem, it is a focus on resistance and resilience that can direct academics and policy makers in the appropriate ways to address these issues. In practice, many aspects of vulnerability cannot be addressed simply at the scale of the community, nation or region. Although there is growing recognition of the need to implement mitigation strategies to reduce the extent of climate change at the global level, this remains a deeply contested process, and the effects will take years or decades to be felt. Likewise, the growing interest in ethical trading networks (Fingal 2008) has, as yet, functioned only as a marginal appendage to the mainstream global economy. Caribbean nations and citizens alike, therefore, while hoping for fundamental changes to global systems, need to identify appropriate methods for resistance and resilience to these challenges.

Mark Pelling identifies resistance as reflecting economic, psychological, and physical health, and their systems of maintenance, while resilience "can be thought of as the ability of an individual or collective actor to cope with or adapt to . . . stress" (Pelling 2001, 182). He takes this forward (Pelling 2003) in the specific context of the vulnerability of Caribbean cities. Following Blaikie et al. (1994), he notes that resistance and resilience are shaped by an actor's access to rights, resources and assets. These, in effect, circumscribe an individual's ability to cope with a crisis and also define the types of coping strategies available to that individual.

Much of the discussion in Blaikie et al. (1994) and in Pelling's work focuses on natural disasters. The context of resistance and resilience in the contemporary Caribbean is wider than, but of course includes, natural disasters. The consistent message, as evidenced by several chapters in this volume, is that the poor (whether rural or urban) are disadvantaged, and community planning needs to look beyond the traditional actors, engaging all stakeholders in searching out holistic solutions. Pelling (2003) argues that the organization of social capital within communities, although not resolving mitigation and response issues on its own, can be a key force in engaging with political and institutional structures, and thus in turn contributing to a more effective political agenda for risk reduction. There remain challenges to this approach, not least of which are entrenched political elites and interface with international agendas.

The ability to bridge scales is an important component of resilience for individuals, communities and nations. In planning and policy terms, while aggregate data are undoubtedly useful, they often mask critical local thresholds of behaviour or activity. Smallholder farmers who are able to tie their agricultural production into the needs of large "international" actors (such as hotels in the Caribbean tourism industry) are at a strong competitive advantage; while poor urban communities can also draw on resources from remittances, aid, and contacts with national and international agencies. Indeed, perhaps the single most important component of enhancing resilience is the ability to remove dependence on any single scale of interaction, and to diversify the sources from which assets (financial or non-financial) can be drawn.

Just as a political ecology approach to vulnerability assesses the inequitable distribution of risk, a political ecology approach to resilience can identify the ultimate causes of risk and point in appropriate directions to reduce this. Within nations, and even within communities, vulnerability to global and local change is unevenly distributed, and a political ecology approach can identify which individuals and groups require particular attention to increase their resilience. Although the importance of gender inequality is increasingly recognized, there are a variety of other inequalities including, although not necessarily limited to, age, family circumstances, and affiliation to ethnic, religious, or political groups (for example, see Verrest, chapter 16). Indeed, throughout the Caribbean it has been shown that young adults are more likely to be unemployed than older individuals, with serious implications for their ability to cope with adverse circumstances.

Economic migration has been a fact of Caribbean life for many decades, and through remittances can be argued to have increased resilience. But this is an uncertain platform on which to cope with the range of natural and human pressures present in the Caribbean today. Remittances are neither stable nor reliable, and indeed opportunities for large-scale economic migration are generally fewer now than in the past. Yet frequently there is a tacit assumption by government that remittances cover at least in part the deficiencies of the states' provision to individuals. Frequently, as shown in the Jamaican case, for example (McGregor, Barker and Campbell, chapter 9), government-funded structures exist but do not provide the required level of resilience at the grassroots level.

In order to identify the best strategies to build resilience, there is a need for more participatory research that raises awareness of the mechanisms already used by individuals and households to reduce their vulnerability, whether it be for environmental or social reasons. Research into community resilience could also be focused on why intervention from the political to the community level has not enhanced resilience to risk in many cases: Why do government and/or aid agency interventions so frequently fail to deliver? More empirical data that seek to isolate the different components of resilience to determine their relative levels of importance are required. This would allow for an evaluation of where aid of whatever nature might be most efficiently delivered. In turn, all this calls for a more integrated research effort that brings the global and the local scales together.

Practical Interventions for Resilience

A frequently recurring theme at the conference Global Change and Caribbean Vulnerability involved the importance of moving beyond academic research to providing practical frameworks for action to reduce vulnerability and increase resilience. One key aspect of this is to take a pragmatic approach, recognizing that ideologically driven approaches to development and disaster management have frequently failed to take into account the complexity of local contexts. This requires a greater degree of flexibility among planners, governments and aid agencies than has frequently been the case.

A pragmatic approach ought also to recognize the already existing presence of indigenous technical knowledge (ITK) that provides a locally generated, context-specific, appropriate means for addressing many of the challenges of vulnerability. Governments, aid agencies and other stakeholders ought to be aware of this pre-existing capacity when devising programmes to reduce vulnerability. Practical solutions to reduce vulnerability that take existing knowledge and skills into account are likely not only to be more cost effective, but also more widely accepted. One example of this was raised in discussion at the Global Change and Caribbean Vulnerability conference by Franklin McDonald, former executive director of Jamaica's National Environment and Planning Agency, who discussed the use of stones on roofs to protect against hurricane wind damage. Not only do these stones act as weight, but also they serve to disrupt the air flow in ways that reduces the pressure differences between the inside and the outside of buildings, thereby helping to prevent the removal of the roof.

There is also the importance of community knowledge as regards the functioning of farming systems, and it is important to examine ways in which this can be accessed. McGregor, Barker and Campbell (chapter 9) show how farmers in semi-arid southern St Elizabeth have adapted to their relatively dry environment to grow a range of crops for the

Jamaican tourist market. Participatory action research is one possible route to examining this issue in more depth, and also one more potential route towards empowerment at the local level.

Another practical intervention for resilience involves awareness-raising among a variety of different populations: at the individual/community level, as well as among planners and decision makers. Whereas individual citizens need to be made aware of the best practices to follow to reduce their vulnerability to natural and social challenges, policy makers also need to be more aware of the challenges facing individuals and the myriad ways that they have already evolved to address these.

Throughout all of these interventions, there ought also to be a strong focus on a participatory approach. Although it has been argued that participatory approaches often shift responsibility for the consequences of projects "away from the agencies and the development workers onto the participating people" (Henkel and Stirrat 2001, 183), these can also have a great effect on improving resilience by providing both *efficiency* benefits, as well as *equity and empowerment* benefits (Cleaver 2001). Particularly in the area of vulnerability and resilience, a participatory approach ensures that the most important decisions are taken by the people who are most at risk, and who therefore have the most to lose from wrong choices, and the most to gain from the right ones.

Concluding Thoughts

This, the fourth in a series of edited collections on aspects of Caribbean environments, may be concluded on a reflective note. The series has evolved from perspectives on environment and development in the Caribbean (1995), through a focus on resource sustainability and Caribbean development (1998), then consideration of resources, planning and environmental management in the region (2003), to the present volume's focus on vulnerability. This parallels a significant shift towards an approach where people, both groups and individuals, become central to issues of environment, resource management, planning, and sustainability in changing environmental and social circumstances. In many ways this mirrors a wider context to geographical research over the period since the early 1990s in better recognition of the importance of social capital in addressing environmental challenges.

Likewise, at the regional scale, Caribbean governments and institutions now have a much greater awareness than in the early 1990s of the environmental challenges facing the region, climatically, environmentally and socially. There is also greater awareness of how the Caribbean is placed within the wider global social and economic context. What has lagged behind this recognition of environmental challenges, be they of natural or human origin, is clear and consistent policy in tackling these challenges.

This present volume has not only pointed towards the complex nature of response to change within the Caribbean Basin, but to the need for greater clarity at the governmental and institutional levels in ways of addressing change. If this collection of research essays assists in the process of reducing vulnerability and increasing people's resilience in the face of vulnerability, then we editors feel that we will have achieved our main objectives. Any credit, however, should properly be attributed to the individual researchers whose work has been reported in this volume and its predecessors.

References

Ahmad, R., ed. 1992. Natural hazards in the Caribbean. Special issue, *Journal of the Geological Society of Jamaica* 12.

———. 1997. *Natural hazards and hazard management in the greater Caribbean and Latin America*. Kingston: Unit for Disaster Studies, University of the West Indies.

Angeles, M.E., J.E. Gonzalez, D.E. Erickson III and J.L. Hernández. 2007. Predictions of future climate change in the Caribbean region using global general circulation models. *International Journal of Climatology* 27: 555–69.

Blaikie, P., T. Cannon, I. Davis and B. Wisner. 1994. *At risk: Natural hazards, people's vulnerability, and disasters*. London: Routledge.

Bryant, R., and S. Bailey. 1997. *Third World Political Ecology*. London: Routledge.

Clarke, C., and D. Howard. 2006. Contradictory socio-economic consequences of structural adjustment in Kingston, Jamaica. *Geographical Journal* 172: 106–29.

Cleaver, F. 2001. Institutions, agency and the limitations of participatory approaches to development. In *Participation: The new tyranny?* ed. B. Cooke and U. Kothari, 36–55. London: Zed Books.

Dodman, D. 2009. Globalization, tourism and local living conditions on Jamaica's north coast. *Singapore Journal of Tropical Geography* 30 (2): 204–19.

Dodman, D., and C. Newstead. 2008. Neoliberalization and Caribbean agriculture: Trends and prospects for livelihoods and national economies. *Caribbean Geography* 15 (2): 91–100.

Emanuel, K. 2005. Increasing destructiveness of tropical cyclones over the past thirty years. *Nature* 436: 686–88.

Fingal, C. 2008. Fair "trade not aid": Globalisation, fair trade and the St Lucian banana industry. *Caribbean Geography* 15 (2): 118–29.

Gordon, D., P. Anderson and D. Robotham. 1997. Jamaica: Urbanization during the years of the crisis. In *The urban Caribbean: Transition to the new global economy*, ed. A. Portes, C. Dore-Cabral and P. Landolt, 190–223. Baltimore: Johns Hopkins University Press.

Henkel, H., and R. Stirrat. 2001. Participation as spiritual duty: Empowerment as secular subjection. In Cooke and Kothari 2001, 168–84.

Henshall, J.D. 1966. The demographic factor in the structure of agriculture in Barbados. *Transactions of the Institute of British Geographers* 38: 183–95.

Huq, S., A. Rahman, M. Konate, Y. Sokona and H. Reid. 2003. *Mainstreaming adaptation to climate change in least developed countries (LDCS)*. London: International Institute for Environment and Development.

Klak, T. 1998. Thirteen theses on globalization and neoliberalism. In *Globalization and neoliberalism: The Caribbean context*, ed. T. Klak, 3–27. Lanham, MD: Rowman and Littlefield.

Lashley, B., and S.B. Bandara, eds. 2001. *Bibliography of natural hazards in the Caribbean*. Kingston: Caribbean Disaster Information Network (CARDIN).

McDonald, F. 2007. Regional perspectives on policy and planning for climate change. Presentation to Conference on Climatic Change Impacts on the Caribbean. Kingston, Jamaica. June.

Mimura, N., L. Nurse, R.F. McLean, J. Agard, L. Briguglio, P. Lefale, R. Payet and G. Sem. 2007. Small islands. In *Climate change 2007: Impacts, adaptation and vulnerability. Contribution of working group II to the fourth assessment report of the Intergovernmental Panel on Climate Change*, ed. M.L. Parry, O.F. Canziani, J.P. Palutikof, P.J. van der Linden and C.E. Hanson, 687–716. Cambridge: Cambridge University Press. Available at http://www.ipcc-wg2.org/

Mistry, J., A. Berardi and D. McGregor. 2009. Natural resource management and development discourses in the Caribbean: Reflections on the Guyanese and Jamaican experience. *Third World Quarterly* 30 (5): 969–89.

Momsen, J.H. 1998. Caribbean tourism and agriculture: New linkages in the global era? In Klak 1998, 115–33.

Momsen, J.H. 2005. Caribbean peasantry revisited: Barbadian farmers over four decades. *Southeastern Geographer* 45 (2): 206–21.

Pattullo, P. 2005. *Last resorts: The cost of tourism in the Caribbean*, 2nd ed. London and New York: Latin America Bureau and Monthly Review Press.

Peck, J., and A. Tickell. 2002. Neoliberalizing space. *Antipode* 34: 380–404.

Pelling, M. 2001. Natural disasters? In *Social nature: Theory, practice and politics*, ed. N. Castree and B. Braun, 170–83. Oxford: Blackwell.

———. 2003. *The vulnerability of cities: Natural disasters and social resilience*. London: Earthscan.

Pelling, M., and C.D. High. 2005. Understanding adaptation: What can social capital offer assessments of adaptive capacity? *Global Environmental Change* 15: 308–19.

Potter, R. 2000. *The urban Caribbean in an era of global change*. Aldershot, UK: Ashgate.

Richardson, B.C. 1992. *The Caribbean in the wider world, 1492–1992*. Cambridge: Cambridge University Press.

Saunders, M.A., and A.S. Lea. 2008. Large contribution of sea surface warming to recent increase in Atlantic hurricane activity. *Nature* 451: 557–60.

Schipper, E.L.F. 2007. *Climate change adaptation and development: Exploring the linkages*. Working Paper no. 107. Norwich, UK: Tyndall Centre for Climate Change Research.

Shepherd, J.M., and T. Knutson. 2007. The current debate on the linkages between global warming and hurricanes. *Geography Compass* 1: 1–24.

Sieber, R. 2006. Public participation geographic information systems: A literature review and framework. *Annals of the Association of American Geographers* 96: 491–507.

Swyngedouw, E. 2004. *Social power and the urbanization of water*. Oxford: Oxford University Press.

Swyngedouw, E., and N. Heynen. 2003. Urban political ecology, justice and the politics of scale. *Antipode* 35: 898–918.

Taylor, M. 2007. Climate change in the future: Results from the PRECIS RCM. Presentation to Conference on Climatic Change Impacts on the Caribbean. Kingston, Jamaica. June.

United Nations Framework Convention on Climate Change (UNFCCC). 2007. Vulnerability and adaptation to climate change in small island developing states. http://unfccc.int/files/adaptation/adverse_effects_and_response_measures_art_48/application/pdf/200702_sids_adaptation_bg.pdf

Walsh, R.P.D. 1980. Runoff processes and models in the humid tropics. *Zeitschrift für Geomorphologie,* Supplement-Band 36: 176–202.

Watts, D. 1987. *The West Indies: Patterns of development, culture and environmental change since 1492.* Cambridge: Cambridge University Press.

Witter, M. 2007. Climate change and Jamaica: Sustainable development as strategic adaptation. Third Annual Public Lecture, Environmental Foundation of Jamaica. Kingston, Jamaica.

Contributors

Duncan McGregor is Senior Lecturer in the Department of Geography, Royal Holloway, University of London, and Fellow of the Royal Geographical Society. His principal research interests include soil erosion, land degradation, land use change, and watershed management strategies in the humid and semi-arid tropics. He has recently worked on watershed management in Kumasi, Ghana, and is currently researching on the effects of environmental change on farming systems in Jamaica and semi-arid West Pokot, Kenya. He has published widely in scholarly publications and has edited a number of tertiary-level books.

David Dodman is Researcher, Human Settlements and Climate Change at the International Institute for Environment and Development in London. Prior to this, he was a lecturer in the Department of Geography and Geology, University of the West Indies, Mona, Jamaica. His primary research interests are in urban governance and the management of urban environments, particularly in relation to climate change adaptation in low- and middle-income countries. His publications include *Adapting Cities to Climate Change* (co-edited with Jane Bicknell and David Satterthwaite).

David Barker is Professor and Head, Department of Geography and Geology, University of the West Indies, Mona, Jamaica. He is co-founder and editor of the journal *Caribbean Geography*. He has published widely on the Caribbean, including books for the tertiary and secondary educational levels. His research interests include small-scale agriculture and indigenous farming knowledge, natural hazards and food security in relation to rural sustainable livelihoods, traditional resource management and environmental degradation in the Caribbean. He is currently President of the Jamaican Geographical Society.

Steve Bailey (1976–2006) died tragically in a road accident just a few weeks before the fifth BCGS Conference in 2006. At the time, he was a teaching assistant and PhD student in the Department of Geography and Geology, University of the West Indies, Mona, Jamaica; a resident advisor at Chancellor Hall; and membership secretary of the Jamaican Geographical Society. His doctoral research was on land degradation and farming systems in the Pindars River Basin, Clarendon, Jamaica. Steve presented preliminary results of this research at the fourth BCGS held in Belize in 2003, but, sadly, was unable to present his definitive paper on his research at the 2006 meeting.

Clinton Beckford is Associate Professor of Geography Education and Associate Dean of Pre-service Education at the Faculty of Education, University of Windsor, Canada. He obtained his PhD in geography from the University of the West Indies, Mona. He has researched and written several scholarly articles on Jamaican small-scale farming systems, their role in local food security and the influence of local knowledge in small-scale farmers' decision-making and survival. He is currently involved in a research project on the experiences of migrant Jamaican farm workers in southern Ontario, and in an international research and development project on the educational experiences of vulnerable children and orphans in Tanzania.

Donovan Campbell is a Jamaican student studying for a PhD at the University of the West Indies, Mona, Jamaica, and is a resident advisor at Irvine Hall. His topic of research is "Domestic Food Production and Hazard Vulnerability in Jamaica: Adaptation and Change in Southern St Elizabeth".

Shauna-Lee Chai is a 2007 Gates scholar and graduate student of Cambridge University, studying deforestation in tropical rainforests. She is a graduate of the University of the West Indies, and has worked for three years as a conservation science officer for the Blue and John Crow Mountains National Park.

Sean Chenoweth is Assistant Professor of Geography, University of Louisiana–Monroe. He teaches GIS and remote sensing and has been studying karst landscapes since 1991, beginning with the Everglades of Florida, then the Ozarks of Arkansas and most recently cockpit karst in the Caribbean.

Colin Clarke is Emeritus Professor of Geography, Oxford University and Emeritus Fellow of Jesus College. He has carried out many research projects in Mexico and the Caribbean. His principal research interests are in urbanization in developing countries; race, ethnicity and class in urban and rural social structures; peasant transformations; and the problems of small states.

Mick Day has been Professor of Geography, University of Wisconsin–Milwaukee since 1977. Originally from the United Kingdom, Mick has conducted research on Caribbean karst since 1974. His initial focus on geomorphology has since broadened to incorporate human land use issues, particularly natural hazards and the designation of protected areas within the Caribbean karst.

Douglas Gamble is Associate Professor of Geography and Director of the Laboratory for Applied Climate Research, Department of Geography and Geology, University of North Carolina–Wilmington. His research expertise includes applied climatology, hydrology, and coastal and island environments of the Caribbean.

Kemi George is a PhD candidate at the University of Massachusetts, Amherst. His research focuses on the production and use of knowledge by networks of scientists to influence environmental policy. He is currently completing a dissertation comparing biodiversity management in Jamaica, Mexico and Egypt.

David Howard is based in the Institute of Geography at the University of Edinburgh, and has research interests in contemporary social and urban geographies of the Caribbean and Latin America. He is currently CNRS Associate at the Centre d'Étude d'Afrique Noire, Université de Bordeaux IV; coordinating editor for the *Bulletin of Latin American Research*; and chair of the Society for Caribbean Studies.

Rivke Jaffe is Lecturer, Department of Cultural Anthropology and Development Sociology, Leiden University, the Netherlands. Her research interests include urban anthropology and sociology, the urban environment, popular culture and violence. She has lived and worked in Jamaica, Curaçao and Suriname and her publications include the edited volume *The Caribbean City*.

Adonna Jardine-Comrie is completing her PhD in environmental management at the University of the West Indies, Mona, Jamaica. In 2007–2008, she was the recipient of a Commonwealth Split-site Award, tenable at the Department of Geography, University of Liverpool.

Linda Johnson-Bhola is Lecturer, School of Earth and Environmental Sciences, University of Guyana. She is an MSc degree in Urban and Regional Planning at the University of the West Indies, St Augustine, Trinidad and Tobago. Her research interests are coastal flooding, geomorphological issues and physical planning; areas in which she has undertaken several research projects.

Shakira Khan is completing a PhD in coastal sediment dynamics in the Department of Geography and Geology, University of the West Indies, Mona, Jamaica and has focused much of her research on coastal processes and their impacts. She is a currently a research associate in the Marine Geology Unit at the University of the West Indies, Mona.

Susan Otuokon is Executive Director of the Jamaica Conservation and Development Trust, a non-government organization that manages the Blue and John Crow Mountains National Park. She is also pursuing her PhD with the Environmental Management Unit at the University of the West Indies, Mona.

Kevon Rhiney is a PhD candidate in the Department of Geography and Geology, University of the West Indies, Mona, Jamaica, and temporary assistant lecturer in the department. He is a resident advisor at Taylor Hall and vice-president of the Jamaican Geographical Society. He is currently completing his doctoral research on the linkages between the local food-supply network and the tourism industry in Jamaica.

Edward Robinson is Director of the Marine Geology Unit, University of the West Indies, Mona, Jamaica. He works on coastline problems and is author/ co-author of more than 140 professional papers. His many awards include the Norman Manley Award for Excellence in Science and Technology in 1978, the National Medal for Science and Technology in 2006, and the Order of Merit in 2008.

Deborah-Ann Rowe is a research associate of the Marine Geology Unit, with a BSc and MPhil in Geology from the University of the West Indies and 10 years experience in the investigation of natural hazards. Supported by a Gladstone Fellowship, she is presently reading for a PhD entitled "Tsunami Hazard Mapping in Southwest Jamaica" at the Universities of Chester and Liverpool, United Kingdom.

Elizabeth Thomas-Hope is the James Seivright Moss-Solomon (Snr.) Chair of Environmental Management, University of the West Indies, Mona Jamaica. She is a member of several international advisory boards, including the EU–UN Joint Migration and Development Initiative, and the UNDP Human Development Report Committee (2009), and ecretary of the Commonwealth Geographical Bureau. Her research interests include issues of environment and sustainable development, as well as international migration, with particular reference to the Caribbean.

Hebe Verrest is Lecturer, Department of Methodology and Statistics, Utrecht University, The Netherlands. Her research interests include urban livelihoods, geography and planning, social stratification and inequality, mixed methods, entrepreneurship, development policies. She has conducted extensive research in Suriname and Trinidad and Tobago.

Patrick Williams is Senior Lecturer, School of Earth and Environmental Sciences, University of Guyana, and Country Manager, World Wildlife Fund. He has undertaken several consultancies and published extensively in the areas of land management and planning, urban and regional planning and environment.

CPSIA information can be obtained
at www.ICGtesting.com
Printed in the USA
LVOW01s1708100816
499850LV00013B/44/P

9 789766 402211